Chemistry and Physics for Nurse Anesthesia

David Shubert, PhD, obtained a bachelor's degree in chemistry at Fort Hays State University in 1983 and then went on to the University of Colorado to obtain a PhD in organic chemistry in 1987 as a National Science Foundation predoctoral fellow. He joined the faculty at Newman University in 1987. Although his area of expertise lies in organometallic chemistry, Dr. Shubert has taught introductory, general, organic, analytical, and instrumental analytical chemistry and biochemistry, in addition to earth and space science, general physical science, and chemistry/physics for nurse anesthesia students. Dr. Shubert has received numerous awards for teaching and service, including the Newman Teaching Excellence Award and the didactic instructor award for the NU NAN program. He and Dr. Leyba were invited to teach chemistry and physics for nurse anesthesia online for the Medical University of South Carolina. Although teaching remains his primary interest, Dr. Shubert has also worked as a chemical consultant with several local industries and has offered numerous workshops that provided training and access to modern instrumentation to high school chemistry teachers. Dr. Shubert now serves as dean of undergraduate studies, and serves the Higher Learning Commission of the North Central Association as a peer reviewer. He is currently working on an electronic organic chemistry textbook.

John Leyba, PhD, received a bachelor's degree in chemistry from Northeast Missouri State University (now Truman State University) in 1986 and obtained a PhD in nuclear chemistry from the University of California, Berkeley, in 1990. He held the positions of senior scientist, senior scientist A, and principal scientist with Westinghouse Savannah River Company at the Department of Energy's Savannah River site. In addition, he was the radiochemistry group leader for Rust Federal Services' Clemson Technical Center located in Anderson, South Carolina. He also held an appointment as a visiting assistant professor in the Department of Chemistry and as an adjunct assistant professor in the Department of Environmental Engineering and Science at Clemson University. Just prior to joining the faculty at Newman, Dr. Leyba was the Denver Area Director of Operations for Canberra Industries. Dr. Leyba joined the Newman faculty in 2002. He is presently a professor of chemistry and chair of the Division of Science and Mathematics at Newman, where he teaches courses in physics, earth and space science, chemistry, and sonography. Dr. Leyba's research interests involve fast chemical separations and detection of radioactive materials. He is the recipient of multiple awards, including the AAUP Academic Freedom Award (2006), the Newman University Scholarship and Publication Award (2006 and 2009), the Newman University Teaching Excellence Award (2005), the Newman University Special Institution Achievement Award (2010), the Newman University Nurse Anesthesia Program Outstanding Didactic Instructor Award (2009 and 2011), and the Newman University Academic Advising Excellence Award (2012). He has published 29 peer-reviewed articles (research) and 22 technical reports (industry). He is a member of the American Nuclear Society and the American Chemical Society.

Chemistry and Physics for Nurse Anesthesia

Second Edition

A Student-Centered Approach

David Shubert, PhD
John Leyba, PhD

SPRINGER PUBLISHING COMPANY
NEW YORK

Springer Publishing Company, LLC
11 West 42nd Street
New York, NY 10036
www.springerpub.com

Acquisitions Editor: Margaret Zuccarini
Composition: Newgen Imaging

ISBN: 978-0-8261-1043-5
Ebook ISBN: 978-0-8261-1044-2

Instructors' Materials ISBN: 978-08261-3483-7
Videos can be found on YouTube through the links provided in this text.
Qualified instructors may request the PowerPoint slide presentation by emailing textbook@springer.pub.com

13 14 15 / 5 4 3 2 1

The author and the publisher of this Work have made every effort to use sources believed to be reliable to provide information that is accurate and compatible with the standards generally accepted at the time of publication. The author and publisher shall not be liable for any special, consequential, or exemplary damages resulting, in whole or in part, from the readers' use of, or reliance on, the information contained in this book. The publisher has no responsibility for the persistence or accuracy of URLs for external or third-party Internet websites referred to in this publication and does not guarantee that any content on such websites is, or will remain, accurate or appropriate.

Library of Congress Cataloging-in-Publication Data

Shubert, David.
  Chemistry and physics for nurse anesthesia : a student-centered approach / David Shubert, John Leyba. — 2nd ed.
    p. ; cm.
  Includes bibliographical references and index.
  ISBN 978-0-8261-1043-5 — ISBN 978-0-8261-1044-2 (e-ISBN)
  1. Chemistry. 2. Physics. 3. Nursing. 4. Anesthesia. I. Leyba, John. II. Title.
  [DNLM: 1. Chemistry—Nurses' Instruction. 2. Anesthetics—chemistry—Nurses' Instruction. 3. Nurse Anesthetists. 4. Physics—Nurses' Instruction. QD 31.3]
  RT69.S56 2013
  617.9'60231—dc23   2013002613

Printed in the United States of America by Courier.

Contents

Videos

Note: The URL links to the videos can also be found as a footnote on those pages in text where the video icons appear.

Foreword

Chemistry and physics are the building blocks on which our anesthesia practice is based. These two areas affect us, and our practice, in ways we don't consciously think about. These are essential elements in the understanding of pharmacology, physiology, and the anesthesia gas machine. Therefore it is important for our students to have a good grasp of these basic principles.

In the first edition of the textbook *Chemistry and Physics for Nurse Anesthesia: A Student-Centered Approach* by David Shubert, PhD and John Leyba, PhD, the authors built on their vast experience as nurse anesthesia instructors to help develop a resource for nurse anesthesia programs that made teaching and understanding these concepts easy and useful. In the second edition, they are expanding the chapters by adding more pictures and graphs, a chapter summary, and end-of-chapter problems to help our students clarify critical points. The authors, through the publisher's website, also provide 15 videos covering such topics as gas laws, Poiseuille's Law, and electrical safety.

This is a text that is easy to read and easily incorporated into any nurse anesthesia chemistry and physics course. I recommend this textbook to any program for incorporation into its courses.

Anthony Chipas, PhD, CRNA
Assistant Division Director
Anesthesia for Nurses Program
Medical University of South Carolina

Preface

We have had the pleasure of teaching chemistry and physics for anesthesia for nearly 13 years to over 200 nurse anesthesia students. The importance of the physical sciences in anesthesia goes without saying, and every one of our students knows this. We are continually impressed by the ability and motivation of our students, but we understand that they generally have only a minimal background in chemistry and almost no background in physics. They are bright and capable, but their math skills are often rusty. Our students are graduate students in nurse anesthesia, not *graduate students in chemistry or in physics*. So, it makes no sense to us to teach this course as though they were. Hence, this text attempts to bridge that gap. We cover material ranging from the most basic introductory chemistry concepts to topics treated in physical chemistry. This text provides a somewhat uneven treatment of a vast array of material that emphasizes the specific content in chemistry and physics that relates to anesthesia and offers illustrations that directly relate to anesthesia. The text has sufficient mathematical rigor to deal with the material, but no more than that.

Our intent in writing this book is to facilitate a conversation about chemistry and physics concepts that are essential to understanding anesthesia, and especially the behavior of gases. Our writing is intentionally conversational and encouraging. We try to make the story as engaging as possible. Some might observe that our writing is *unacceptably casual*, but we believe it is more important for these students to master this material, rather than to learn how to sound like a stuffy college professor.

In writing this second edition, we listened to our students, to your students, and to you, our fellow instructors. We have incorporated many excellent suggestions. Every chapter has an increased focus on clinical stories and applications. The end-of-chapter problems have been expanded, edited, and reorganized. We have included end-of-chapter summaries and an end-of-book Glossary. **Most excitingly (to us anyway) are the demonstration videos available on YouTube, courtesy of Springer Publishing Company, and can be viewed through links to the Springer website.** There are 15 videos illustrating concepts ranging from the various gas laws to fluid dynamics. They might look like home videos (mostly because they are) but the science is real, and some of the demonstrations are original.

In addition to this text, we have provided instructors' material that includes chapter-based PowerPoint slides. **To receive these electronic files, faculty should please contact Springer Publishing Company at textbook@springerpub.com.**

And (of course) we think we have eliminated every typo from the text. If (when) you find one, please let us know. Thank you for your support and ENJOY CHEMISTRY AND PHYSICS!

David Shubert
John Leyba

Measurement

A REVIEW OF SOME BASIC MATHEMATICAL SKILLS

Chemistry and physics are both very logical sciences, but both depend on math to translate concepts into application. Unfortunately, many students who struggle in these disciplines have more trouble with the mathematics than with the scientific concepts. Therefore, let us begin our exploration of chemistry and physics with a review of some basic math skills and concepts that you will use throughout this course. While this chapter reviews basic math skills, it cannot replace a basic understanding of college-level algebra.

You will need a calculator for this course. Some of you may have fancy graphing calculators, but the capacity of these instruments far exceeds your mathematical needs for this course. Any calculator that is labeled as a "scientific calculator" will more than suffice. You should be able to buy an adequate calculator for less than $20. When dealing with a math problem, don't reach for your calculator first. Whether or not the calculator gives you the correct answer depends on your ability to give it the correct information with which to work. Think about what operations you need to do first. See how far you can get with just a paper and pencil. You don't have to do *arithmetic* computations, such as long division, by hand, but you should be able to perform the *mathematical* manipulations (e.g., solve the equation for x) without a calculator. In fact, it is typically best to solve the equation first for the variable of interest *before* plugging in numbers. As you become more adept at thinking your way through math problems, you are more likely to get your calculator to give you the right answer.

Order of Operations

There are four principal arithmetic operations: addition, subtraction, multiplication, and division. When an equation involves multiplication and/or division as well as addition and/or subtraction, you need to do the multiplication and division operations *before* doing the addition and subtraction operations. For example, what does 12 plus 3 times 10 equal?

$$x = 12 + 3 \cdot 10$$

Since this equation mixes addition and multiplication, we need to do the multiplication first:

$$x = 12 + 30$$
$$x = 42$$

Now, you evaluate this expression:

$$x = \frac{12}{4} - 3 \cdot 2 + 4$$

Answer: $x = 1$

Parentheses and division bars (fraction bars) are both symbols of enclosure. Whenever there are symbols of enclosure, execute any arithmetic operations inside the symbol of enclosure first. The general rule is *to work from the innermost symbol of enclosure to the outermost, multiplying and dividing, then adding and subtracting.* Consider this example, which involves both kinds of symbols of enclosure and order of operations.

$$x = \frac{12 + 3 \cdot (4 + 2)}{3 \cdot 5}$$

To solve this example, we need to first evaluate inside the parentheses, then multiply and finally add:

$$x = \frac{12 + 3 \cdot (4 + 2)}{3 \cdot 5} = \frac{12 + 3 \cdot (6)}{3 \cdot 5} = \frac{30}{3 \cdot 5}$$

Here we come to a very common error. Many students are tempted to divide 30 by 3 first, and then *multiply* by 5. But the division bar is a symbol of enclosure, so multiply 3 times 5 first and then divide 30 by 15.

$$x = \frac{30}{15} = 2$$

Let's evaluate this expression:

$$x = \frac{\frac{10}{5} + (6 + 4) \cdot (5 - 3)}{\frac{8}{4} + 9}$$

Answer: $x = 2$

Negative numbers are common sources of confusion. Just remember that multiplying (or dividing) two positive numbers or two negative numbers gives a positive number. Multiplication or division involving a positive number and a negative number gives a negative number. Multiplying a number by negative 1 changes the sign of the number. Likewise, a negative number can be expressed as a positive number times –1. Subtracting a negative number is the same as adding a positive number.

Some examples are shown below:

$$4 \cdot (-3) = -12$$
$$(-4) \cdot (-3) = +12$$
$$-4x - (-3x) = -4x + 3x = -x$$

Algebra: Solving Equations for an Unknown Quantity

Addition and subtraction are inverse operators of each other, as are multiplication and division. Usually, a problem will involve solving an equation for some variable. You want to convert the equation into the form $x = some\ number$ (or some expression). Therefore, when you want to isolate a variable (i.e., "solve the equation"), you need to undo any operations that are tying up the variable by applying the inverse operation. That is, if the variable is multiplied by a number, you need to divide *both sides* of the equation by that number. If a variable is divided by some number, you need to multiply *both sides* of the equation by that number. This eliminates that number from the variable by converting it into a 1. Since 1 is the "multiplicative identity element," you can ignore 1's that are involved in multiplication or division. If some number is added to the variable, you need to subtract that number from *both sides* of the equation. If some number is subtracted from the variable, you need to add that number to *both sides* of the equation. This eliminates that number from the variable by converting it into a zero. Since zero is the "additive identity element," you can ignore zeros that are involved in addition or subtraction.

The critical thing to remember, though, is that whatever you do to one side of the equation, you have to do the other side. Think of it as being fair. If you add a number to one side, you have to add the same number to the other side; otherwise the equation would no longer be balanced.

For example, if $12x = 180$, what's x?

$$12x = 180$$

x is multiplied by a factor of 12: let's divide both sides by 12:

$$\frac{12x}{12} = \frac{180}{12}$$

12 divided by 12 equals 1, and we can ignore it.

$$x = \frac{180}{12} = 15$$

On your calculator, first enter the number on top (the numerator) and then divide by the number on the bottom (the denominator). Division operators are not commutative (i.e., $a \div b \neq b \div a$), so the order in which you enter the numbers is critically important. On the other hand, multiplication and division are associative, so you can regroup fractions multiplied by a quantity. For example:

$$6 \cdot \frac{12}{3} = \frac{6 \cdot 12}{3} = 12 \cdot \frac{6}{3} = 24$$

Division signs ($\div$) are rarely used in the sciences. Usually, a division operation is signaled by a fractional notation. Just think of the horizontal bar as the division sign. For example, if x divided by 15 equals 180, what does x equal?

$$\frac{x}{15} = 180$$

x is divided by a factor of 15: let's multiply both sides by 15:

$$15 \cdot \frac{x}{15} = 180 \cdot 15$$

15 divided by 15 equals 1, and on the left side of the equation it drops out.

$$x = 180 \cdot 15 = 2700$$

Sometimes the variable is in the denominator. Since anything in the denominator denotes the division operation, you can undo the division by multiplying, which is the inverse operation. For example:

$$\frac{15}{x} = 180$$

x is tied up in a division, so let's multiply both sides by x:

$$x \cdot \frac{15}{x} = 180 \cdot x$$

On the left side, x divided by x is 1, and we can ignore it.

$$15 = 180 \cdot x$$

Dividing both sides by 180 gives

$$\frac{15}{180} = x = 0.083$$

For example, Charles's Law is shown below and describes the effect of temperature on the volume of a gas. Solve this equation for T_2:

$$\frac{V_1}{T_1} = \frac{V_2}{T_2}$$

Answer: $T_2 = \dfrac{V_2 \cdot T_1}{V_1}$

When you need to solve an equation that involves addition and/or subtraction *as well as* multiplication and/or division, the order of operations reverses. First, you need to clear the addition and/or subtraction operations and then the multiplication and/or division. For example, the equation that relates Fahrenheit temperature to Celsius temperature is

$$°F = \frac{9}{5}(°C) + 32$$

To solve this equation for °C, we need to remove the addition term (32), the multiplication factor (9), and the division factor (5). Remember that, since the 5 is below the division (or fraction) bar, it is included in the equation by a division operation. We have to start with clearing the addition operation:

$$°F - 32 = \frac{9}{5} \cdot °C + 32 - 32$$

This converts the 32 on the right into zero, which we can ignore.

$$°F - 32 = \frac{9}{5} \cdot °C$$

Now, the °C is multiplied by 9 and divided by 5, so we need to divide by 9 and multiply by 5.

$$(°F - 32) \cdot \frac{5}{9} = \frac{9}{5} \cdot °C \cdot \frac{5}{9}$$

Nine divided by 9 is 1, as is 5 divided by 5. This clears the right side of the equation, and we finally have:

$$(°F - 32) \cdot \frac{5}{9} = °C$$

First, notice the °F – 32 term is contained in parentheses. When we multiplied and divided by 5 and 9, respectively, these operations applied to the *entire* left side of the equation, not just the °F term. Second, notice that you do not need to memorize two equations for converting between °F and °C. The first equation, $°F = 9/5 °C + 32$, converts °C into °F. If you need to convert °F into °C, you don't need to memorize $°C = 5/9(°F – 32)$. You just need to solve the first equation for °C, and then plug in your numbers and evaluate the expression. This strategy will serve you well throughout your studies of both physics and chemistry.

How can we solve this expression for *x*?

$$\frac{(x + 5)}{(x - 5)} = 5$$

Answer: 7.5

Two other functions you're likely to encounter are square roots and logarithms. You can review your college algebra book for the nuances of these functions. For our purposes, however, it is more important that you are able to use these functions to generate numerically correct answers. Your calculator should have an "x^2" button and a "$\sqrt{x}$" button. These operators are inverses of each other. That is, if you enter 10 on your calculator and then press x^2, you get 100. Now if you press $\sqrt{x}$, you get 10 back. Each operator undoes the effect of the other.

Let's try another example. To solve this equation for x, we clear the square root operator by squaring both sides, and then solve for x.

$$\sqrt{x + 3} = 10$$
$$(\sqrt{x + 3})^2 = 10^2$$
$$x + 3 = 100$$
$$x = 97$$

In extracting square roots, remember that you always obtain two answers: a positive root and a negative root. This is a mathematical reality, but in science typically only one of these answers has a physical reality. For example, starting from rest, the distance (x) an object travels, when undergoing constant acceleration (a) for a time interval (t), is given by $x = 1/2 \cdot at^2$. Let's solve this equation for t when $x = 6$ and $a = 3$:

$$6 = \frac{1}{2} \cdot 3t^2$$

Dividing both sides by 3 and multiplying both sides by 2 gives $t^2 = 4$. If we take the square root of both sides, t can equal either $+2$ or $–2$. However, the concept of negative time makes no physical sense, so we would report that answer *in a physics class* as $t = 2$ seconds.

Logarithm functions come up fairly frequently in chemistry and physics, so you need to be able to handle them. Logarithms are mixed up exponents. You know $10^2 = 100$. In this expression, 10 is the base, 2 is the power, and 100 is the result. In other words, you need 2 factors of 10 to get to 100, 3 to reach 1000, and so forth. A logarithm is another way to express this exponential expression. So, the log expression for $10^2 = 100$ is:

$$\log_{10}(100) = 2$$

Again, 10 is the base of the logarithm, and the 2 tells us that we need two factors of ten to reach 100. This becomes more challenging when we want the log of a number such as 110, which is not an integral power of 10. Fortunately, you do not need to calculate logs by hand. Your calculator should have a "log" button (for base 10, or common, logarithms) and a "10^x" button (for antilog). Unless you are told otherwise, the operator log means base-10 logarithms. The log and antilog operators are inverses of each other. So, if you take the log of a number, and then the antilog of the result, you get the original number. For example, if we take the log of 110, we get 2.041.

$$\log(110) = 2.041$$

Then, if we take the antilog of 2.041, we get 110. Notice the antilog function can be represented as either "antilog" or 10^x:

$$\text{antilog}(2.041) = 10^{2.041} = 110$$

Depending on your calculator, you might have to enter the number first, and then press the operator key, or first press the operator key and then enter the number. You need to explore the operation of your calculator.

One reason logs are so handy is the way they reduce multiplication and division to the simpler operations of addition and subtraction. The log of a product of two numbers is equal to the sum of the logs of those numbers:

$$\log(a \cdot b) = \log(a) + \log(b)$$

And the log of a ratio of two numbers is equal to the difference of the logs of the numbers:

$$\log\left(\frac{a}{b}\right) = \log(a) - \log(b)$$

Perhaps most usefully, the log of a power is equal to the exponent in the power times the log of the base of the power:

$$\log(a^b) = b \cdot \log(a)$$

Use your calculator to verify for yourself that the following statements are true:

$$\log(10 \cdot 5) = \log(10) + \log(5) = 1.699$$
$$\log\left(\frac{10}{5}\right) = \log(10) - \log(5) = 0.3010$$
$$\log(10^5) = 5 \cdot \log(10) = 5$$

There are other bases for logarithms. Your calculator should also be able to handle natural logarithms, which are based on the irrational number "e" rather than 10. The natural log key is "ln" and the natural antilog key is "e^x".

Exponents

Exponents are shorthand for the number of times a quantity is multiplied. We will most commonly encounter exponents when using scientific notation, but units in physics are sometimes exponential, and this gives them a unique meaning. Imagine we measure the edge of a cube and find each edge to be 1 centimeter (cm) in length. What is the volume of the cube? The volume of a cube is given by length × width × height, so we start with:

$$\text{Volume} = 1 \text{ cm} \times 1 \text{ cm} \times 1 \text{ cm}$$

Let's regroup the like terms (multiplication is associative, so it is okay to group the terms together in any order we like).

$$\text{Volume} = 1 \times 1 \times 1 \times \text{cm} \times \text{cm} \times \text{cm}$$

Well, $1 \times 1 \times 1 = 1$, but what about the cm? Since there are three cm terms, the exponential expression of cm $\times$ cm $\times$ cm is cm³. Therefore, the volume of our cube is 1 cm³. Notice that cm connotes "length" but cm³ connotes volume, a different concept than length.

There are some special cases of exponents. Any quantity (except zero) raised to the zero power equals one.

$$5^0 = 1$$
$$cm^0 = 1$$

Any quantity (except zero) to the "–1" power is equal to the reciprocal of the quantity. Moving a quantity across a division bar changes the sign of the exponent. This is true for numbers as well as variables or units.

$$2^{-1} = \frac{1}{2^1} = \frac{1}{2}$$
$$\frac{g}{cm^3} = g \cdot cm^{-3}$$

When powers are multiplied, the exponents are added.

$$10^3 \cdot 10^{-4} = 10^{-1}$$

When powers are divided, the exponents are subtracted.

$$\frac{cm^3}{cm^1} = cm^2$$

Let's evaluate this expression:

$$\frac{10^{-3} \cdot 10^6}{10^{-8}}$$

Answer: 10^{11}

Scientific Notation

Scientific notation uses exponents (powers of 10) for handling very large or very small numbers. A number in scientific notation consists of a number multiplied by a power of 10. The number is called the *mantissa*. In scientific notation, only one digit in the mantissa is to the left of the decimal place. The "order of magnitude" is expressed as a power of 10, and indicates how many places you had to move the decimal point so that only one digit remains to the left of the decimal point.

$$11,000,000 = 1.1 \times 10^7$$
$$0.0000000045 = 4.5 \times 10^{-9}$$

If you move the decimal point to the left, the exponent is positive. If you move the decimal to the right, the exponent is negative. Memorized rules like this are easy to confuse. You might find it more logical to recognize that large numbers (greater than 1) have positive exponents when expressed in scientific notation. Small numbers (less than 1) have negative exponents.

As an example, let's express 1.02×10^{-3} and 4.26×10^4 in decimal form.

Answer: 0.00102 and 42,600

Numbers in proper scientific notation have only one digit in front of the decimal place. Can you convert 12.5×10^{-3} into proper form? You have to move the decimal point one place to the left, but does the exponent increase or decrease? Is the correct answer 1.25×10^{-4} or 1.25×10^{-2}? Rather than memorize a rule, think of it this way. You want your answer to be equal to the initial number. So, if you DECREASE 12.5 to 1.25, wouldn't you need to INCREASE the exponent to cancel out this change? So, the correct answer is 1.25×10^{-2}.

As an other example, let's express 0.0034×10^{-25} in proper scientific notation.

Answer: 3.4×10^{-28}

You will probably use your calculator for most calculations. It is critical that you learn to use the scientific notation feature of your calculator properly. Calculators vary widely, but virtually all "scientific" calculators have either an "EE" key or an "EXP" key that is used for scientific notation. Unfortunately, these calculators also have a "10^x" key, which is the "antilog" key and has nothing to do with scientific notation so don't use the 10^x key when entering numbers in scientific notation. One common mistake is to include the keystrokes " $\times$ 10" when entering numbers in scientific notation. DO NOT TYPE " $\times$ 10"! The calculator knows this is part of the expression, and manually entering " $\times$ 10" will cause your answer to be off by a factor of 10. You should refer to the owner's manual of your calculator for exact instructions. However, most calculators follow this protocol:

1. Enter the "mantissa"
2. Press the "EXP" key
3. Enter the exponent, changing the sign if necessary

Let's evaluate this expression:

$$\frac{(4.8 \times 10^{-6})}{(1.2 \times 10^3)(2.0 \times 10^{-8})}$$

Answer: 2.0×10^{-1}

Remember to follow the rules for symbols of enclosure. The division bar is a symbol of enclosure, so do the operations in the numerator and in the denominator before performing the division. By the way, did you notice you can easily solve this problem without a calculator? Try regrouping the mantissas separately from the powers of 10.

$$\frac{4.8}{(1.2 \cdot 2)} = 2$$

and

$$\frac{10^{-6}}{(10^3 \cdot 10^{-8})} = \frac{10^{-6}}{10^{3+(-8)}} = \frac{10^{-6}}{10^{-5}} = 10^{-1}$$

Therefore, the answer is 2.0×10^{-1}.

When representing a negative number using scientific notation, the negative sign goes in front of the mantissa. This negative sign has nothing to do with the sign in front of the exponent. For example, -0.0054 would be represented in scientific notation as -5.4×10^{-3}.

Graphing

It is frequently very helpful to plot data. The data is often easier to analyze when it is presented so as to obtain a linear relationship. One of the most useful forms of graphing data is to use the "slope-intercept" equation for a straight line:

$$y = mx + b$$

The x variable appears on the horizontal axis and represents how you are manipulating the system. For this reason, it is called the *independent variable*, because it is not free to change as the system is manipulated. The y variable appears on the vertical axis and represents the measured response of a system. For this reason, it is called the *dependent variable*, because its value depends on the way in which you change the system. In an anesthesia setting, you would control the amount of anesthetic administered to a patient. This would be the independent variable. The amount of time a patient remains under anesthesia depends on the dose, which is, therefore, the dependent variable. If you were conducting a study and wanted to graph the relationship between the dose and time of anesthesia, the dose should be graphed on the horizontal axis and the time on the vertical axis. It may not always be clear in every case which variable is independent and which is dependent.

The slope of the line (m) relates how strongly, and in which sense, the dependent variable changes as you change the independent variable. If the line rises from left to right, the slope is positive. If the line falls from left to right, the slope is negative. The slope can be calculated for a line by the equation:

$$m = \frac{y_2 - y_1}{x_2 - x_1}$$

The intercept (b) mathematically describes the value of the function where it crosses the y-axis. In a physical sense, however, the intercept represents a correcting or offset value that relates the dependent variable to the independent variable. That is, when your controlled (independent) variable is zero, the experimental (dependent) variable may have an inherently nonzero value.

Consider the simple example of relating temperatures on the Fahrenheit scale to the Celsius scale. Imagine a container of water at its freezing point. A Celsius thermometer registers 0°C, and a Fahrenheit thermometer registers 32°F. Similarly, when the water is boiling, the two thermometers register 100°C and 212°F, respectively. If you graph the two data pairs, and we consider the Fahrenheit temperature to be the dependent variable (although this assignment is somewhat arbitrary), our graph would look like Figure 1.1.

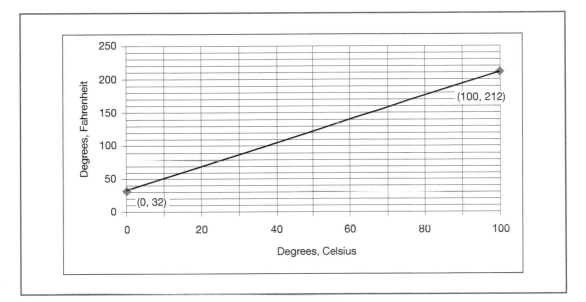

Figure 1.1 Fahrenheit versus Celsius Temperatures

The line has a value of 32°F where it crosses the y-axis, so $b = 32$. The slope of the line is:

$$m = \frac{y_2 - y_1}{x_2 - x_1} = \frac{212 - 32}{100 - 0} = \frac{180}{100} = 1.8$$

So the equation of our line is:

$$y = 1.8x + 32$$

But the y variable is °F and the x variable is °C, so we can also express this equation as:

$$°F = 1.8°C + 32$$

Of course, this is the formula most of us memorized for converting between Fahrenheit and Celsius temperatures (notice that 1.8 is just the decimal equivalent of 9/5). One important lesson from this example is that formulas don't just "fall from the sky." They are experimentally (empirically) or theoretically derived. Second, it is a poor strategy to get through chemistry and physics by trying to just memorize formulas. You will have a much better understanding of the concepts if you focus on how the formulas are derived and the underlying reasons for the relationship between various quantities.

MEASUREMENTS AND SIGNIFICANT FIGURES

Science is based on observation and measurement. All measurements have an inherent error or uncertainty. Therefore, our scientific conclusions can't be more reliable than the least reliable measurement. One tool to assess the reliability of a measurement is the number of digits that are reported. Significant figures are digits in a measured value that have a physical meaning and can be reproducibly determined. For example, what is the pressure reading in the pressure gauge in Figure 1.2?

Figure 1.2 Pressure Gauge/Significant Figures

Could you report the pressure as 3.8945 psi? Well, the needle is clearly between 3 and 4; there's no doubt about that. If we want another decimal place, we have to estimate. The next decimal place looks to be about 0.8, although some people might estimate it to be 0.7 or 0.9. This is where the uncertainty enters. So we report the pressure as 3.8 psi. Any more decimal places would be pure guesses with absolutely no reproducibility. Therefore, this measuring device is capable of delivering a maximum of two significant figures. As a matter of habit, you should always push measuring devices to their limits for delivering good data. That is, you should always estimate one decimal place past the smallest division on the scale.

The concept of significant figures applies only to measured quantities, which always have some inherent uncertainty. Numbers not obtained using measuring devices are called *exact numbers*. That is, they are easily countable, and there is absolutely no question as to the value. For example, most hands have four fingers and one thumb, and there is no question regarding the four or the one. Quantities that are defined are another example of exact numbers. For example, there are 12 inches in 1 foot and 2.54 centimeters in 1 inch. The 12 and 2.54 are definitions; therefore they are exact quantities. You can consider these kinds of measurements to have an infinite number of significant figures. Not all counted quantities are exact numbers. For example, counting very large numbers (e.g., the number of red blood cells under a microscope) will have some uncertainty, because the result is difficult to reproduce.

In order to deal with measurements reported to you by other investigators, you need to recognize how many significant figures a given measurement contains. Nonzero digits are always significant and never seem to confuse anyone. Zeros are more problematic. For example, if you measure your height as 1.7 m, does the reliability of this measurement change if you report it as 0.0017 km or 1700 mm? Of course not. These zeros are "placeholders" and separate the decimal place from the nonzero digits. Zeros are *not significant* in the following cases:

- A single zero to the left of the decimal point is a visual emphasis for the decimal point and is not significant, for example, 0.25.
- Zeros to the right of a decimal point and preceding nonzero numbers are not significant, for example, 0.001.
- Zeros to the left of a decimal point that follow nonzero numbers, for example, 100, are usually not significant, unless one of the zeros is identified by the reporter as significant. This is done by underlining the zero or placing a bar over that zero, for example, $1\bar{0}$.

Zeros are significant when:

- they occur between significant numbers, for example, 101 or $10\bar{0}$
- they occur to the right of the decimal point and also to the right of nonzero numbers, for example, 0.10 or 1.0

One operational way to identify significant zeros is to convert the number into scientific notation. If the decimal point does not cross a zero when converting a number into scientific notation, the zero is always significant. Some additional examples are given below (significant figures are in bold).

VALUE	NUMBER OF SIGNIFICANT FIGURES
2.0	2
20.00	4
0.00**2**	1
0.0**20**	2
200	1
2.30 × 10^3	3
2020	3
20000	1

SIGNIFICANT FIGURES IN CALCULATIONS

The result of a calculation cannot be more reliable than any of the measurements on which the calculation is based. The statistical way in which the inherent uncertainties in measurements are propagated through addition and subtraction operations is different from the way they are propagated in multiplication and division operations. A rigorous treatment of this difference is beyond the scope of our purposes, although certainly not beyond your capacity to understand it. You should examine a statistics text if you are interested in learning more. Fortunately, there are some simple rules for determining the number of significant figures to retain when performing calculations involving measured quantities.

- When adding or subtracting, keep the smaller number of decimal places.
- When multiplying or dividing, keep the smaller number of significant figures.
- Apply these rules according to the algebraic hierarchy of calculations (consider symbols of enclosure first, then multiplication/division operations, and lastly addition/subtraction operations).

For example, how should we report the answer in the problem below to the proper number of significant figures?

$$4.023 + 102.3 =$$

When adding, the answer should have the same number of decimal places as the addend with the smallest number of decimal places. The second number has only one decimal place. Therefore the answer can have only one decimal place and is correctly reported as 106.3.

Let's try a multiplication example. Evaluate the following expression and report the answer to the proper number of significant figures.

$$4.023 \times 102.3 = ?$$

Each number has four significant figures. Therefore the correct answer is 411.6 (four significant figures).

We'll crank it up a notch and try a mixed operation example next. Evaluate the following example and report the answer to the proper number of significant figures.

$$\frac{4.023 - 3.954}{5.564} =$$

A division bar is a symbol of enclosure: do the subtraction first, keeping three decimal places:

$$\frac{0.069}{5.564} = 0.012$$

The numerator has two significant figures, while the denominator has four. Therefore the answer should have only two significant figures.

ACCURACY AND PRECISION

When reporting a measurement, a reasonable question is "how good is the measurement?" There are two ways to answer this question: accuracy and precision. Accuracy is the agreement between experimental data and the "true" value, while precision is the agreement between replicate measurements. For example, imagine shooting two arrows at a target. If both arrows strike the bull's-eye, your data are both accurate and precise. If both arrows land close together, but miss the bull's-eye, the data will be precise but not accurate. If one arrow hits 5 in to the right of the bull's-eye, and the other arrow hits 5 in to the left, then the average position of the arrows does hit the bull's-eye. These data are accurate but not precise. Finally, if one arrow hits to the left of the bull's-eye and the other arrow misses the target completely, the data are neither precise nor accurate.

Which estimate of "goodness" is *more* important? Well, both of them are important. It is important to you and your patient that the reading on a pulse oximeter *accurately* represents the patient's blood P_{O2}. Likewise, you won't have confidence in a pulse oximeter if it gives a different reading every time you look at it (assuming the patient's condition has not changed).

The accuracy of a measurement can be improved by making replicate measurements and taking the average. All measurements have some inherent uncertainty. But if the uncertainties are random, then half of the measurements should be too big and half too small. The errors will cancel each other out. Accuracy is assessed by calculating the percent error, which is given by the formula:

$$\% error = \frac{\text{measured value} - \text{"true" value}}{\text{"true" value}} \times 100\%$$

Notice the quotation marks around "true" value. A method is verified by testing it against known standards, and in this case, you do know the "true" value, because you prepared the system. However, in most cases, you will never know what the true value is. The patient certainly has some value for serum cholesterol level, but you don't know what that is.

The precision of a measurement is improved by careful lab technique and/or using instruments capable of yielding greater precision (signaled by showing more significant figures). Repeating a measurement doesn't necessarily improve precision (except in that your technique might improve), but it does allow you to determine whether or not the measurement is, in fact, reproducible. In any case, a greater number of significant figures implies greater precision. Precision can be quantified by standard deviation. Unfortunately, we do not have time for a rigorous exploration of statistics, but you probably had a statistics course sometime during your undergraduate nursing program. So the equation for calculating averages and standard deviations should be familiar.

$$\text{average} = \frac{\Sigma \text{ measurements}}{\text{number of measurements}}$$

$$\text{standard deviation} = \sqrt{\frac{\Sigma(\text{measurement} - \text{average of measurements})^2}{\text{number of measurements} - 1}}$$

The smaller the ratio of the standard deviation to the average value, the better the precision. However, there is no absolute standard that defines good precision or bad precision. In general, you should become increasingly suspicious of measurements as the ratio of standard deviation to the average value increases. For an experienced analytical chemist, working with a well-known system, this ratio will typically be less than 0.1%, but this level of precision is unlikely in a clinical setting.

For example, imagine measuring the (arterial) P_{O2} of a patient three times and obtaining these values: 105, 96.0, and 102 mmHg (in that order). What is the average value? Does the data seem precise? What is the standard deviation of the measurements? What is the percent error?

First, notice that the values are not "trending." That is, they seem to bounce around the average value, and this should be reassuring. If each successive measurement is smaller or larger than the previous measurement, you should be concerned that there is a problem with the measurement system, instrument, or method. The average value is 101 mmHg and the standard deviation is 4.6 mmHg. Since the standard deviation is much smaller than the average value (4.6/101 = 0.045 or 4.5%) these data are fairly precise. You can't determine the percent error (and thereby assess the accuracy) because you don't know the "true" value. If these measurements were obtained from an artificial system in which you *knew* the P_{O_2} was equal to $10\overline{0}$ mmHg (three significant figures), the percent error is 1%.

SI (METRIC SYSTEM)

The metric system consists of a base unit and (sometimes) a prefix multiplier. Most scientists and health care providers use the metric system, and you are probably familiar with the common base units and prefix multipliers. The base units describe the type of quantity measured: length, mass, or time. The SI system is sometimes called the MKS (*meter, kilogram, second*) system, because these are the standard units of length, mass, and time upon which derived quantities, such as energy, pressure, and force, are based. An older system is called the CGS (*centimeter, gram, second*) system. The derived CGS units are becoming extinct. Therefore, we will focus on the MKS units.

The prefix multipliers increase or decrease the size of the base unit, so that it more conveniently describes the system being measured. The base units that we will be using are listed in Table 1.1. Chemists frequently express mass in units of grams, which is derived from the kilogram. There are three other SI base units (the mole, the candela, and the ampere). We will consider moles (amount of material) and amperes (electric current) in subsequent chapters. The candela is a unit of light intensity or luminosity and does not concern us at this point.

Table 1.1 SI BASE UNITS

BASE UNIT	QUANTITY MEASURED	ABBREVIATION
meter	length	m
kilogram	mass	kg
kelvin	temperature	K
mole	amount of material	mol

The prefix multipliers are given in Table 1.2. It is crucial that you learn all the multipliers, as well as their numerical meanings and abbreviations. Notice some abbreviations are capital letters, while others are lowercase letters. It is important that you not confuse these. The letter m

Table 1.2 SI PREFIX MULTIPLIERS

PREFIX MULTIPLIER	NUMERICAL MEANING	ABBREVIATION
giga	1×10^9	G
mega	1×10^6	M
kilo	1×10^3	k
deci	1×10^{-1}	d
centi	1×10^{-2}	c
milli	1×10^{-3}	m
micro	1×10^{-6}	μ
nano	1×10^{-9}	n

is particularly overused as an abbreviation. Notice that the abbreviation for micro is the Greek letter mu (μ), which is equivalent to the English letter m.

Table 1.3 gives some useful relationships between metric and English measurements for length, mass, and volume. There is a common error in this table. A pound is a unit of weight, not mass. So it is not correct to equate pounds (a weight unit) and kilograms (a mass unit). However, in a clinical setting, this distinction is almost never made. Weight is related to mass through Newton's Second Law, a concept we will explore in Chapter 3.

Table 1.3 COMMON CONVERSION FACTORS

LENGTH	MASS	VOLUME
1 in = 2.54 cm (exactly)	1 lb = 0.454 kg*	$1 L = 1 dm^3$
1 mile = 1.609 km	1 lb = 16 oz	1 gal = 3.79 L
1 mile = 5280 ft (exactly)	1 oz = 28.35 g**	$1 mL = 1 cm^3$
	1 grain = 64.80 mg	$1 cc = 1 cm^3$

*This is correctly stated as 1 lb = 4.45 N.
**This is commonly cited, even though it equates a mass quantity with a weight quantity.

ABSOLUTE ZERO AND THE KELVIN SCALE

Much of physics and physical chemistry is based on the Kelvin temperature scale, which begins at absolute zero. Absolute zero is the coldest possible temperature. Therefore, there is no temperature lower than 0K. Notice that it is not zero "degrees kelvin." It is simply zero kelvins. On the Celsius temperature scale, 0K is equivalent to −273.15°C. Since the temperature difference represented by 1K is the same as the temperature difference represented by 1 degree Celsius, converting between the two scales is easy:

$$K = °C + 273.15$$

As an example, what is body temperature (98.6°F) on the Kelvin scale?

We don't have an equation to convert between K and °F. It would be a waste of time for you memorize one, because we can simply use the two relationships we already know. First, let's convert 98.6°F into degrees Celsius:

$$°F = 1.8°C + 32$$
$$°C = (98.6°F - 32)/1.8 = 37.0°C$$

Then we use the second relationship to convert °C into kelvins:

$$K = °C + 273.15$$
$$K = 37.0°C + 273.15 = 310.2K$$

Now, you should be able to confirm that absolute zero is equivalent to –459.67°F

CONVERSION FACTORS

The factor-label method, or using conversion factors, is arguably the single most important skill for physical science students to master. Most problems in chemistry and physics can be solved by allowing the units associated with the measurements to walk you through the calculations. If you manipulate the units so that the undesired units drop out of the calculation, your numerical answer is almost sure to be correct. The basic strategy is to identify a beginning quantity and desired final quantity. You multiply a conversion factor times the beginning quantity so that the units of the beginning quantity are mathematically replaced by the units of the desired quantity. Notice $unit_1$ divided by $unit_1$ equals 1 and we can ignore it.

$$\text{beginning quantity } unit_1 \cdot \left(\text{conversion factor } \frac{unit_2}{unit_1}\right) = \text{desired quantity } unit_2$$

A conversion factor is a statement of equality between two measurements of the same object or property. As a beginning example, let's explore converting between metric quantities. For example, a length of 1 m obviously has a length of 1 m:

$$1 \text{ m} = 1 \text{ m}$$

Of course, we are free to mathematically manipulate this equation to suit our needs. So it is perfectly legitimate to divide both sides of the equation by 100, which gives:

$$1 \times 10^{-2} \text{ m} = 1 \times 10^{-2} \text{ m}$$

Recall the numerical meaning of "centi" is 1×10^{-2}, so we can substitute the prefix multiplier "c" for its numerical equivalent. We don't have to do this substitution to both sides because in trading "c" for its numerical equivalent, we haven't changed the value on the right side.

$$1 \times 10^{-2} \text{ m} = 1 \text{ cm}$$

Notice that our conversion factor between metric quantities has an interesting form. One side gets the prefix multiplier letter (in this case, the "c"), and the other side gets the numerical equivalent of the multiplier (in this case, 1×10^{-2}). Students sometimes get the number on the wrong side of the conversion factor. If you stick with the definition of the prefix multipliers given in Table 1.2, your conversion factors will always have a prefix multiplier on one side and a number on the other side. Seems fairer that way, doesn't it?

Of course, we could divide both sides of our conversion factor equation by 1×10^{-2}, which would give:

$$1 \text{ m} = 100 \text{ cm}$$

You might have learned this form in a previous class. Of course, either form of our conversion factor can be used. However, if you have ever had trouble converting between metric units, you might want to stick with the first version. Learn the prefix multipliers and their numerical meanings. Then, when you write a conversion factor expression, make sure one side gets the letter and the other side gets the number.

Let's explore a couple of other manipulations of our conversion factor. If we divide both sides by 1 cm, our conversion factor would look like:

$$\frac{1 \times 10^{-2} \text{ m}}{1 \text{ cm}} = \frac{1 \text{ cm}}{1 \text{ cm}}$$

Notice that, in doing this, we have 1 cm divided by 1 cm. Any quantity (except zero) divided by itself equals 1.

$$\frac{1 \times 10^{-2} \text{ m}}{1 \text{ cm}} = \frac{1 \text{ cm}}{1 \text{ cm}} = 1$$

Notice further that the reciprocal of 1 is 1. Therefore our conversion factor is equal to its own reciprocal.

$$\frac{1 \times 10^{-2} \text{ m}}{1 \text{ cm}} = \frac{1 \text{ cm}}{1 \times 10^{-2} \text{ m}}$$

This is important because 1 is the multiplicative identity element. This means we are free to multiply any beginning quantity times our conversion factor, and we will always get an equivalent quantity.

Can you write a conversion factor expression for converting bytes into gigabytes?

Answer: $1 \text{ Gbyte} = 1 \times 10^{9} \text{ bytes}$

$$\text{or} \quad \frac{1 \text{ Gbyte}}{1 \times 10^{9} \text{ bytes}}$$

$$\text{or} \quad \frac{1 \times 10^{9} \text{ bytes}}{1 \text{ Gbyte}}$$

The trick, of course, is knowing which form of the conversion factor we need to use. Fortunately, we can always let the units tell us how to solve the problem. Now, let's try a concrete example: How many inches are equal to 25.0 cm? First, we need to recognize the beginning quantity is 25.0 cm. The beginning quantity is the one about which you have all the information (i.e., both the number and the unit). Next we need a conversion factor that relates centimeters and inches. Conversion factors always involve (at least) two units. From Table 1.3 we see that 1 in = 2.54 cm. We want the cm unit in the conversion factor in the denominator, so that it will divide out the cm unit in the beginning quantity. This will leave units of inches in the numerator, which are the units of the desired quantity. Now, if we do the arithmetic and divide 25.0 by 2.54, we obtain 9.84 as the numerical portion of the answer. Notice we have kept three significant figures, because 25.0 cm has three significant figures. More importantly, the units that remain are inches, which are what we wanted to end up with.

$$25.0 \; \cancel{cm} \left(\frac{1 \; in}{2.54 \; \cancel{cm}} \right) = 9.84 \; in$$

When using conversion factors, there are two key points to remember. First, this method works only if you are meticulous in including units with all numerical data. Second, if your final answer has the correct units, the answer is *probably* correct. If your final answer has the wrong units, the answer is *almost certainly* incorrect.

Now let's try an example that requires more than one conversion factor. A red blood cell is about 6.0 μm in diameter. How many nanometers is this? In this case, we don't have a conversion factor that directly relates nanometers to micrometers. Fortunately, we can relate micrometers to meters, and we can relate meters to nanometers:

$$1 \; \mu m = 1 \times 10^{-6} \; m$$
$$1 \; nm = 1 \times 10^{-9} \; m$$

If you are very proficient in using conversion factors, you can "stack them up" and do all the conversions in one step. However, we will generally proceed stepwise, so that we are less likely to make an error. First, we need to convert micrometers into meters:

$$6.0 \; \cancel{\mu m} \left(\frac{1 \times 10^{-6} \; m}{1 \; \cancel{\mu m}} \right) = 6.0 \times 10^{-6} \; m$$

Our second conversion factor changes meters into nanometers:

$$6.0 \times 10^{-6} \; \cancel{m} \left(\frac{1 \; nm}{1 \times 10^{-9} \; \cancel{m}} \right) = 6.0 \times 10^{3} \; nm$$

If a patient receives 8.0 L of oxygen per minute, how many milliliters per second is this? Notice we need to divide the units of L from the numerator and units of minutes from the denominator. You know that 60 s = 1 min and 1 mL = 1 × 10⁻³ L. Be careful to apply the conversion factor so that the unit we need to eliminate is on the opposite side of the division bar from the same unit in the conversion factor. Let's change minutes into seconds first, and then change liters into milliliters. Of course, you can apply the conversion factors in any order you like.

$$\frac{8.0 \text{ L}}{\text{min}}\left(\frac{1 \text{ min}}{60 \text{ s}}\right) = \frac{0.13 \text{ L}}{\text{s}}$$

$$\frac{0.13 \text{ L}}{\text{s}}\left(\frac{1 \text{ mL}}{1 \times 10^{-3} \text{ L}}\right) = \frac{130 \text{ mL}}{\text{s}}$$

DENSITY

Density is the ratio of the mass of an object to its volume. Chemists usually abbreviate density with a "d" while physicists typically use the Greek letter ρ (rho). Since the course is chemistry and physics for nurse anesthesia, you will need to recognize both abbreviations.

$$\rho = d = \frac{\text{mass}}{\text{volume}}$$

Notice that density will always have two units: a mass unit divided by a volume unit. For example, water has a density of 1.0 g/mL. Density can have a different combination of units, though. For example, water also has a density of 1000 kg/m^3 and 8.33 pounds/gallon. By the way, the density of water as 1.0 g/mL is a useful quantity to remember. Since the mass of 1 mL of water has a mass of 1 g, we can interchange grams and milliliters of water (as long as the temperature is close to 25°C).

If 28.0 g of nitrogen gas occupies 22.4 L at 0°C and 1 atm of pressure, what is the density of N_2 in grams per milliliter? By now, you should easily be able to convert 22.4 L into 22,400 mL, so we can plug the numerical values into the definition of density and find the density of nitrogen under these conditions is 0.00125 g/mL.

$$d = \frac{28.0 \text{ g}}{22,400 \text{ mL}} = 1.25 \times 10^{-3} \frac{\text{g}}{\text{mL}}$$

Since density always has two units and conversion factors have two units, guess what? Density is a conversion factor between mass and volume. Don't memorize when you need to multiply by density or divide by density to achieve a required conversion. Let the units tell you what to do.

What is the edge length of a wooden cube if the mass of the cube is 145 g? The density of this sample of wood is 0.79 g/cm^3. Using density, we can convert the mass of the wooden cube into its volume.

$$145 \text{ g}\left(\frac{1 \text{ cm}^3}{0.79 \text{ g}}\right) = 184 \text{ cm}^3$$

Hold on a minute. The density is reported to only two significant figures, so we are entitled to retain only two significant figures in our final answer, but you see this intermediate answer with three significant figures. If you round off to the correct number of significant figures after each calculation, your final answer can be significantly different from the correct answer. This is called a rounding error. It is a good idea to carry a "guard digit" through a multistep calculation, and only round the *final* answer to the correct number of significant figures. You can also prevent rounding errors by not clearing your calculator after each step and re-entering the number. Just leave each intermediate answer in the calculator's memory and proceed with the next computation.

To complete this problem, you need to know a little geometry. The volume of a cube is equal to the length of its edge cubed. So, if we take the cube root of the volume, we will obtain the length of the edge.

$$\text{volume of a cube} = \text{edge length}^3$$
$$\text{edge length} = \sqrt[3]{184 \text{ cm}^3} = 5.7 \text{ cm}$$

Now, it's time to round the final answer to two significant figures. Notice the units also work out in this calculation. The cube root of cm^3 is cm.

DENSITY AND SPECIFIC GRAVITY

Video 1.1

Specific gravity is the ratio between an object's density and the density of water. Specific gravity, is, therefore, dimensionless and does not depend on the units. Specific gravity will always have the same numerical value, regardless of the units of density you choose to employ in determining the specific gravity. Any sample with a specific gravity greater than 1 is denser than water. Any sample that has a specific gravity less than 1 is less dense than water.

There is often some confusion between specific gravity and density, because the density of water is 1.00 g/mL. This means the density expressed as grams per milliliter and specific gravity of a sample are *numerically* equal. For example, let's calculate the specific gravity of mercury, which has a density of 13.6 g/mL.

$$sg = \frac{13.6 \frac{g}{mL}}{1.00 \frac{g}{mL}} = 13.6$$

It appears that density and specific gravity are the same, but they aren't. They are numerically equal when the units of density are expressed as grams per milliliter (g/mL). Specific gravity equals density only when the units of measure are grams per milliliter, because the density of water is 1 g/mL. To illustrate this difference, let's start with different density units. One gallon of mercury has a mass of 113 lb (i.e., the density of mercury is 113 lb/gal). The density of water is 8.33 lb/gal. If we calculate the specific gravity using these values, we find the same value for the specific gravity.

$$sg = \frac{113 \frac{lb}{gal}}{8.33 \frac{lb}{gal}} = 13.6$$

Summary

■ Laboratory and clinical measurements all require measurements, and all measurements have an inherent uncertainty.

■ Scientific notation is a tool to deal with very large or very small numbers. A number expressed in scientific notation consists of a number having only one digit to the left of the decimal point, multiplied by some power of 10.

■ Significant digits represent measurements that can be replicated or determined with a reasonable level of certainty. A digit is not significant if it represents a random guess in a measurement. By convention, all of the nonzero digits in a number are significant. Zeros are significant when they are located anywhere between two significant digits. Zeros are significant when they are separated from a significant digit by the decimal place (except for the optional leading zero in the one's place in a number that is less than one.) Otherwise, zeros are not significant. The zeros in boldface below are all significant.

$$1.0 \qquad 0.010 \qquad 100.0$$

These zeros in boldface are not significant:

$$0.01 \qquad 10$$

■ When adding or subtracting measurements, the final answer should have the same number of decimal places as the smallest number of decimal places in the measurements. When multiplying or dividing measurements, the final answer should have the same number of significant digits as the measurement with the smallest number of significant digits.

■ The conversion factor method is a tool to assist you in converting one unit of measurement into another. A conversion factor is an equality statement that relates the same quantity in two different units (e.g., 1 in = 2.54 cm). Conversion factors are applied as a ratio so that the old units cancel out, and only the desired units remain.

■ Accuracy represents agreement of a measurement with the true or accepted value, although the true or accepted value is typically not known in a clinical setting. Accuracy can be evaluated through a percent error calculation:

$$\%\,error = \frac{measured\ value - true\ value}{true\ value}\,100\%$$

■ Precision represents agreement among repeated measurements. A larger number of significant digits in a measurement imply greater precision. Precision can be evaluated with the standard deviation of the replicate measurements.

■ The SI, or metric, system of measurements consists of base units, plus prefix multipliers. The SI unit of mass is the kilogram, the SI unit of length is the meter, and the SI unit of time is the second. Clinical data contains several derived units of measurement, such as volume, with is the cube of length. One liter is defined as $0.001 m^3$.

The most commonly encountered prefix multipliers include kilo (10^3), milli (10^{-3}), and centi (10^{-2}), however, you should be familiar with all of the multipliers listed in Table 1.2.

■ The SI unit of temperature is the kelvin. The Kelvin temperature scale begins at absolute zero, the coldest possible temperature. The change in temperature represented by one kelvin is equivalent to a change in temperature of 1°C. The relationships between the three most common temperature scales (Kelvin, Celsius, and Farenheit) are:

$$°F = 1.8°C + 32$$
$$K = °C + 273.15$$

■ Density is mass divided by volume. The SI unit of density is kg/m^3, although clinical densities are more likely to be reported as g/cm^3. Density, like all measurements that have units in terms of the ratio of two other units, can be used as a conversion factor between mass and volume.

$$\text{density} = \frac{\text{mass}}{\text{volume}}$$

■ Specific gravity is defined as the density of an object divided by the density of water. The density of both substances must be expressed in the same units. Since the density of water is approximately 1.0 g/cm^3, the specific gravity of an object is approximately equal to the density of the object, expressed as g/cm^3. Specific gravity has no units and will have the same numerical value, regardless of the units of density used to determine it. Objects with a specific gravity less than one will float on water; objects with a specific gravity greater than one will sink.

$$\text{specific gravity} = \frac{\text{density of object}}{\text{density of water}}$$

Review Questions for Measurement

1. Evaluate the following exponential expressions. You should not need to use your calculator.
   a. $10^3 \cdot 10^{-4}/10^{-8}$
   b. $(10^3)^{-1}$
   c. $(10^{450})/10^{-253}$
   d. $(cm^2)(cm)$

2. Put the following numbers into scientific notation.
   a. 23.4
   b. 26.4×10^{-4}

3. Use your calculator to simplify these expressions.
   a. $8.46 \times 10^5/3.25 \times 10^{-6}$
   b. $8.02 \times 10^{-2}/6.02 \times 10^{23}$

4. What is a significant figure?

5. What is an exact number? How many significant figures are in an exact number?

6. Differentiate between accuracy and precision and describe how to assess and improve each.

7. If lug nuts on your car should be tightened to a torque of $15\overline{0}$ ft-lb, what is the corresponding metric torque in units of N-m? (1 lb = 4.45 N and 1 in = 2.54 cm)

8. What is the volume of 1.0 cm^3 in units of cubic inches?

9. How is density different from specific gravity?

10. The density of water is 62 lb/ft³. The specific gravity of gold is 19.3. What is the density of gold in lb/ft³?

11. The density of gold is 19.3 g/cm³. What is the mass in kilograms of a gold brick measuring 5.00 in by 3.00 in by 12.0 in?

12. If $PV = nRT$, what does $R = $?

13. The Clausius–Clapeyron equation states:

$$\ln\left(\frac{P_1}{P_2}\right) = \frac{\Delta H}{R}\left(\frac{1}{T_2} - \frac{1}{T_1}\right)$$

   What does T_1 equal?

14. Evaluate the value of k, which, though of no particular importance to you at this time, is the famous Boltzmann constant.

$$k = \frac{(7.60 \times 10^2) \cdot (2.24 \times 10^4)}{(273.15) \cdot (6.02 \times 10^{23})}$$

15. Complete the following table of the metric prefix multipliers.

MULTIPLIER	ABBREVIATION	MATHEMATICAL MEANING
micro		
	m	
		10^{-2}
kilo		
	n	
		10^{6}

16. This graph shows how the volume of a sample of gas responds to changing temperature. Estimate the volume of the gas at a temperature of 0°C.

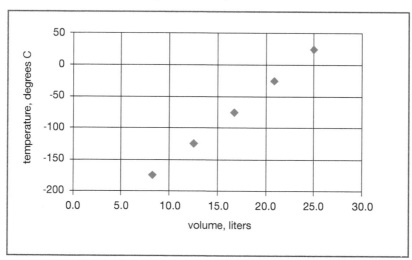

17. Consider the syringe pictured at the right. Each mark on the syringe represents 0.1 mL. What is the correct reading of the material in the syringe and the maximum number of significant figures that can be reported?
    a. One significant figure, and the reading is 0.7
    b. Two significant figures, and the reading is 0.7
    c. Two significant figures, and the reading is 0.68
    d. Three significant figures, and the reading is 0.687

18. Which of these statements is correct?
    a. The precision of a data set is improved by averaging multiple measurements.
    b. Standard deviation is a way of quantifying the precision of data.
    c. Accuracy refers to the reproducibility of a measurement.
    d. All of the above statements are correct.

19. Dr. Kyotay has ordered a new operating table from the ACME surgical supply. The table will support a total of 50 kg. What is the corresponding weight of the largest patient you can put on this table in pounds?
    a. 0.11 lb
    b. 110 lb
    c. 23 lb
    d. 230 lb

20. Which of these quantities represents the greatest mass?
    a. 10^4 ng
    b. 10^{-1} mg
    c. 10^{-8} kg
    d. All of these represent equivalent masses.

21. A patient has a temperature of $31\overline{0}$ K. Should you be concerned? Or, rather, what is the patient's temperature on the Fahrenheit scale?
    a. No, it's 98.6 degrees
    b. Not terribly, it's 97.0 degrees
    c. Not really, it's 101 degrees

22. A syringe has a stated capacity of 5 cc (or 5 mL). What is the volume of the syringe in microliters?
    a. 5000 μL
    b. 0.005 μL
    c. 0.05 μL
    d. 5×10^6 μL

23. Ethyl alcohol has a density of 0.79 g/mL. What volume of ethyl alcohol is needed to treat methanol poisoning, if the physician orders $2\overline{0}0$ g of ethanol?
    a. 160 mL
    b. 250 mL
    c. $2\overline{0}0$ mL
    d. There is insufficient information to solve this question.

24. Water has a density of 8.3 lb/gal. What is the density of water in units of stones per bushel? A stone is 14 lb and there are 8 gal per bushel.
    a. 8.3 stones/bushel
    b. 15 stones/bushel
    c. 4.7 stones/bushel
    d. 0.074 stones/bushel

25. The reason specific gravity is often cited rather than density is that:
    a. Specific gravity has units that are consistent with the SI (metric) system.
    b. The numerical value of density depends on the mass and volume units chosen, whereas specific gravity is unit-independent.
    c. The specific gravity of bodily fluids is always close to 1.
    d. If the specific gravity of a substance is greater than 1, you know it will float on water.

26. Charles's Law relates the volume of a gas to its temperature. The table below gives the temperature of a gas as a function of volume (when the pressure is 1 atm). Treat the volume as the dependent variable. Graph the data and determine the equation of the line that relates these variables.

VOLUME, L	TEMPERATURE, °C
20.0	−29
30.0	93
40.0	215
50.0	336

27. Using conversion factors found in this chapter, convert each of these measurements into the indicated units.
    a. 45 cm into meters
    b. 36 m into kilometers
    c. 0.45 L into milliliters
    d. 0.97 ms into seconds
    e. 17 µg into grams

28. Using conversion factors found in this chapter, convert each of these measurements into the indicated units.
    a. 26 Mg into mg
    b. 0.26 nm into mm
    c. 38 dL into cL
    d. 72 µm into nm
    e. 55 Mb into Gb

29. Using conversion factors found in this chapter, convert each of these measurements into the indicated units.
    a. 12 in into centimeters
    b. 38 gal into liters
    c. 55 lb into kilograms
    d. 85 km into miles
    e. 59 mg into grains

30. Using conversion factors found in this chapter, convert each of these measurements into the indicated units.
    a. 12.0 cu in into cubic centimeters
    b. 67 km^2 into square miles
    c. 9.8 m/s^2 into $feet/min^2$
    d. 67 $watts/mi^2$ into watts/square kilometers
    e. 86 lb/ft^3 into g/cm^3

31. Use the conversion factors provided to perform each of these transformations.
    a. A urine sample has a concentration of 1.9 mEq sodium/mL. How many mEq of sodium are in 55 mL of urine?
    b. A car is traveling at 75 mi/hr. How far will the car go in 89 hours?
    c. A motorcycle has a mileage of 89 mi/gal. How many gallons of gasoline are required for a 45-mile trip?
    d. An IV has a drip rate of 52 mL/min. How long will it take to deliver 12 L of fluid?
    e. If the Kessel run represents a distance of 18 parsecs, and the Millennium Falcon can travel at a maximum velocity of 2.6×10^8 m/s (just less than the speed of light), how long will this journey take? 1 parsec = 3.26 light-years and 1 light-year = 9.5×10^{15} m.

32. The temperature of a sample is measured using both a Celsius thermometer and a Fahrenheit thermometer, and both thermometers register the same value. What is the temperature of the sample? Start with °F = 1.8°C + 32 and use the requirement in this example that °F = °C.

33. Can an object have a temperature measured on the Kelvin scale that is equal to the same temperature on the Fahrenheit scale? Explain.

34. A cube has an edge length of 1.5 cm and a mass of 15 g. What is the density of the cube in units of g/cm³?

35. A cylinder has radius (R) of 5.0 in and a height (H) of 12 in. If the cylinder has a mass of 1.5 lb, what is the density of the cylinder in lb/ft³? The volume of a cylinder is given by $V_{cylinder} = \pi R^2 H$.

36. A sphere has a diameter of 34 m and a mass of 75 kg. What is the density of the sphere in units of kg/m³? The volume of a sphere is given by $V_{sphere} = 4/3 \pi R^3$, and the radius of a sphere is one-half of the diameter.

A Review of Some Chemistry Basics

WHAT IS CHEMISTRY?

Chemistry is the study of matter and the changes it undergoes. Every aspect of your patients' physiology is a proper study of chemistry. Even such abstract processes as thought and emotion are believed to be based on biochemical processes in the brain. Therefore, it is appropriate that we review some essential concepts upon which chemistry is founded.

Chemistry is a vast and complex field, so don't fall into the common trap of believing you know nothing about chemistry because you don't know everything about chemistry. Generally, even PhD-level chemists are considered to be experts in only one or two of the five classic areas of chemistry. Analytical chemists study the composition of samples, both in terms of what is present in the sample, and the percent composition of each component. Physical chemists strive to discern models needed to understand chemical systems from a theoretical framework. Inorganic chemists study substances that are derived from all elements except carbon. Organic chemists study compounds based on carbon, while biochemists study the chemistry that occurs in living systems. In its common use, the term *organic* connotes derivation from natural (and presumably) healthy sources, but this is not how chemists understand the term. Virtually all of the substances studied by biochemists are organic compounds, while a huge number of organic compounds are synthetic and not found in living systems. There are also several cross-disciplinary areas of chemistry, including nuclear chemistry, polymer chemistry, nanochemistry, and material sciences. You no doubt took one (or more) chemistry courses as a part of your undergraduate studies in nursing. In this chapter, we will review some of the basics of chemistry that you might have gotten rusty on. Your previous experience with chemistry notwithstanding, chemistry is not just an applied algebra course. Many of the essential chemistry concepts are not quintessentially mathematical in nature. In fact, it is sometimes more useful to gain an understanding of these ideas at a qualitative level. Therefore, we will focus on conceptual and holistic topics. The major goal of this chapter is for you to correctly and precisely use the language of chemistry.

Matter

Chemistry studies matter. Let's consider further what that means and the kinds of matter that exist. Matter is formally defined as anything that has mass and occupies space. Matter doesn't have to be visible. For example, air and nitrous oxide are matter. While you might argue that you can see light, light is not considered to be matter because it does not have mass. Atoms are the basic building blocks of matter, so under ordinary conditions of temperature and pressure, essentially all matter is comprised of atoms.[1] Sometimes atoms join to form molecules. An element contains only a single kind of atom, whereas compounds contain two or more kinds of atoms. Descriptions for these terms are given below.

Atoms are the fundamental building blocks of matter. Atoms themselves are comprised of three simpler particles: protons, neutrons, and electrons.

Protons are positively charged and have a mass of approximately 1 atomic mass unit (amu). One amu is equal to 1.66×10^{-27} kg. The number of protons, also called the atomic number (Z), determines the identity of the atom. Neutrons are electrically neutral and have a mass of approximately 1 amu. Electrons are negatively charged and have a much, much smaller mass than either protons or neutrons. Chemists typically ignore the mass of an electron.

Ions are atoms or groups of atoms bonded together that have a net electrical charge. This charge is attained by adding or removing electrons. Cations have a positive electrical charge. Anions have a negative electrical charge.

Elements are comprised of only one kind of atom.

Compounds are comprised of more than one kind of atom in a fixed ratio by mass.

Molecules (or molecular compounds) are groups of atoms chemically bonded together into a discrete unit by covalent bonds. Molecules are electrically neutral.

Ionic Compounds contain positively charged ions and negatively charged ions. There are no identifiable discrete units in an ionic compound. All positively charged ions are attracted to all of the negatively charged ions (and vice versa). Therefore, ionic compounds are not molecules.

Question: Can a substance be both a molecule and an element?

Answer: Yes. Oxygen (O_2) consists of two oxygen atoms bonded together to form an oxygen molecule. But oxygen is also an element, because it contains only oxygen atoms.

---

[1] The Sun, for instance, is a very massive body that is composed mostly of a plasma. A plasma is a distinct phase of matter in which the nuclei of atoms and electrons move about at extremely high speeds.

Physical and Chemical Properties and Changes

Chemistry also studies the *properties* of matter, as well as the changes that matter undergoes. There are two categories of change. Physical changes occur without changing the chemical makeup of the substance undergoing the changes. Chemical changes always result in the formation of chemically different substances. When an ice cube melts, it changes from *water* in the solid form into *water* in the liquid form. However, it is still water. Therefore, melting is an example of a physical change. If we run an electric current through water (containing a small amount of an electrolyte, such as sodium sulfate), the water will break down into hydrogen gas and oxygen gas. We started with one chemical substance, water, and ended up with new chemical substances, hydrogen and oxygen. This is an example of a chemical change.

A physical property can be observed or measured without changing the chemical makeup of the substance. Physical properties fall into two categories. An intensive physical property is integral to the material, regardless of how much material there is. Color is a good example of an intensive physical property. Extensive physical properties depend on the sample size. Volume and mass are good examples of extensive physical properties. A chemical property describes the type of chemical changes the material tends to undergo. For example, a chemical property of some anesthetic gases is that they are flammable.

Pure Substances and Mixtures

Pure substances are materials that cannot be physically separated into simpler components. The chemical and physical properties of a substance are uniform through all samples of that substance. A substance can be either a compound or an element. If the substance is a compound, it can be chemically separated into its elemental components.

Mixtures are comprised of two or more pure substances. Mixtures can be resolved into simpler components through physical processes. Homogeneous mixtures are uniform in chemical and physical properties throughout the sample. Normal saline (like all solutions) is an example of a homogenous mixture. Air is another good example of a homogeneous mixture. A heterogeneous mixture exhibits distinct phase boundaries between its components. A phase boundary is a demarcation where the chemical and/or physical properties of the sample change. Emesis is one example of a heterogeneous mixture. The relationships between the various kinds of matter are summarized in Figure 2.1.

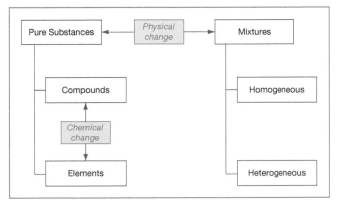

Figure 2.1 Kinds of Matter

ATOMIC STRUCTURE AND DIMENSION

Atoms are incredibly small. As you no doubt already know, the protons and neutrons are bound together by the *strong nuclear force* into an incredibly dense structure called the nucleus at the center of the atom. This strong nuclear force results from conversion of a part of the proton and neutron mass into energy, according to the famous Einstein relationship $E = mc^2$. Electrons are bound to the nucleus by the electromagnetic force (attraction of opposite charges). However, you should abandon the "solar system" picture of electrons orbiting the nucleus in well-defined paths, like the planets revolve around the sun. Electrons surround the nucleus in a nebulous cloud that is described by quantum-mechanical rules. When dealing with quantities on the atomic scale, everyday, common-sense concepts like determinism (if *a* is true, then *b* must be true) and geometry no longer apply. We are forced to adopt quantum mechanical models that give us only probabilities as to where we are *likely* to find the electrons. Furthermore, subatomic "particles" behave as point-localized particles when an observer interacts with them in a manner that predicts they will be point-localized particles. Absent an observer, protons, neutrons, and electrons can be described only by wave-like probability functions, whose energy and momentum emerge by application of an appropriate mathematical operator. Wave–Particle Duality and uncertainty are the heart of quantum theory and underlie the stark contrast between the world at the atomic scale and our everyday experiences.

Atomic radii are on the order of 100 pm (picometers) (100×10^{-12} m = 10^{-10} m), whereas the radius of an atomic nucleus is on the order of femto meters (10^{-15} m). Thus, to a first approximation, the nucleus of an atom is about 100,000 times smaller than the atom. Put into terms of volume, the volume of the nucleus is smaller than the volume of the atom by a factor of about 1×10^{15}.

Atomic Number and Mass Number

All atoms contain protons and electrons. With the exception of ordinary hydrogen, all atoms contain neutrons. The atomic number (Z) of an atom is the number of protons in the nucleus. The identity of an atom is determined exclusively by the atomic number. For example, carbon has atomic number 6. All carbon atoms have six protons, and all atoms having six protons are carbon atoms. The neutron number (N) is the number of neutrons in the nucleus. The mass number (A) of an atom is the sum of the atomic number (or proton number) and the neutron number and is an integer. The unit for atomic mass is the atomic mass unit, or amu. One amu is defined to be exactly 1/12 the mass of one ^{12}C atom. Protons and neutrons each have a mass of about 1 amu. In more precise terms, neutrons have a mass of 1.0087 amu or 1.6749×10^{-27} kg. Protons have a mass of 1.0073 amu or 1.6726×10^{-27} kg. When protons and neutrons combine to form an atomic nucleus, there is a mass deficit, because some of the mass is converted into a binding energy. In other words, the mass of the nucleus is slightly less than the sum of the individual masses of the protons and neutrons. However, for our purposes, it is adequate to consider the masses of the proton and neutron to be equal and to ignore the mass deficit.

You can easily tell the atomic number from the mass number, because the mass number can never be smaller than the atomic number. The mass number is frequently written as a superscript and the atomic number as a subscript (^A_Z Atomic Symbol). Sometimes, the element name is simply followed by the mass number.

Technetium-99 ($^{99}_{43}$Tc) is used as a radioactive tracer in nuclear medicine. How many protons and neutrons does an atom of this material have? The atomic number (the smaller of the two numbers) is 43. Therefore, a technetium atom has 43 protons. The mass number is 99. The mass number is equal to the sum of the number of protons plus the number of neutrons. Therefore, if we subtract the atomic number from the mass number, we find the number of neutrons. In this case the neutron number is 99 − 43 = 56.

Elements are always electrically neutral, and, therefore, each atom must have equal numbers of positively charged protons and negatively charged electrons. The number of electrons is equal to the number of protons *in an electrically neutral atom*. To emphasize the importance of electrical neutrality, imagine a sample of graphite (elemental carbon) with a mass of 1 g. That is about the amount of graphite in a pencil. Carbon is element number 6, and each carbon atom has six protons as well as six electrons. If each carbon atom were missing one electron, the amount of charge on this quantity of carbon ions would amount to about 8,000 C (coulombs) of charge.[2] You might not be familiar with coulombs as a quantity of charge, but consider that an average lightning bolt has somewhere between 5 and 20 C of charge.

Isotopes and Mass Spectroscopy

In a modern laboratory, it is possible to determine the mass of an atom or molecule using a mass spectrometer. Figure 2.2 illustrates how a mass spectrometer works. The sample (A) is introduced into the instrument. The injection port is maintained under conditions of high temperature and low pressure, which causes the sample to (at least partially) vaporize into

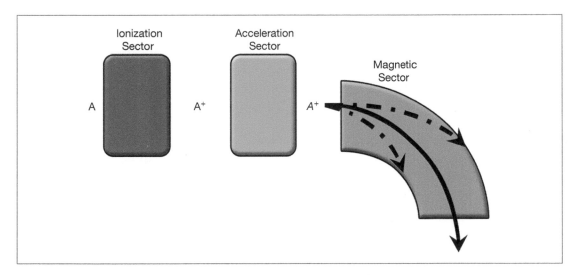

Figure 2.2 Schematic Diagram of Mass Spectrometer

---

[2] Charge = 1 g C (1 mol/12 g)(1 mol e⁻/mol C)(1 Faraday/mole e⁻)(96,500 C/Faraday) = 8000 C.

the gas phase. The sample diffuses into the ionization sector. Here the sample is bombarded with high-energy electrons (an "electron gun"), and this electronic attack knocks one electron off of A. Removing a negatively charged electron from an electrically neutral particle affects introduction of a positive charge onto A, and this gives a molecular ion A^+. It is necessary to place a charge onto the sample because, at the molecular level, it is very difficult to manipulate electrically neutral particles. Charged particles, however, are easy to manipulate. First, they are attracted by oppositely charged particles. The acceleration sector consists of a series of charged metal plates. The molecular ion is attracted to the negatively charged plates and repelled by the positively charged plates. This series of pushes and pulls accelerates the molecular ion, and sends it hurtling into the magnetic sector.

A very useful property of charged particles is that, once they are moving, they behave like little magnets. These little magnets can be given a push or a pull by other magnets. In the magnetic sector, external magnets are arranged so as to push the molecular ion in a curved path. The amount of curvature in the molecular ion's path depends on the mass of the molecular ion.[3] If the molecular ion is too light or too heavy, the ion is either deflected too much or too little and crashes into the wall. Only ions of the correct mass curve around the bend and reach a detector. By manipulating the strength of the external magnets, you can scan to find the atomic (or molecular) mass of the sample.

The mass spectrometer was developed by J.J. Thompson. In obtaining the mass spectrum of neon, he found a large signal at 20 amu, but also a small signal at 22 amu. He first assumed it was an impurity. So, he kept repeating the experiment, each time more and more carefully purifying the sample, but the small signal at 22 amu remained constant. What Thompson discovered was the existence of isotopes. Isotopes have the same atomic number but different mass number (same Z, different A), or the same number of protons and different number of neutrons.

The discovery of isotopes illustrates an important characteristic of a good scientist and, by extension, a good clinician. Many students, when confronted with unexpected information that does not conform to a predicted model, are usually tempted to ignore that information. They will assume that the instrument was not working properly, or that the experiment was flawed, or that they made an error. You should always be cognizant of this fact. Many important scientific discoveries were made when experimental data was obtained that did not conform to an initial hypothesis. Of course, good science requires replication of the measurement by a variety of experiments in order to validate that the unexpected result is reproducible. When you encounter unexpected results, don't be afraid to consider the possibility that you have encountered something really new and that existing theories might have to be modified, expanded, or even discarded in order to accommodate new information.

DALTON'S ATOMIC THEORY

The idea that all matter is made of atoms is a familiar concept that that can be recited by the average first grader. However, the general acceptance of the atomic theory dates to the

---

3 Actually, it depends on the mass-to-charge ratio of the ion. However, it is unusual for the ionization sector to remove more than one electron. Therefore, the charge on the molecular ion is almost certain to be +1. Therefore, the mass/charge ratio is equal to the mass/1, which is numerically equal to the mass of the particle.

early part of the 20th century, when Albert Einstein published a paper explaining Brownian motion. The notion of atoms dates back to ancient times, but these models are largely based on philosophical arguments, rather than empirical evidence. John Dalton first formulated an atomic theory in the early nineteenth century that was founded on experimental evidence. The Postulates of Dalton's Atomic Theory include the following:

1. Elements are composed of tiny, indivisible particles called atoms. All atoms of a given element are identical and unique to that element.
2. Compounds are formed by bonding atoms together in a fixed ratio.
3. Chemical reactions do not create, destroy, or change atoms into atoms of other elements. Chemical reactions cause atoms to recombine into new substances.

Dalton based his theory on two important laws: the Law of Conservation of Mass and the Law of Definite Proportions. Antoine Lavoisier proposed the Law of Conservation of Mass at the end of the 18th century. This law observes that no *detectable* change in the total mass occurs during a chemical reaction. In other words, during a chemical change, the components of a system are neither created nor destroyed. They simply recombine into new substances. Another French chemist, Joseph Proust, proposed the Law of Definite Proportions, which states that different samples of a pure compound always contain the same elements in the same proportion by mass. For example, a sample of water taken from any source always contains 11.2% hydrogen and 88.8% oxygen by mass. Dalton discerned a third law, the Law of Multiple Proportions. Some elements can combine to give more than one compound. For example, carbon can burn in oxygen to give carbon dioxide as well as carbon monoxide. Likewise, nitrogen and oxygen combine to give several oxides, including nitrous oxide (N_2O: an anesthetic gas) and nitrogen dioxide (NO_2: a component of air pollution). The ratios of the weights of the elements in these compounds is always a ratio of small, whole numbers, which supports the notion that the elements must be delivered as particles (i.e., atoms) rather than a continuous substance.

Dalton's Atomic Theory was an important milestone in the development of chemistry, but modern chemistry students will correctly note that it was incomplete, and in some cases, just plain wrong. For example, not all atoms of a given element are identical, because Dalton did not know about the existence of isotopes. Likewise, we now know that atoms are comprised of still smaller particles and that nuclear processes convert atoms of one element into atoms of other elements. By the very nature of science, when a hypothesis, law, or model—no matter how dearly held—fails to make correct predictions, it must be discarded or modified. So, significant portions of Dalton's original theory have been modified. However, the importance of Dalton's theory can hardly be understated and should not be assessed by whether or not it was correct in the finest details, but in how it provided a working foundation that guided current and future scientists in their quest to understand the physical world.

THE PERIODIC TABLE OF THE ELEMENTS

One of the most important achievements of 19th-century chemists was recognizing that the chemical and physical properties of the known elements repeat in a regular or periodic fashion. The Periodic Law states the properties of elements are periodic functions of their atomic

IA	IIA	IIIB	IVB	VB	VIB	VIIB	VIIIB	VIIIB	VIIIB	IB	IIB	IIIA	IVA	VA	VIA	VIIA	VIIIA
1 H 1.008																	2 He 4.003
3 Li 6.941	4 Be 9.012											5 B 10.81	6 C 12.01	7 N 14.01	8 O 16.00	9 F 19.00	10 Ne 20.18
11 Na 22.99	12 Mg 24.31											13 Al 26.98	14 Si 28.09	15 P 30.97	16 S 32.07	17 Cl 35.45	18 Ar 39.95
19 K 39.10	20 Ca 40.08	21 Sc 44.96	22 Ti 47.88	23 V 50.94	24 Cr 52.00	25 Mn 54.94	26 Fe 55.85	27 Co 58.93	28 Ni 58.69	29 Cu 63.55	30 Zn 65.39	31 Ga 69.72	32 Ge 72.64	33 As 74.92	34 Se 78.96	35 Br 79.90	36 Kr 83.80
37 Rb 85.47	38 Sr 87.62	39 Y 88.91	40 Zr 91.22	41 Nb 92.91	42 Mo 95.94	43 Tc (98)	44 Ru 101.1	45 Rh 102.9	46 Pd 106.4	47 Ag 107.9	48 Cd 112.4	49 In 114.8	50 Sn 118.7	51 Sb 121.8	52 Te 127.6	53 I 126.9	54 Xe 131.3
55 Cs 132.9	56 Ba 137.3	57 La 138.9	72 Hf 178.5	73 Ta 180.9	74 W 183.9	75 Re 186.2	76 Os 190.2	77 Ir 192.2	78 Pt 195.1	79 Au 197.0	80 Hg 200.6	81 Tl 204.4	82 Pb 207.2	83 Bi 209.0	84 Po (209)	85 At (210)	86 Rn (222)
87 Fr (223)	88 Ra (226)	89 Ac (227)	104 Rf (263)	105 Db (268)	106 Sg (266)	107 Bh (270)	108 Hs (270)	109 Mt (278)	110 Ds (281)	111 Rg (281)	112 Cn (285)	113 Uut (286)	114 Fl (289)	115 Uup (289)	116 Lv (293)	117 Uuh (294)	118 Uuo (294)

58 Ce 140.1	59 Pr 140.9	60 Nd 144.2	61 Pm (145)	62 Sm 150.4	63 Eu 152.0	64 Gd 157.3	65 Tb 158.9	66 Dy 162.5	67 Ho 164.9	68 Er 167.3	69 Tm 168.9	70 Yb 173.0	71 Lu 175.0
90 Th 232.0	91 Pa 231.0	92 U 238.0	93 Np (237)	94 Pu (244)	95 Am (243)	96 Cm (247)	97 Bk (247)	98 Cf (251)	99 Es (252)	100 Fm (257)	101 Md (258)	102 No (259)	103 Lr (262)

Figure 2.3 Periodic Table of the Elements

numbers. The Russian chemist Dmitri Mendeleev is generally credited with organizing the first periodic table. To demonstrate the utility of Mendeleev's periodic table, three elements, namely, gallium, scandium, and germanium, had not been observed and hence were unknown when the table was first published in 1872. Using information contained in his periodic table along with holes in the table, Mendeleev was able to accurately infer the existence and chemical properties of these three missing elements. Even today, chemists and physicists rely on the immense predictive power of the periodic law.

The modern periodic table looks very different from Mendeleev's, but it is still organized according to the periodic repetition of chemical and physical properties. Figure 2.3 illustrates a modern version of the periodic table. Each box contains the chemical symbol of an element, along with its atomic number and its average atomic mass. The elements are listed in order of increasing atomic number, so each successive element has one additional proton. The vertical columns are called groups or families. Elements in a given group have similar chemical and physical properties. Unfortunately, there is more than one numbering system for the groups. The simplest system numbers the groups from 1 to 18. While this is the simplest and the officially accepted system, the authors strongly feel the older system of numbering groups by a number from 1 to 8, along with a letter (A or B), is a much more useful system. For example, the elements listed in the column on the far right of the table (Group 8A or Group 18, depending on which version of the periodic table you are working with) are called the noble gases. They are all colorless, odorless gases, and they all are extremely reluctant to combine with other elements to form compounds.

Each row on the periodic table is called a period. The first period contains hydrogen and helium. The second period begins with lithium and ends with neon. Because each successive element has one additional proton, each successive element also has one additional electron. In fact, most chemists tend to focus on the electrons, rather than the protons as a means of predicting chemical and physical proclivities of the elements. Periods represent adding electrons to quantum energy levels in the atom, which are called *electron shells*. Atoms at the end of a period each have an electron shell filled to its capacity with electrons.

Although a detailed treatment of electronic configuration of atoms is beyond the scope of this text, and is not necessary to understand the remaining material, it is instructive to consider how the number of electrons in an atom relates to electron shells. As shown in Figure 2.4, a hydrogen atom has a single electron in the first electron shell, and a helium

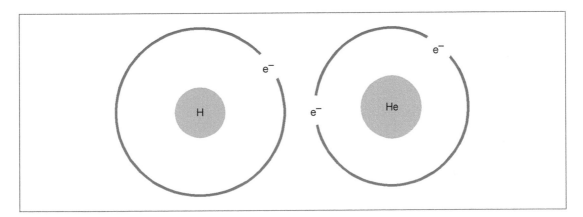

Figure 2.4 The First Electron Energy Shell

atom has two electrons, enough to fill the first energy shell. The light blue circle with the atomic symbol represents the atomic nucleus, and the darker blue path represents the region in space where you're likely to find an electron in the first quantum state of the atom (i.e., the first shell).

Lithium, in addition to the two electrons that fill its first electron shell, has one electron in the second electron shell, as illustrated in Figure 2.5. The region encompassed by the second shell is shown by a larger dark blue circle. Similarly, beryllium has two electrons in the second shell, boron has three, and so forth.

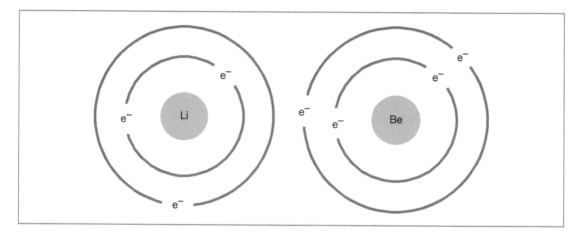

Figure 2.5 The Second Electron Energy Shell

Neon, a noble gas located at the end of the second period, has enough electrons to fill the second electron shell. Atoms (or ions) with filled electron shells are especially stable, and there is a strong energetic tendency for atoms to react in a way so as to acquire as many electrons as a noble gas. Therefore, elements that have the highest electron shell nearly filled to capacity tend to accept additional electrons, forming negatively charged anions. Elements with nearly vacant electron shells tend to surrender the electrons from partially filled shells, forming positively charged cations.

Fluorine is element 9 and is located in Group 7A. How many electrons are in the second electron shell of a fluorine atom? How many electrons does a fluorine atom need in order to have as many as the nearest noble gas? What charge do you expect on a fluoride anion? Fluorine is in Group 7A; therefore, fluorine has seven valence electrons. Since addition of one electron would give fluorine as many electrons as the nearest noble gas (Ne), fluorine forms an ion with a charge of negative 1.

Average Atomic Weights

The atomic weights listed in the periodic table are weighted averages of the atomic masses of the naturally occurring isotopes of that element. For example, consider the element chlorine. The atomic number is 17, and an *average* atomic mass of 35.45 is listed on the periodic table. Yet no chlorine atom has a mass of 35.45 amu, because that would require a fractional part of a neutron. The solution to this conundrum comes from considering that about 76% of all naturally occuring chlorine atoms have a mass of ~35 amu, while 24% of all naturally occuring chlorine atoms have a mass of ~37 amu. The value reported on the periodic table is a weighted average of these two exact masses.

To calculate the average atomic mass, we begin with the precise values of the atomic masses and percent abundance of each isotope. We see that 75.77% of Cl has a mass of 34.97 amu and 24.23% of Cl has a mass of 36.97 amu.

You probably learned to calculate an average by adding up all the values and dividing by the number of data points. A mathematically equivalent calculation of an average is to multiply each value times the percent of the population that value represents. Of course, the sum of the percent occurrences must total 100%.

$$\bar{x} = \Sigma(\text{occurrence})(\text{value})$$

The average is equal to the total of these products. This method is more convenient when the sample is very large and cannot be easily counted. Applying this strategy to chlorine, we obtain:

$$\bar{x} = (0.7577)(34.97 \text{ amu}) + (0.2423)(36.97 \text{ amu})$$
$$\bar{x} = 35.45 \text{ amu}$$

Classifying Elements on the Periodic Table

Classifying elements as representative, transition, or inner transition elements is one of the most useful distinctions on the periodic table. The representative elements are contained in the "high-rise" portions located at the left and right extremes of the periodic table. In Figure 2.6 the representative elements are contained in the gray boxes. In the older numbering system, representative elements have a group number with an A. Transition elements, shown in blue boxes, have a B designation in their group number and form the connection between the representative high-rise towers. The inner transition elements (shown in green) are the "footnotes" located at the bottom of the periodic table.

Most of the elements on the periodic table are metals and are listed on the left side of the table. Metals have a characteristically shiny luster. They tend to be ductile (able to be drawn into wires) and malleable (able to be beaten into thin sheets). Metals are good conductors of both heat and electricity. They tend to react chemically to form cations by giving away electrons from partially filled electron shells. Figure 2.7 lists the metallic elements in gray boxes.

Nonmetals are located on the right side of the periodic table and are in dark blue boxes. Nonmetals may be solids, liquids, or gases. The solid nonmetals tend to be brittle. With the exception of carbon in the form of graphite, they do not conduct electricity. If they form ions, nonmetals tend to form anions.

Metalloids (or semimetals) have properties intermediate between the metals and nonmetals and are listed in light blue boxes in Figure 2.7. Metalloids have a shiny luster, but they are less malleable and ductile than metals. They conduct electricity but not nearly as well as metals. For this reason, semimetals are called semiconductors. Silicon is the most well known of the semimetals because of its use in constructing semiconductor computer chips.

IA	IIA	IIIB	IVB	VB	VIB	VIIB	VIIIB	VIIIB	VIIIB	IB	IIB	IIIA	IVA	VA	VIA	VIIA	VIIIA
1 H 1.008																	2 He 4.003
3 Li 6.941	4 Be 9.012											5 B 10.81	6 C 12.01	7 N 14.01	8 O 16.00	9 F 19.00	10 Ne 20.18
11 Na 22.99	12 Mg 24.31											13 Al 26.98	14 Si 28.09	15 P 30.97	16 S 32.07	17 Cl 35.45	18 Ar 39.95
19 K 39.10	20 Ca 40.08	21 Sc 44.96	22 Ti 47.88	23 V 50.94	24 Cr 52.00	25 Mn 54.94	26 Fe 55.85	27 Co 58.93	28 Ni 58.69	29 Cu 63.55	30 Zn 65.39	31 Ga 69.72	32 Ge 72.64	33 As 74.92	34 Se 78.96	35 Br 79.90	36 Kr 83.80
37 Rb 85.47	38 Sr 87.62	39 Y 88.91	40 Zr 91.22	41 Nb 92.91	42 Mo 95.94	43 Tc (98)	44 Ru 101.1	45 Rh 102.9	46 Pd 106.4	47 Ag 107.9	48 Cd 112.4	49 In 114.8	50 Sn 118.7	51 Sb 121.8	52 Te 127.6	53 I 126.9	54 Xe 131.3
55 Cs 132.9	56 Ba 137.3	57 La 138.9	72 Hf 178.5	73 Ta 180.9	74 W 183.9	75 Re 186.2	76 Os 190.2	77 Ir 192.2	78 Pt 195.1	79 Au 197.0	80 Hg 200.6	81 Tl 204.4	82 Pb 207.2	83 Bi 209.0	84 Po (209)	85 At (210)	86 Rn (222)
87 Fr (223)	88 Ra (226)	89 Ac (227)	104 Rf (263)	105 Db (268)	106 Sg (266)	107 Bh (270)	108 Hs (270)	109 Mt (278)	110 Ds (281)	111 Rg (281)	112 Cn (285)	113 Uut (286)	114 Fl (289)	115 Uup (289)	116 Lv (293)	117 Uuh (294)	118 Uuo (294)

58 Ce 140.1	59 Pr 140.9	60 Nd 144.2	61 Pm (145)	62 Sm 150.4	63 Eu 152.0	64 Gd 157.3	65 Tb 158.9	66 Dy 162.5	67 Ho 164.9	68 Er 167.3	69 Tm 168.9	70 Yb 173.0	71 Lu 175.0
90 Th 232.0	91 Pa 231.0	92 U 238.0	93 Np (237)	94 Pu (244)	95 Am (243)	96 Cm (247)	97 Bk (247)	98 Cf (251)	99 Es (252)	100 Fm (257)	101 Md (258)	102 No (259)	103 Lr (262)

Figure 2.6 Representative and Transition Elements

Figure 2.7 Metals, Nonmetals, and Metalloids

Figure 2.8 Solid, Liquid, and Gaseous Elements

Most of the elements are solids under normal conditions. Only two (mercury and bromine) are liquids, although gallium has a melting point of around 30°C and will liquefy in your hand. Hydrogen, nitrogen, oxygen, fluorine, chlorine, and all of the noble gases are gases under normal conditions. Figure 2.8 illustrates this information. Solids are listed in gray boxes, gases in blue boxes, and liquids in red boxes.

SOME COMMON ELEMENTS

Every element has a chemical symbol. In many cases, the chemical symbol is the first two letters of the element's name. While 118 elements have been reported in the chemical literature, you are likely to encounter only 30 to 40 of these in your nursing studies. The following paragraphs outline a few chemical and physical properties of some common elements as well as some medical applications for the most common elements. This information is not presented for you to memorize. Rather, it is presented for you to gain an appreciation of the myriad of important roles these elements play in your careers in the medical field.

Aluminum (Al) Aluminum is a silvery white metal with a relatively low density of 2.7 g/cm^3. For this reason, aluminum is a popular material for construction. Alloying aluminum with other metals improves its strength. Elemental aluminum does not occur in nature, and must be synthesized by the energy-intensive Hall process. Compounds containing aluminum are found in antiperspirants (aluminum chloride) and antacids (aluminum hydroxide).

Barium (Ba) Barium sulfate ($BaSO_4$) is given orally or as an enema to patients who are undergoing radiographic procedures involving the digestive tract. Because barium is a heavy metal, it is relatively opaque to X-rays and thus improves the definition of the gastrointestinal (GI) tract. Although all compounds containing barium are toxic, the solubility of barium sulfate is very low (2.5 mg/L of pure water), and patients excrete the material before they absorb a fatal dose.

Bromine (Br) Elemental bromine exists as a diatomic molecule Br_2 and is a reddish-orange, fuming, highly toxic liquid.

Calcium (Ca) Calcium is a silvery white metal. It does not occur in nature in its elemental form. However, compounds containing calcium are the inorganic components of bone.

Carbon (C) Carbon occurs in nature as two common varieties: graphite and diamond. You are probably familiar with diamonds, and you encounter graphite as pencil lead and charcoal. Highly purified and anhydrous graphite is known as *activated charcoal*, which can be administered orally to adsorb certain poisonous materials that a patient might have ingested. Different forms of the same element are called *allotropes*. Carbon is arguably the most versatile of all chemical elements, at least in terms of its ability to combine with other elements to form a staggering number of different compounds. In fact, an entire branch of chemistry (organic chemistry) is devoted to studying the compounds of carbon.

Chlorine (Cl) Chlorine, like bromine, is a diatomic molecule, Cl_2. Chlorine is a toxic green gas that has excellent disinfectant properties. Chlorine gas dissolves in sodium hydroxide to give sodium hypoclorite (NaOCl), which you probably know as Clorox®.

Chromium (Cr) Chromium is a silvery white metal. It forms many highly colored compounds, hence the name *chromium* (Greek, khromo, color). Chromium is an essential component of the iron alloy known as stainless steel.

Cobalt (Co) Cobalt is a silvery white metal. Cobalt chloride absorbed onto a piece of felt can measure the relative humidity in air. When the air is humid, cobalt chloride picks up water from the air to form a reddish purple hydrate. When the air is dry, cobalt chloride releases the water of hydration and turns blue.

Copper (Cu) Copper is a reddish metal. All metals are variations of silvery white except for copper and gold. Copper is an excellent electrical conductor. The most important application of copper in a clinical setting is as wires in electronic instruments.

Fluorine (F) Fluorine is a yellowish, poisonous gas. In its natural state it is a diatomic molecule, F_2. Sodium fluoride is added to drinking water to strengthen teeth. Other fluoridation sources include sodium monofluorophosphate and stannous fluoride. The polymer of tetrafluoroethylene is a strong, slippery solid known as Teflon®.

Gold (Au) The chemical symbol for gold is Au, after *aurum* (Latin for sun). Gold is often used for electrical contacts because of its excellent electrical conductivity, combined with its virtual imperviousness to corrosion.

Helium (He) Helium is a colorless inert gas named after the sun (Greek, *helios*), because that is where it was first discovered. Liquid helium is used as a coolant in MRI instruments.

Hydrogen (H) Hydrogen is the most common atom in the universe, and accounts for more than 95% of all known ordinary matter. On Earth, hydrogen is a colorless flammable gas that is found as a diatomic molecule, H_2. The name "hydrogen" literally means "water forming," and hydrogen reacts violently with oxygen to give water. Hydrogen represents a tantalizing energy alternative to fossil fuels, but current methods of producing hydrogen consume significant quantities of fossil fuel.

Iodine (I) Iodine is a purplish-black solid. Like its vertical neighbors in the periodic table, bromine and chlorine, iodine is a diatomic molecule, I_2. The topical antiseptics, tincture of iodine and betadine, both contain iodine.

Iron (Fe) Iron is arguably the most important construction metal. In addition to buildings and cars, iron alloyed with chromium, molybdenum, nickel, and carbon form surgical stainless steel. Each hemoglobin molecule, the protein that transports oxygen in the blood, contains four ferrous (Fe^{+2}) ions, each of which serves as a point of attachment for an oxygen molecule.

Lead (Pb) The chemical symbol for lead comes from the Latin *plumbum*, which also gives us the word "plumber." It is highly malleable and ductile and has been known since antiquity. One ancient, albeit poorly chosen, application of lead was to fashion water pipes, and, hence the connection to the word "plumber."

Lithium (Li) Lithium is a silvery, highly reactive metal. Lithium is used in batteries, and compounds containing lithium are used to treat bipolar disorder.

Magnesium (Mg) Magnesium is a silvery white metal. Magnesium sulfate (Epsom salts) is used to slow down uterine contractions in preterm labor.

Mercury (Hg) Mercury is a silvery white liquid metal. Because of its high density, mercury is used in sphygmomanometers. It is also found in some thermometers. Before its toxic nature was fully understood, mercury compounds had been used in medical applications ranging from treatment of syphilis to constipation. Mercury alloys well with many other metals, and alloys containing mercury are known as *amalgams*.

Neon (Ne) Neon is a colorless inert gas that is used in some lighting applications.

Nickel (Ni) Nickel is a silvery white metal that plays more of a supporting role in medicine. For example, alloys containing nickel are components in such diverse applications as stainless steel and magnets.

Nitrogen (N) Nitrogen is an odorless and colorless gas that occurs as a diatomic molecule, N_2. Nitrogen gas comprises about 80% of air. Nitrous oxide (N_2O) is an important anesthetic gas.

Oxygen (O) Oxygen is an odorless and colorless reactive gas. Diatomic oxygen (O_2) comprises roughly 20% of air. Oxygen is a very strong oxidizing agent. That is, oxygen has a strong tendency to accept electrons from other chemical species. This chemical property finds application in the electron transport chain (the biochemical pathway that is coupled to adenosine triphosphate (ATP) synthesis). Triatomic oxygen (O_3) is called ozone, which is a highly toxic gas with a characteristic sharp odor. Ozone is produced by electrical discharges. In the upper atmosphere, ozone absorbs ultraviolet radiation and provides us some protection from skin cancers.

Phosphorus (P) Red phosphorus is used to make matches. The white allotrope of phosphorus is a much more dangerous material. White phosphorus causes horrific burns. In biological systems, phosphorus is found in genetic materials (RNA and DNA) and in high-energy molecules such as ATP.

Potassium (K) Potassium is a silvery white metal. Because of its high reactivity, elemental potassium is not found in nature. Potassium chloride is a component in "Lite Salt." Potassium and sodium ions are required for muscle contractions.

Silicon (Si) Silicon is a lustrous silvery gray material. Because silicon conducts electricity, but not as well as a metal, silicon is classified as a semimetal. Crystals of pure silicon that have been doped with arsenic or gallium are known as *semiconductors* and are used to fabricate computer chips. Silicone rubbers are polymers containing silicon, oxygen, and various hydrocarbon groups, and are used in applications ranging from sealants to breast implants.

Silver (Ag) The chemical symbol for silver comes from the Greek *argentum*, for white or shining. Silver is the best of all electrical conductors. Alloys of silver and mercury were once used in dental fillings, although the toxicity of each of these metals has detracted from this application. Silver also finds application in photography (including X-rays). When exposed to light, compounds such as silver chloride are more easily reduced to metallic silver. X-rays that are not blocked by the patient activate the silver emulsion. Developing the image reduces the activated silver chloride to finely divided silver metal, which appears as the black portions of the X-ray image. The unexposed silver compounds are washed away in the fixing process, leaving the white areas of the X-ray image.

Sodium (Na) The chemical symbol for sodium comes from the Latin *natrium*, which means swimmer. When a small piece of sodium is placed in water, it skitters around the surface, like a swimmer in a pool. It reacts with water to give sodium hydroxide and hydrogen gas. The hydrogen sometimes ignites. Larger samples of sodium simply explode. Sodium ions, along with potassium ions, are required for muscle contractions.

Sulfur (S) Sulfur is a yellow solid. Iron–sulfur clusters are found in cytochrome enzymes. The sulfur-containing amino acid cysteine is common in hair. Cysteine residues can connect to each other via disulfide bridges, giving hair a natural curl. Permanent waves are achieved by artificially removing and then re-forming these disulfide bridges.

Tin (Sn) Tin is a silvery white metal. Solder is an alloy of tin and lead and is used in electrical connections. Stannous fluoride (SnF_2) was once a common ingredient in toothpaste, but has largely been replaced by sodium monofluorophosphate.

Titanium (Ti) Titanium is a grayish metal often used in the manufacture of prosthetic implants because of its light weight, low toxicity, and high strength. Titanium oxide is used in white pigments, especially in paints.

Zinc (Zn) Zinc is a bluish silver metal. Medical applications of zinc compounds include calamine lotion and zinc oxide (sun block). Galvanized steel is resistant to corrosion and is prepared by coating iron with a thin layer of zinc.

CHEMICAL NOMENCLATURE

The rules for naming chemical compounds depend on whether the substance is a molecular substance or an ionic substance. Molecular compounds are comprised only of nonmetals. Ionic compounds are almost always comprised of a metal and a nonmetal. If a compound contains one of the polyatomic ions listed in Table 2.1, it is an ionic compound.

Naming Molecular Compounds

Molecular compounds are easier to name than ionic compounds, so let's begin there. The molecular formula of a substance gives the number of each kind of atom in the molecule. To name a molecular substance:

- Name each element
- Indicate how many of each element is present with a prefix multiplier (mono = 1; di = 2; tri = 3; tetra = 4; penta = 5; hexa = 6; hepta = 7; octa = 8)
- Add the suffix "ide" to the last element name

Table 2.1 SOME COMMON POLYATOMIC IONS

CHARGE	NAME AND FORMULA	
+1	NH_4^+ ammonium ion	H_3O^+ hydronium ion
−1	HCO_3^- bicarbonate ion or hydrogencarbonate ion NO_2^- nitrite ion NO_3^- nitrate ion HSO_3^- bisulfite or hydrogensulfite ion HSO_4^- bisulfate or hydrogensulfate ion	OH^- hydroxide ion $C_2H_3O_2^-$ acetate ion ClO^- hypochlorite ion CN^- cyanide ion $H_2PO_4^-$ dihydrogenphosphate ion
−2	CO_3^{2-} carbonate ion HPO_4^{2-} hydrogenphosphate ion	SO_3^{2-} sulfite ion SO_4^{2-} sulfate ion
−3	PO_4^{3-} phosphate ion	

Notice that chemical names are not proper nouns and should not be automatically capitalized.

FORMULA	NAME
N_2O	Dinitrogen monoxide (commonly: nitrous oxide)
NO	Nitrogen monoxide (commonly: nitric oxide)
SCl_2	Sulfur dichloride
P_2O_5	Diphosphorus pentoxide
CCl_4	Carbon tetrachloride
SiO_2	Silicon dioxide

Several molecular compounds have common (nonsystematic) names (see table below), and must be memorized. Sorry.

FORMULA	NAME	COMMON USES
H_2O	Water	Drinking; chemical solvent
NH_3	Ammonia	Fertilizer; window cleaner
CH_4	Methane (natural gas)	Heating
C_3H_8	Propane	LPG fuel gas used in rural areas
N_2O	Nitrous oxide	Laughing gas; anesthetic

Naming Ions and Ionic Compounds

An ion is an atom or group of atoms with a charge. Cations are positively charged and anions are negatively charged. Ionic compounds consist of ions and are held together by *ionic bonds*, or the attraction of the oppositely charged ions. In the solid state, ionic compounds form *crystalline lattices* in which all the cations are attracted to all the neighboring anions, and vice versa. Since you cannot identify any anion that is associated with any particular cation, there are no discrete ionic "molecules."

Ionic compounds are sometimes referred to as salts. You are no doubt familiar with common table salt, NaCl. While NaCl is certainly a salt, there are many other salts besides NaCl. Technically, a salt is produced by the reaction of an acid and a base, which is a subject for a later chapter. For now, you can consider all ionic compounds as being salts.

Monatomic Cations of Representative Metals

Representative metals almost always form cations in which the ionic charge equals the group number. This is because the group number is equal to the number of electrons in the highest energy, partially filled electron shell. By giving these electrons away, a representative metal is able to obtain an electron configuration that is identical with the closest noble gas. All of the metals in group 1A form cations with a charge of +1. All of the metals in group 2A form cations with a charge of +2. Aluminum forms a cation with a charge of +3. To name a representative cation, name the element and add "cation" or simply "ion." There are no simple cations with charges of +4.

For example, sodium is in group 1A, and forms a cation with a charge of +1. Sodium metal has 11 electrons, and one of these electrons is in the third electron shell, all by itself. By giving away that electron, the resulting sodium cation has the exact number of electrons as neon, and all of its electrons are in filled shells.

Predict the charge on the ions that are formed from Ba, Rb, and Tl.

Answer: Ba^{2+}; Rb^+; Tl^{3+}

Monatomic Anions of Representative Nonmetals

For anions of representative elements, the ionic charge is based on the number of electrons the nonmetal needs to gain in order to have as many electrons as the nearest noble gas. For example, let's look at oxygen on the periodic table. Oxygen is two boxes away from neon, the nearest noble gas. Therefore, when forming an anion, oxygen will acquire two electrons and gain a charge of –2. Nitrogen forms an anion with a charge of –3, while fluorine forms an anion with a charge of –1. Again, however, it is virtually impossible to form an anion with a charge greater than –3. To name monatomic anions, add the suffix "ide" to the stem name.

Cl^- chloride ion
S^{2-} sulfide ion
P^{3-} phosphide ion

Transition-Metal Cations

Most transition metals form more than one cation. For example, iron commonly forms two different cations: Fe^{2+} and Fe^{3+}. It is not as easy to predict the charges on the cations that transition metals are likely to form, and most beginning students usually wind up memorizing a list. A few examples of some common transition metal cations are given below.

Ag	+1 only	Ag^+ silver cation
Zn	+2 only	Zn^{2+} zinc cation
Cu	+1 and +2	Cu^+ copper (I) cation or cuprous cation Cu^{2+} copper (II) cation or cupric cation
Fe	+2 and +3	Fe^{2+} iron (II) cation or ferrous cation Fe^{3+} iron (III) cation or ferric cation

If the transition metal forms only one cation, you name it like a representative metal cation: name the element and call it a cation. If the transition metal forms more than one cation, you need to name the metal and then indicate the charge on the cation with Roman numerals in parentheses.

$$Fe^{3+} \text{ is the iron (III) cation}$$
$$Sn^{2+} \text{ is the tin (II) cation}$$

An older system for naming transition metals is to name the lower charged ion as the "ous" ion and the higher charged ion as the "ic" ion. This system is becoming less common in general academic use, but is still frequently encountered in medical and industrial settings.

$$Cr^{2+} \text{ is the chromous ion}$$
$$Cr^{3+} \text{ is the chromic ion}$$

If the chemical symbol is based on the Latin or Greek name of the element (i.e., the chemical symbol isn't the first two letters of the element's name), you need to use the Latin or Greek stem name.

$$Fe^{2+} \text{ is the ferrous ion}$$
$$Fe^{3+} \text{ is the ferric ion}$$

Polyatomic Ions

Polyatomic ions are formed from two or more nonmetal atoms that are bonded together in a way that results in a net electrical charge. The subtleties of predicting the charge and formula of these ions is fairly involved, and you should content yourself with learning the ions listed in Table 2.1.

Some of these ions merit special comment. Some ions, like SO_4^{-2} and SO_3^{-2}, differ only in a single oxygen atom. In these cases, the ion with the larger number of oxygen atoms is given the "ate" suffix, while the ion with the smaller number of oxygen atoms is given the "ite" suffix.

$$SO_3^{2-} \text{ is the sulfite ion}$$
$$SO_4^{2-} \text{ is the sulfate ion}$$

Some of the ions, for example, HCO_3^-, contain a hydrogen atom. They are formed by combining the parent ion (CO_3^{2-} in this case) with an acid (H^+ ion).

$$CO_3^{2-} + H^+ \rightarrow HCO_3^-$$

These "half acid salts" can be named in two ways. The systematic way is to simply name both components: HCO_3^- is formed from a hydrogen ion plus a carbonate ion, and is, therefore, named hydrogencarbonate. The common, if less systematic, approach is to call this the bicarbonate ion.

Formulas of Ionic Compounds

As we mentioned earlier, compounds (like atoms) must be electrically neutral. The net charge on an ionic compound must be zero. When a metal "gives" its electrons away to form a cation, there has to be some other species, often a nonmetal, present to accept the electrons. Thus, whenever a cation is forming, there is also a concurrent formation of an anion. Chemical systems have to remain electrically neutral.

As you recall, molecules are comprised of atoms chemically bonded into a discrete and identifiable unit. There are no ionic molecules, because every cation is attracted to every anion, so there are no identifiable ion pairs that belong exclusively to each other. Therefore, the formula of an ionic compound is an *empirical formula*. That is, the formula of an ionic compound lists the simplest ratio of cations to anions necessary to achieve electrical neutrality. For example, calcium oxide contains Ca^{2+} ions and O^{2-} ions. While the formulas CaO and Ca_2O_2 both represent electrically neutral combination of ions, the subscripts in the latter formula have a common factor of 2. Therefore, Ca_2O_2 is not the simplest ratio of calcium cations to oxide anions, and so the formulation of calcium oxide as Ca_2O_2 is incorrect. The correct empirical formula of calcium oxide is CaO.

When forming an ionic compound, the total positive charge contributed by the cation(s) must be equal and opposite to the total negative charge contributed by the anion(s). For ion pairs with equal (but opposite) charges, this ratio is obviously 1:1.

IONS	FORMULA OF IONIC COMPOUND
Na^+ and Cl^-	NaCl
Ca^{2+} and O^{2-}	CaO

What about when the ionic charges aren't equal to each other? The most mathematically sophisticated way to determine this ratio is to seek the least common multiple between the absolute values of the charges on the cation and anion. Most beginning students, however, simply use the "criss-cross" method, as shown in Figure 2.9. That is, the number of cations in the formula is the absolute value of the charge on the anion. The number of anions in the formula is the absolute value of the charge on the cation. For example, as shown above, the formula of the ionic compound containing aluminum ions and oxide ions is Al_2O_3.

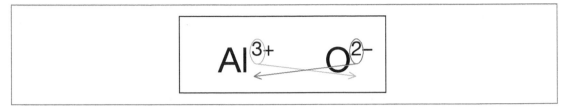

Figure 2.9 Formula of an Ionic Compound

Here are some more examples:

IONS	FORMULA OF IONIC COMPOUND
Na^+ and CO_3^{2-}	Na_2CO_3
Mg^{2+} and PO_4^{3-}	$Mg_3(PO_4)_2$
NH_4^+ and SO_4^{2-}	$(NH_4)_2SO_4$

Ionic Compounds

To name an ionic compound, you just name the cation and then the anion. There is a crucial difference between naming ionic compounds and molecular compounds. In molecular compounds you *must* include prefix multipliers (di, tri, etc.) to indicate the number of each kind of atom in the molecule. In ionic compounds you *must not* include prefix multipliers, because the number of each ion in the formula unit is controlled by the charges on the ions. If the cation is a representative element, it is not necessary to indicate the charge, because (with few exceptions) these metals form cations with an ionic charge equal to the group number.

FORMULA	NAME
K_2SO_4	Potassium sulfate
Na_2O	Sodium oxide
Li_2CO_3	Lithium carbonate
$FeSO_4$	Iron (II) sulfate or ferrous sulfate
$Fe_2(SO_4)_3$	Iron (III) sulfate or ferric sulfate

When naming ionic compounds that contain transition-metal cations, you need to indicate the charge on the cation, either by including Roman numerals to indicate the charge or by using the "ic/ous" suffixes. But how do you determine the charge on the cation from the formula? You have to learn the charges on the anions and assign a charge to the cation so that the formula achieves electrical neutrality. For example, let's consider the last two examples.

$FeSO_4$ contains an iron cation and sulfate anion. Since the sulfate anion has a charge of negative two, the iron must have a charge of positive two. Therefore, $FeSO_4$ is iron (II) sulfate.

$Fe_2(SO_4)_3$ contains three sulfate anions. Since each sulfate carries a -2 charge, there are a total of six negative charges contributed by the sulfate anions. Therefore, there must be six positive charges distributed between the two iron cations. Thus, the charge on the iron cations must be $+3$, and $Fe_2(SO_4)_3$ is iron (III) sulfate.

$FeSO_4$	Iron (II) sulfate or ferrous sulfate
$Fe_2(SO_4)_3$	Iron (III) sulfate or ferric sulfate

Hydrates

Some ionic compounds incorporate a fixed number of water molecules into their formula unit. The compound that contains the water is called a hydrate, and removal of the water affords the anhydrous salt. Compounds that have a strong tendency to absorb water are called *hygroscopic*. To name a hydrate, you simply name the ions and then add the appendage hydrate, along with a multiplier to indicate the number of water molecules in the formula.

$CuSO_4 \cdot 5\ H_2O$	Copper (II) sulfate pentahydrate
$MgSO_4 \cdot 7\ H_2O$	Magnesium sulfate heptahydrate

Naming hydrates only makes sense when you are dealing with solid substances, because when a hydrate is dissolved in water, both the ions and the waters of hydration separate and mix uniformly with the water solvent.

The *anhydrous* form of a compound that has a strong tendency to absorb water can be used as a desiccant. Desiccants scavenge the last traces of water from a system. One of the most commonly used desiccants is silica gel (SiO_2). Addition of water to a desiccant is a reversible process, so hydrated forms of a desiccant can be used as moisturizers.

Formulas From Names

To assign a chemical formula of a compound from its name, you first must recognize whether the compound is ionic or molecular. For ionic compounds, you must also know the charges on the ions.

ELECTROLYTES

Video 2.1

We most commonly encounter electricity as a flow of electrons through copper wiring in our homes and offices. However, electricity is more generally described as the flow of charged particles under the influence of an electric field. We can't have an electric current in a system unless there are charged particles that are free to move around. That brings us to an important concept in both physiology and in chemistry: electrolytes. An electrolyte is a substance that dissolves in water to give a solution that conducts electricity. A nonelectrolyte is a substance that dissolves in water to give a solution that does

COMPOUND	IONIC OR MOLECULAR?	FORMULA
Calcium sulfide	Ionic	CaS
Chromium (III) acetate	Ionic	$Cr(C_2H_3O_2)_3$
Cupric hydroxide	Ionic	$Cu(OH)_2$
Nitrogen triiodide	Molecular	NI_3
Sulfur trioxide	Molecular	SO_3

not conduct electricity. While most ionic compounds are only sparingly soluble in water, those few ionic compounds that readily dissolve in water are electrolytes. Molecular compounds are nonelectrolytes, unless they have acid or base properties. We will explore acids and bases in a later chapter. For now, let's just consider molecular compounds to be nonelectrolytes.

It is a remarkable thing for an ionic compound to dissolve in water. You probably learned at some point that opposite charges attract each other. The energy cost of separating positively charged cations from negatively charged anions is immense. Dissolution occurs only because water interacts very effectively with ions. We will explore this phenomenon more fully in Chapter 8. For now, however, you just need to accept that when ionic compounds dissolve in water, they (mostly) separate into ions that freely and independently move around in the solution. Since these ions are free to move around in the solution, the solution conducts electricity.

Most beginning chemistry students believe water conducts electricity, because of the (very reasonable) prohibition against using a hair dryer in the bathtub. The fact is, water is a nonelectrolyte. Now, let's be clear: There is a very good probability that you will electrocute yourself if you use an electrical device in the bathtub, but that is because water in the bathtub comes from the tap which contains a fair concentration of several electrolytes, such as NaCl. Water is a molecular substance, because it is a compound of nonmetals. *Pure* water is a nonelectrolyte and is a very poor conductor of electricity.

STOICHIOMETRY

Stoichiometry includes the calculations that relate amounts of reactants and/or products in a chemical reaction. You, no doubt, covered this topic in depth during your undergraduate chemistry course. For our purposes here, we need only the basics.

Moles

Moles are a quantity of material, analogous to a dozen. In the case of moles, however, the number is much larger. A mole is an amount of substance that contains exactly as many particles as there are in exactly 12 g of carbon-12. This number is often called Avogadro's number, and it is equal to 6.02×10^{23} particles/mol. "Particles" could mean molecules, atoms, ions, or electrons. Notice Avogadro's number has two units: particles and moles. Therefore, Avogadro's number is a conversion factor between particles and moles.

Molar Mass (Molecular Weight)

Chemists use moles mainly because molecules are just too small to deal with on an individual basis. So, we gather 6.02×10^{23} of them together into a mole, and we use that as our working unit in chemistry. So what is the mass of one mole of carbon? Well, the system is set up so that the molar mass (expressed in g/mol) of an element is equal to its atomic mass (expressed in amu).

For molecules, the molar mass is equal to the sum of the masses of the component atoms. Most chemists use the terms *molar mass* and *molecular weight* interchangeably, even though mass and weight are not the same concept. So, for example, the molar mass of magnesium sulfate ($MgSO_4$) is:

$$1 \text{ Mg} \times (24.3 \text{ g/mol}) + 1 \text{ S} \times (32.1 \text{ g/mol}) + 4 \text{ O} \times (16.0 \text{ g/mol}) = 120.4 \text{ g/mol}$$

Since molar masses have units of grams and moles, they are conversion factors between grams and moles.

Summary

- Chemistry studies matter and the changes it undergoes.
- Changes can be physical or chemical. A chemical change produces a new substance. A physical change transforms a substance into a different state, but it's still the same substance.
- Chemical properties describe the chemical changes that a sample undergoes (e.g., flammability).
- Physical properties can be observed without changing the chemical nature of the sample (e.g., mass, volume, color, density).
- Matter is anything that occupies space and has mass.
- Atoms are the basic building blocks of all matter.
- Atoms are comprised of protons (mass ≈ 1 amu; charge = +1), neutrons (mass ≈ 1 amu; charge = 0), and electrons (mass ≈ 0 amu; charge = −1).
- The number of protons is the atomic number of an atom and determines the identity of the atom.
- The number of protons plus the number of neutrons is the mass number of the atom.
- Isotopes contain the same number of protons, but a different number of neutrons (i.e., same atomic number but different mass number).
- In a neutral atom, the number of electrons equals the number of protons.
- If there are more electrons than protons, the net charge on the atom is negative, and it is called an anion.
- If there are fewer electrons than protons, the net charge on the atom is positive, and it is called a cation.
- A pure substance cannot be separated into its components through physical changes.
- A pure substance that consists of only one kind of atom is called an element. There are about 118 known elements. A pure substance that is comprised of more than one kind of atom is called a compound. There are millions of different compounds reported in the chemical literature.
- Molecules consist of two or more atoms chemically bonded into a discrete unit called a molecule. Molecules that consist of only one kind of atom (e.g., O_2) are elements. However, most molecules are compounds (e.g., CO_2).
- Ionic compounds consist of anions and cations. Ionic compounds are not molecules because every anion is attracted to every cation, so none of the ions are bound into a discrete unit.
- John Dalton is credited with the atomic theory that postulates all matter is comprised of indestructible atoms. While this theory contains some inaccuracies in its details, it still forms the foundation upon which modern chemistry is based.
- Dmitri Mendeleev first organized the known elements into a periodic table demonstrating the chemical and physical properties of elements repeated in a predictable, periodic fashion. The modern periodic table organizes elements by increasing atomic number. Rows of elements are called periods and represent electron energy shells filling. Columns are called groups or families. Elements in a given family have similar chemical and physical properties.

- The atomic weights on the periodic table represent average atomic weights, based on the relative abundance of the naturally occurring isotopes of an element.
- Representative elements are located in the "high-rise" ends of the periodic table. The transition elements occupy the center of the periodic table. The inner transition elements are the "footnotes" at the bottom of the periodic table.
- Nonmetals are located on the right-hand side of the periodic table. Nonmetals may be solids, liquids, or gases. Nonmetals tend to form anions and to be nonconductors of electricity and heat.
- Metals are located on the left side through the center of the periodic table. Metals are almost always solids and tend to form cations. Metals are malleable, ductile, and good conductors of both heat and electricity.
- Molecular compounds are comprised of two or more nonmetals. To name a molecular compound, name each element and add the suffix "ide." If there is more than one atom of a particular element, include a prefix multiplier (di = 2, tri = 3, etc.).
- Cations are named as the metal. If the metal can form more than one charge on a cation, the charge of the cation is included as a Roman numeral in parentheses. Representative metals form cations with charges equal to the group number.
- Simple monoatomic anions are named by adding the suffix "ide" to the stem name of the nonmetal (e.g., chloride). Representative nonmetals form anions with charges equal to the group number minus 8.
- Polyatomic ions have more than one atom in the ionic unit. The names of common polyatomic ions are given in Table 2.1.
- Ionic compounds are named by first naming the cation and then naming the anion. Prefix multipliers are not needed because the ratio of cations to anions is dictated by the charge on each ion.
- Electrolytes are compounds that dissolve in water to give a solution that conducts electricity. Ionic compounds that dissolve in water are strong electrolytes. Molecular compounds that dissolve in water are normally nonelectrolytes.

Review Questions for Chemistry Basics

1. List and define the five major areas of chemistry.

2. Classify and describe types of matter.

3. What is the difference between physical and chemical changes?

4. Give one physical property of water and one chemical property of water.

5. What is the difference between intensive and extensive physical properties?

6. Give one intensive physical property of water and one extensive physical property of water.

7. Density is calculated from two extensive physical properties, namely, mass and volume. Is density an intensive or an extensive physical property?

8. Define and describe molecules, atoms, compounds, and mixtures.

9. What are isotopes?

10. Can a molecule be an element? Explain.

11. Give the mass number, atomic number (Z), proton number (P), neutron number (N), and electron number (E) for the following: ^1H, ^4He, ^6Li and ^8Be.

12. What is the difference between mass number and average atomic mass?

13. Naturally occurring bromine consists of two isotopes, ^{79}Br and ^{81}Br, in roughly equal amounts. What is the average atomic mass of bromine?

14. Describe how a mass spectrometer works.

15. Describe the organization of the periodic table (periods, groups, group number).

16. What is the difference between metals, nonmetals, and metalloids?

17. What is an ion? A cation? An anion? How do ions form?

18. Is there a trend in the periodic table about which kind of ion an element is likely to form?

19. How do you predict the charge of the ions derived from the representative elements?

20. What's a hydrate?

21. How do you decide whether a compound is molecular or ionic?

22. What charge ion is sodium likely to form? How about a phosphide ion? How about a barium ion?

23. What are the representative elements? The transition elements?

24. "Lactated ringers" contain sodium lactate. If the formula of sodium lactate is $NaC_3H_5O_3$, what is the charge on the lactate ion? (lactate = $C_3H_5O_3$ with some unspecified charge).

25. Is glucose an ionic or molecular compound? The molecular formula of glucose is $C_6H_{12}O_6$.

26. What is the molar mass of nitrous oxide?

27. If a patient is given $2\overline{0}$ g of nitrous oxide, how many molecules of nitrous oxide have been given? HINT: Use the molar mass to convert grams into moles, and then use Avogadro's number to convert moles into molecules.

28. What is the difference between a period and a group on the periodic table?

29. Provide a name or chemical formula for each of the following compounds or ions, as required.

   a. Sodium bicarbonate   i. $CuCO_3$ q. Dinitrogen monoxide
   b. Potassium chloride   j. PO_4^{3-} r. Lithium dihydrogen phosphate
   c. Ammonia   k. Fe^{3+} s. Nitrogen dioxide
   d. $MgSO_4$ l. Na_2HPO_4 t. $AlCl_3$
   e. Carbon tetrachloride m. NH_4^+ u. Calcium carbonate
   f. NH_4NO_3 n. Carbon monoxide   v. Sodium fluoride
   g. Ferrous sulfate   o. PCl_3 w. Ozone
   h. $Mg_3(PO_4)_2$ p. KCl   x. $BaSO_4$

30. Which of these is a chemical property of nitrous oxide gas, rather than a physical property?
   a. Nitrous oxide is less dense than oxygen gas.
   b. Nitrous oxide is flammable.
   c. Nitrous oxide is colorless.
   d. All of these are physical properties.
   e. All of these are chemical properties

31. You give a glass of (pure) water and a hamburger to a severely dehydrated patient, who then immediately throws up. The vomit is an example of a ___, while the water is an example of ___.
   a. homogeneous mixture; element
   b. homogeneous mixture; compound
   c. heterogeneous mixture; element
   d. heterogeneous mixture; compound

32. Based on ions that the elements we discussed in class typically form, which of these is an UNLIKELY compound?
   a. Al_3O_2
   b. BeO
   c. NaF
   d. FeS

33. Which of these is a common property of metals?
    a. Ductility
    b. Good electrical conductivity
    c. Tendency to form cations
    d. Shiny and lustrous surface
    e. All of the above

34. Which of these statements is true of the elements symbolized by $^{14}_{6}X$ and $^{14}_{7}Y$?
    a. X and Y are isotopes of each other.
    b. X and Y are in the same chemical family.
    c. Y has more neutrons but fewer protons than X.
    d. X and Y have the same atomic number.
    e. None of these statements is correct.

35. Which of these lists only representative metals?
    a. Na, B, C, N, O
    b. C, Si, Ga, Sn
    c. Li, Na, K, Rb
    d. Fe, Co, Ni, Cu
    e. Ce, Pr, Ne, Pm

36. What cation is each of these metals likely to form?
    a. Al
    b. Rb
    c. Mg
    d. Li
    e. Ga

37. What anion is each of these nonmetals likely to form?
    a. S
    b. I
    c. N
    d. P
    e. Se

38. Based on the trends discussed in this chapter, give the formula and name of the ionic compound formed between these pairs of elements.
    a. Mg and N
    b. Al and S
    c. Li and I
    d. K and O
    e. Ca and C

39. Give the name of each of these ions.
    a. CO_3^- d. OH^- g. PO_4^{3-}
    b. SO_4^{2-} e. NH_4^+ h. HSO_3^-
    c. NO_2^- f. Fe^{3+} i. $H_2PO_4^-$

40. Give the formula of each of these ions.
    a. Ferrous ion   d. Sulfate   g. Sulfite
    b. Hydroxide   e. Bicarbonate   h. Hydrogen sulfate
    c. Phosphate   f. Carbonate   i. Dihydrogen phosphate

41. Give the formula of each of these ionic compounds.
    a. Sodium bicarbonate   f. Copper(II) acetate
    b. Potassium phosphate   g. Iron(II) sulfate
    c. Lithium carbonate   h. Nickel(II) oxide
    d. Ammonium chloride   i. Ammonium nitrate
    e. Barium oxide   j. Silver(I) chloride

42. Give the formula of each of these molecular compounds.
    a. Carbon disulfide   f. Oxygen difluoride
    b. Iodine trichloride   g. Nitrogen dioxide
    c. Nitrogen triiodide   h. Nitric oxide
    d. Diphosphorus pentoxide   i. Nitrous oxide
    e. Ammonia   j. Sulfur tetrachloride

43. Sodium dodecyl sulfate (SDS) is the surfactant in Tide®. If the formula of this ionic compound contains one sodium ion and one dodecyl sulfate ion, what is the charge on the dodecyl sulfate anion?

44. Soap scum is largely comprised of magnesium stearate. If magnesium stearate has a formula $Mg(C_{18}H_{36}O_2)_2$, what is the charge on the stearate anion? What is the formula of iron(III) stearate?

45. Give the name of the element that fits these criteria.
    a. The lightest element
    b. A noble gas in the third period
    c. The halogen that is a liquid at room temperature and pressure
    d. The alkali metal that is in the second period
    e. The semimetal in the fourth period that has the larger atomic number
    f. The element whose allotropes include diamond and graphite
    g. The fourth period element whose properties are most similar to nitrogen

46. The permanganate ion has the formula MnO_4^-. What is the formula of the pertechnetate ion?

47. If the chromate ion has the formula CrO_4^{2-}, what is the formula of the tungstate ion?

48. Aqueous solutions of which of these substances do you expect to be a strong electrolyte?
    a. NaCl
    b. Ethanol, C_2H_6O
    c. Glycerin, $C_3H_8O_3$
    d. Lithium pyruvate, $LiC_3H_3O_3$
    e. Ammonium carbonate

49. Lidocaine $(C_{14}H_{22}N_2O)$ is not very soluble in water. Mixing lidocaine with hydrogen chloride gives lidocaine hydrochloride, which is very soluble in water. A solution of lidocaine hydrochloride is a good conductor of electricity. Which of these statements is true? If the statement is false, explain why it is false.
    a. The solubility of lidocaine is a physical property.
    b. Conversion of lidocaine into lidocaine hydrochloride is a physical change.
    c. Lidocaine is a strong electrolyte.
    d. Lidocaine hydrochlroide is an ionic compound.

50. Which of the tenets of Dalton's atomic theory are still considered correct today? If there is a problem with one of the tenets, cite an example that contradicts that tenet. (N.B., All of these counterexamples are founded on technologies that were not available to Dalton.)
    a. Atoms of a given element are all identical.
    b. Atoms are neither created nor destroyed in a chemical reaction.
    c. Matter is comprised of tiny particles called atoms.
    d. Atoms are indivisible.
    e. Compounds are formed by combining atoms in a fixed ratio by mass.

Basics of Physics Part 1 (Force and Pressure)

The word "physics" often provokes negative responses from people, not unlike a cat coughing up a hair ball. Irrespective of first impressions, a fundamental grasp of physics is vital to a solid understanding and appreciation of our universe. When all is said and done, physics is a beautiful fundamental science that uses mathematics as its tool set. Physics finds a myriad of applications in a medical setting. The regulation of gases, the breathing process, and the flow of energy are just a few examples of clinical care that are founded on basic physics concepts. One of the major strengths of physics is its ability to take a fundamental set of concepts and apply them to virtually every aspect of our universe. Therefore, let's start with some of the foundational concepts in physics. As with much of science, every concept needs to be considered first, so you will probably want to read this chapter a couple of times.

NEWTON'S LAWS

Much of classical physics is based on the work of Isaac Newton, who formulated three laws of motion. Because the extant mathematics of his day was not adequate to formulate these laws, he also invented a new branch of mathematics called calculus.[1] Incidentally, Newton accomplished most of this work in the 18 months after he graduated from college.

Newton's First Law states an object at rest or moving at constant speed in a straight line will continue in that state until a net external force acts upon it. Newton's first law is simply a summary of everyday observations that we all make. An example of Newton's first law is a coasting bicycle on a long, level, straight stretch of highway. If

---

[1] Newton is not the undisputed inventor of calculus. Gottfried Leibniz independently and nearly simultaneously developed calculus in the seventeenth century.

the bicyclist is moving with constant velocity and happens to ride through a large mud puddle on the road, the bicycle will slow down or even stop. The external frictional force provided by the mud puddle had the effect of changing the velocity of the bicycle and rider.

Newton's Second Law is usually stated as a deceptively simple equation: Force is equal to mass times acceleration, or $\vec{F} = m\vec{a}$. In one sense, the second law is a quantitative extension of the first law. For example, you have to push harder to get a bigger object to start moving.

Newton's Third Law states that, for every action, there is an equal and opposite reaction. If you push on an object, the object is pushing back on you.

MASS

Mass can be defined as the amount of matter in an object. In our experience of the macroscopic world, all ordinary objects have mass. Electromagnetic radiation, such as visible light, does not have mass. From a physics perspective, mass can be defined as the resistance of an object to acceleration. That is, if we want an object, which has a mass, to start moving, stop moving, or change direction, we have to give it a push. Thus, by simply rearranging Newton's second law we get:

$$m = \frac{\vec{F}}{\vec{a}}$$

$\vec{F}$ is the force in newtons (N), m is the mass in kilograms, and $\vec{a}$ is the acceleration. Acceleration has the somewhat unusual units of meters per second squared (m/s²). We will explore acceleration more fully in a later section. A larger mass requires a greater force than a smaller mass if we wish to achieve the same acceleration in both bodies.

VELOCITY

This brings us to the subject of velocity, a word that is frequently misused. Let's imagine rolling a patient toward the operating room on a gurney. How fast are you moving? This is where the distinct concepts of velocity and speed become confused. Average velocity is formally defined as the displacement divided by the time it takes to make that trip.

$$\text{average velocity} = \frac{\text{displacement}}{\text{time}}$$

In mathematical symbols, it is

$$\vec{v}_{\text{ave}} = \frac{\Delta \vec{r}}{\Delta t}$$

Displacement ($\Delta \vec{r}$) is the net change in the position of an object, while Δt is the time interval over which the change takes place. The capital delta (Δ) means "change in" (Δ = final − initial).

Note this is the average velocity over a finite time interval and does not take into account how the velocity might have been changing over that time interval. If we want to know the instantaneous velocity, that is, the velocity at a given instant (i.e., $\Delta t \rightarrow 0s$), we have to resort to calculus. Also note the small arrows over the velocity and displacement symbols. These arrows are used to indicate vector values. We will discuss more on vectors in the next section. The symbol $\Delta \vec{r}$ is used as a general variable for displacement, rather than, say, x, because the motion can have components in the x, y, and/or z directions in a Cartesian coordinate system.

Velocity is frequently, and incorrectly, interchanged with speed. Speed is not a vector value, but rather a scalar value and does not specify any particular direction of motion. Average speed is defined as:

$$\text{average speed} = \frac{\text{distance}}{\text{time}}$$

Using mathematical symbols, we get:

$$v_{\text{ave}} = \frac{\Delta x}{\Delta t}$$

Speed involves *distance*, whereas velocity involves *displacement*. Distance is simply the total length of travel, which is always going to be a positive number. Displacement is the net change in position with respect to some specified beginning point or origin in a coordinate system. Displacement can be positive, negative, or zero. Therefore, displacement, unlike distance, necessarily has directionality. Consider a 500-mile car race in which a vehicle travels 250 times around a 2-mile oval track in 3 hours and then stops at the finish line, which was also the starting point in the race. The average *speed* of the vehicle is 500 miles/3 h = 167 mph. However, the average *velocity* is 0 mph, because the car returned to exactly the same point from where it started, and so its displacement is 0 miles.

Vectors

Scalar quantities have magnitude only. Distance, height, mass, and age are examples of scalar quantities. Note that units typically need to be specified with scalar quantities. When indicating one's age, for example, we usually add the word "years" after the numerical value. Vector quantities have magnitude *and* direction. Once again the proper units need to be specified for vectors, and, in order to gain a complete understanding, the direction of the vector must also be specified. Vectors are prevalent in the world around us. Examples of vectors include velocity, weight, and the force applied to a syringe in order to inject a medication. Magnetic resonance imaging (MRI) relies on a field of vectors, known as a magnetic field, to orient the spin of protons contained in the hydrogen atoms of water molecules. Finally, the movement of a limb with a muscle involves a force applied at a certain distance from an axis of rotation, such as the elbow, to produce a torque.

Vector Addition

From a mathematical perspective, vectors are manipulated differently than scalar quantities. Since scalar quantities only have magnitude, to add scalar quantities, you just add the numerical values of the scalars. If you work 5 days this week and 5 days next week, you worked a total of 10 days. When adding vectors, however, we need to take into account both their magnitudes and their directions. When two or more vectors are added together, the sum is called the resultant. For example, think about giving someone a high five. Your hand is moving toward your friend and slightly upward. The motion of your hand, therefore, is a vector quantity. Your friend's hand is moving toward you and slightly upward. Now, when your hands collide the vectors are added, and where do your hands go? Don't they move (more or less) straight up? That is the resultant vector, which has a magnitude and direction that are based on the magnitude and direction of the motions of your hands before they collided.

One technique used to add vectors is called the graphical or head-to-tail method. Vectors are represented graphically by arrows of a given length, pointing in a specific direction. The length represents the magnitude or size of the vector quantity, while the direction of the arrow is an indication of the directionality of the vector quantity. The end of the vector arrow with the tip is called the head, while the other end is called the tail. In the graphical method, one simply slides the head of the first vector so that it touches the tail of the second vector. The resultant vector is then drawn from the tail of the first vector to the head of the second vector. When sliding vectors in order to put them head to tail, you are not allowed to change the length or the direction of any vector.

For example, suppose you are walking at 5 mph east on a train that is going 30 mph east. What is your resultant velocity? Let's represent your velocity and that of the train using arrows of appropriate length and direction. We will assume that east is to the right:

$$v_{you} = 5 \text{ mph} \qquad v_{train} = 30 \text{ mph}$$

Now, if we move the tip of one vector to the tail of the second vector,

and then draw the resultant by connecting the tail of the first vector to the head of the second vector,

we find the length of the resultant is 35 mph, while the direction indicates east.

Suppose you are walking at 5 mph west on a train that is going 30 mph east. What is your resultant velocity (relative to the ground)?

Answer: 25 mph, east

You are in a boat traveling east at 3.0 m/s and the current is going north at 1.0 m/s, as shown in Figure 3.1. What is your resultant velocity? This situation is more complicated because the two vectors are no longer parallel (pointing in the same direction) or antiparallel (pointing in the opposite directions). In this case they happen to be perpendicular. First, we'll draw the two velocity vectors, connect the head of one to the tail of the other, and draw the resultant (in blue). We will assume north is up and east is to the right.

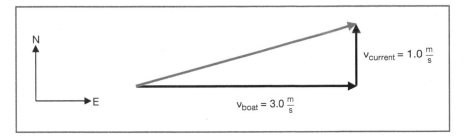

Figure 3.1 Displacement Diagram

Since, the three vectors form a right triangle, the length (magnitude) of the resultant can readily be calculated using the Pythagorean Theorem:

$$v_{resultant} = \sqrt{(v_{boat})^2 + (v_{current})^2}$$

$$v_{resultant} = \sqrt{(3.0)^2 + (1.0)^2} = \sqrt{10} = 3.2\,\frac{m}{s}$$

ACCELERATION

Acceleration is what gives thrill rides at amusement parks their "thrill." Newton's second law ensures that you will feel a force anytime you experience an acceleration. Average acceleration is defined as:

$$\text{average accelaration} = \frac{\Delta \text{ velocity}}{\Delta \text{ time}}$$

Mathematically, we can write average acceleration as:

$$\vec{a}_{ave} = \frac{\Delta \vec{v}}{\Delta t}$$

We thus see that acceleration is a vector and describes how velocity changes with time. Let's take a more careful look at the units on acceleration. In the official metric units, velocity has units of m/s. If we divide m/s by s, we get m/s². These are the units for acceleration, but you might find it useful to think of them in terms of "meters per second" *per second*. That is, how velocity (m/s) changes with each second of time.

Most people associate acceleration with "speeding up" or, more correctly, increasing the magnitude of velocity. However, one can have acceleration by changing the speed, the direction, or both. As an example, imagine a ball thrown straight upward with an initial velocity of 10.0 m/s. How does the position of the ball change with time? How does the velocity of the ball change with time? The position y and the velocity v are governed by the following two equations:

$$y = y_0 + v_0 t + \tfrac{1}{2}at^2$$

$$v = v_0 + at$$

where y is the position at time t; y_0 is the initial position at time $t = 0$; v is the velocity at time t; a is the acceleration in the y direction; and v_0 is the velocity at time $t = 0$. Note that when vector equations are written for one-dimensional motion, no arrows are used over the symbols representing the vectors.

For objects like a ball thrown in the air, a is equal to the acceleration due to gravity (9.8 m/s^2) and is pointed down toward the ground. Both equations above are for the y direction. Analogous equations can be written for the x and z directions. In order to use the above equations, we must first specify a coordinate system. In this case we will use the y-axis with the positive y direction pointing up and the negative y direction pointing down.

Because we decided that the $+y$ direction points up and gravity pulls objects down, the acceleration vector has a negative sign: $a = -9.8$ m/s^2 We will assign the starting position (y_0) as 0 m. The next step is to substitute the numbers into the two equations. For example, at $t = 0.20$ s, we have:

$$y = y_0 + v_0 t + \tfrac{1}{2}at^2 = 0 \text{ m} + \left(10 \tfrac{m}{s}\right) \cdot (0.20 \text{ s}) + \tfrac{1}{2} \cdot \left(-9.8 \tfrac{m}{s^2}\right) \cdot (0.20 \text{ s})^2$$
$$y = 0 \text{ m} + 2.0 \text{ m} - 0.196 \text{ m} = 1.804 = 1.80 \text{ m}$$

Similarly, the velocity at $t = 0.20$ s is:

$$v = v_0 + at = 10 \tfrac{m}{s} - \left(9.8 \tfrac{m}{s^2}\right) \cdot (0.2 \text{ s})$$
$$= 10 \tfrac{m}{s} - 1.96 \tfrac{m}{s}$$
$$= 8.04 \tfrac{m}{s}$$

The values for y and v as a function of time for the ball are given in Table 3.1.

Table 3.1 DISPLACEMENT AND VELOCITY AS A FUNCTION OF TIME

TIME (s)	POSITION (m)	VELOCITY (m/s)
0.00	0.00	10.0
0.20	1.80	8.04
0.40	3.22	6.08
0.60	4.24	4.12
0.80	4.86	2.16
1.00	5.10	0.20
1.0205	5.102	0.00
1.20	4.94	−1.76
1.40	4.40	−3.72
1.60	3.46	−5.70
1.80	2.12	−7.64
2.041	0.00	−10.0

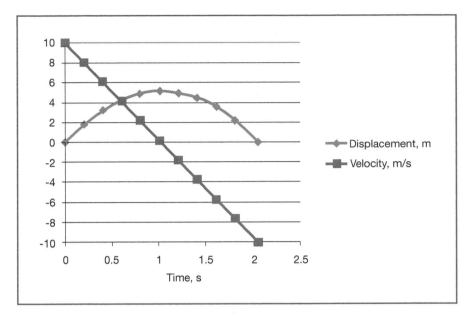

Notice how the velocity gets smaller and smaller as the time gets bigger and bigger, reaching 0 m/s at 1.0205 s after release. This makes sense, because you know that, if you throw a ball straight up in the air, it will slow, eventually stop, and then fall back toward you. The negative signs in front of the velocities past $t = 1.0205$ s simply indicate that the ball is headed in the $-y$ direction. The initial and final positions of the ball are the same. If we ignore air resistance, the only force acting on the ball once it leaves your hand is gravity. Notice also that the velocity of the ball when it returns to the starting point is equal in magnitude to the initial velocity, but in the opposite direction. Such motion is called *projectile motion*.

FORCE

Generally speaking, human beings are naturally curious. Using our curiosity, we find that a quick survey of the world around us may lead to the conclusion that things are hopelessly complicated. However, as far as we currently know, all of the interactions in our physical world are dominated by only four fundamental forces. In order of strongest to weakest, these four forces are: (1) the strong nuclear force; (2) the electromagnetic force; (3) the weak nuclear force; and (4) the gravitational force. Of course, it is easy to grasp the concept of gravity, but what exactly is a force?

Simply put, a force is a push or a pull. Pushing a plunger in a syringe, pulling a needle out of a patient's arm, and dripping IV fluid from a suspended bottle are all examples of forces. Newton's second law mathematically describes the relationship between force and mass:

$$\vec{F} = m\vec{a}$$

Newton's second law tells us that a force is required to produce an acceleration. For a given mass, a larger force produces a greater acceleration. A ride on a roller coaster is a wonderful way to experience forces. All of the exciting parts of the roller coaster involve acceleration, whether on a steep hill or on a sharp bank. In these situations our bodies feel the acceleration and we respond accordingly. Remember to eat that funnel cake after the roller coaster ride!

Gravity

The force of gravity dominates our macroscopic world. Gravity can be described as the universal *attraction* between all objects with mass. Even though gravity is the weakest of the four fundamental forces, it is ultimately responsible for perhaps the most violent of all objects in the universe, black holes. The *magnitude* form of Newton's Law of Universal Gravitation gives us the mathematical description of the attractive gravitational force between two point objects of mass m_1 and m_2:

$$F = G\frac{m_1 m_2}{r^2}$$

where G is the universal gravitational constant and r is the distance between the two masses. Note that the above equation gives the magnitude of the gravitational force. The direction of the force is attractive and directed along the imaginary line that connects the two objects.

In modern terms, we say the Earth creates a gravitational field around it, and the gravitational field is what exerts the force on any object that happens to be in the field. Thus, an object near the Earth, such as the Moon, is gravitationally attracted to it. Conversely, the Moon attracts the Earth. These two forces are equal in magnitude and oppositely directed. This is a clear example of Newton's third law: *For every action there is an equal and opposite reaction.*

Weight

At or near the Earth's surface, objects experience the attractive force of the Earth's gravitational field, the result of which is represented by the symbol g. Thus, at or near the Earth's surface, the magnitude of the acceleration due to gravity is:

$$g = 9.8 \,\tfrac{m}{s^2} = 32.2 \,\tfrac{ft}{s^2}$$

Weight is simply the gravitational force exerted on an object by a much larger object such as the Earth. Starting with the magnitude form of Newton's second law, $F = ma$, we can use W to represent the magnitude of the weight and g to represent the magnitude of the acceleration due to gravity to give:

$$W = mg$$

If using SI units, W will be in newtons, m in kilograms, and g will be in m/s². If one is using the British system, W will be in pounds, m in slugs, and g will be in ft/s². Notice that weight is a vector and is always directed downward, toward the center of the Earth.

Units of Mass and Weight

Some formal discussion is in order regarding mass and weight, since most of us use these terms loosely and frequently interchange them. Strictly speaking, mass and weight are two separate concepts. Mass is the amount of matter contained in an object, whereas weight is the gravitational force exerted on an object by a much larger object. If a patient were taken from the Earth's surface to the Moon, her mass would remain constant but her weight would change.

As previously discussed, the SI unit of mass is the kilogram while the British unit of mass is the slug. The SI unit of force is the newton (N), given by Newton's second law. One newton of force results from accelerating a mass of 1 kg by 1 m/s².

$$\vec{F} = m\vec{a}$$

$$1\text{N} = (1 \text{ kg}) \cdot \left(1 \tfrac{\text{m}}{\text{s}^2}\right) = 1 \tfrac{\text{kg} \cdot \text{m}}{\text{s}^2}$$

An object having a *mass* of 1 kg would thus have a *weight* of 9.8 N.

$$9.8 \text{ N} = (1 \text{ kg}) \cdot \left(9.8 \tfrac{\text{m}}{\text{s}^2}\right)$$

The British unit of force is the pound (lb), once again given by Newton's second law:

$$1 \text{ lb} = (1 \text{ slug}) \cdot \left(1 \tfrac{\text{ft}}{\text{s}^2}\right) = 1 \tfrac{\text{slug} \cdot \text{ft}}{\text{s}^2}$$

The weight of a 1 slug mass on the Earth's surface is therefore 32.2 lb.

$$32.2 \text{ lb} = (1 \text{ slug}) \cdot \left(32.2 \tfrac{\text{ft}}{\text{s}^2}\right)$$

The key conversion factors are 1 slug is equal to 14.62 kg and 1 lb is equal to 4.45 N.

For example, let's calculate the mass (in kg) and the weight (in N) of a $15\overline{0}$-lb person.

You probably learned to convert pounds into kilograms by dividing pounds by 2.2. This is technically incorrect because you are converting pounds (a weight quantity) into kilograms (a mass quantity), although it does give the numerically correct answer, at least when the acceleration on the object is 9.8 m/s². The formally correct solution to this question is:

Step 1: Convert the weight in pounds into weight in newtons.

$$\text{Weight} = 150 \text{ lb}\left(\tfrac{4.45 \text{ N}}{\text{lb}}\right) = 667.5 \text{ N}$$

To the correct number of significant figures, the weight is 668 N.

Step 2: Use Newton's second law to calculate mass.

$$\frac{667.5 \text{ N}}{9.8 \tfrac{\text{m}}{\text{s}^2}} = 68.1 \text{ kg}$$

To the correct number of significant figures, the mass is 68.1 kg.

The acceleration due to gravity on the moon is one-sixth that of Earth, 1.63 m/s². Let's calculate the mass and weight of a patient on the moon, if her mass on Earth is 68.1 kg.

Mass is invariant, so the mass of the patient will be 68.1 kg on the moon. However, the weight depends on gravity, and the patient will weigh (68.1 kg) (1.63 m/s²) = 111 N.

PRESSURE

In the world of physics, pressure is defined to be the force per unit area. Mathematically, we can write this as

$$\text{Pressure} = \frac{\text{Force}}{\text{Area}}$$

Pressure can be increased by either *increasing* the applied force or *decreasing* the area over which the force is applied. Conversely, the pressure can be decreased by *decreasing* the applied force or *increasing* the area over which the force is applied. A practical application of this can be demonstrated using a balloon. It is difficult to pop a balloon using a blunt object such as your finger. You must apply a fairly large force due to the large area of contact between your finger and the balloon. However, it is much easier to pop the same balloon using a pin. In this case, a small force applied over an extremely small area produces a very large pressure.

Units of Pressure

The units typically used for pressure are greatly varied, depending on the particular field of science involved. Since pressure is defined as a force per unit area, one would expect the units to be that of force divided by area. In many cases we find this to be true; however, in others we find some very unexpected units.

In the British system, a common set of units for pressure is pounds per square inch (psi). The SI unit of pressure is the pascal, which is the pressure exerted by a force of 1 newton over 1 square meter of area.

$$\frac{1 \text{ N}}{\text{m}^2} = 1 \text{ pascal} = 1 \text{ Pa}$$

The pascal represents a very small pressure, and therefore the most common applications, such as tire pressure, will use kilopascals (kPa) instead of Pa. Other units of pressure include the torr (or millimeter of mercury, mmHg), inches of mercury, the atmosphere (atm), and the bar. A torr is an amount of pressure necessary to support a column of mercury 1 mm in height. One atmosphere of pressure is loosely defined by the atmospheric pressure at sea level, but is more precisely defined as the pressure necessary to support a column of mercury 760 mm in height. One bar of pressure is equal to 100 kPa. The relationships between the various units of pressure are as follows:

$$1 \text{ atm} = 760 \text{ torr} = 760 \text{ mm Hg} = 101{,}325 \text{ Pa}$$
$$1 \text{ atm} = 1.013 \text{ bar}$$
$$1 \text{ bar} = 100{,}000 \text{ Pa} = 10^5 \text{ Pa} = 100 \text{ kPa}$$
$$\frac{1 \text{ lb}}{\text{in}^2} = 6895 \text{ Pa}$$

For example, if a barometer, which is a device used to record atmospheric pressure, reads 735 mmHg, what is this pressure in units of bar, torr, atm, and kPa?

Answer: 0.980 bar, 735 torr, 0.967 atm, 98.0 kPa

Now we'll let you work another example. What is the pressure of 1.00 atm in psi?

Answer: 14.7 psi

Syringes

As a nurse, there is no doubt that you use syringes frequently. Let's take a closer look at how much pressure is created when you apply a 5.00 N force on a syringe plunger that has a diameter of 1.00 cm. Pressure is force per unit area. We have been provided with the force in newtons; however, we need to calculate the cross-sectional area of the plunger in m² in order to obtain units of pascals or kilopascals. The area of a circle is πr^2:

Video 3.1

$$\text{Area} = \pi r^2 = \pi \left(\frac{0.01 \text{ m}}{2}\right)^2 = 7.854 \times 10^{-5} \text{ m}^2$$

Note in the above calculation that the diameter was converted to meters and subsequently divided by 2 in order to get the radius of the circle in meters. We are now ready to calculate the pressure created by the plunger in the syringe:

$$\text{Pressure} = \frac{\text{Force}}{\text{Area}} = \frac{5.00 \text{ N}}{7.854 \times 10^{-5} \text{ m}^2} = 6.37 \times 10^4 \text{ Pa} = 63.7 \text{ kPa}$$

Example: How much pressure is created when you apply a 5.00 N force on a syringe plunger that has a diameter of 2.00 cm?

Answer: 15.9 kPa

Notice that doubling the diameter of the syringe decreases the pressure by a factor of 4!

Table 3.2 shows actual data obtained by applying known forces to real syringes. The pressure in each syringe was measured by connecting the syringe to a computer-controlled pressure sensor. Thus, these data were collected under static conditions. From the data, one can clearly see that the pressure inside a given syringe increases as the magnitude of the applied force is increased. In addition, for a given magnitude of force, as the syringe area decreases, the corresponding pressure increases. Notice that smaller syringes have the capacity to develop very high pressures. As an example, a plugged insulin syringe with a cross-sectional area of 7.07 x 10⁻⁶ m² could develop a pressure greater than 13 atm if a force of just 9.8 N was applied.

A block of wood, having a mass of 16 g and measuring 3.0 cm × 3.0 cm × 3.0 cm, is resting on a table. How much pressure does it exert on the table? To calculate the pressure, we will need the force involved and the area. The force is simply the weight (in newtons) of the block, and the area is the cross-sectional area of one face of the cube of wood.

$$\text{Force} = \text{Weight} = mg = (0.016 \text{ kg}) \cdot \left(9.8 \frac{\text{m}}{\text{s}^2}\right) = 0.1568 \text{ N}$$
$$\text{Area} = (0.030 \text{ m}) \cdot (0.030 \text{ m}) = 9.0 \times 10^{-4} \text{ m}^2$$

Finally,

$$\text{Pressure} = \frac{\text{Force}}{\text{Area}} = \frac{0.1568 \text{ N}}{9.0 \times 10^{-4} \text{ m}^2} = 174.2 = 1.7 \times 10^2 \text{ Pa}$$

Table 3.2 FORCE/PRESSURE DATA FOR REAL SYRINGES

NOMINAL SYRINGE VOLUME (mL)	DIAMETER (mm)	AREA (m²)	APPLIED FORCE (N)	PRESSURE (kPa)
60	26.45	5.49×10^{-4}	4.9	7.7
60	26.45	5.49×10^{-4}	9.8	16.1
60	26.45	5.49×10^{-4}	14.7	25.0
30	23.42	4.31×10^{-4}	4.9	8.5
30	23.42	4.31×10^{-4}	9.8	20.7
30	23.42	4.31×10^{-4}	14.7	31.4
20	19.30	2.93×10^{-4}	4.9	11.0
20	19.30	2.93×10^{-4}	9.8	29.3
20	19.30	2.93×10^{-4}	14.7	45.9
10	14.73	1.70×10^{-4}	4.9	18.9
10	14.73	1.70×10^{-4}	9.8	48.0
10	14.73	1.70×10^{-4}	14.7	82.4
3	8.72	5.97×10^{-5}	0.98	6.5
3	8.72	5.97×10^{-5}	2.0	25.1
3	8.72	5.97×10^{-5}	2.9	40.6
3	8.72	5.97×10^{-5}	3.9	55.9
3	8.72	5.97×10^{-5}	4.9	73.7
3	8.72	5.97×10^{-5}	5.9	94.0
3	8.72	5.97×10^{-5}	6.9	107.5

A cube measuring 10 cm on an edge and a cylinder with a diameter of 10 cm and a height of 10 cm are on a table. Each has a mass of 1 kg. Which exerts a greater pressure on the table?

Answer: $P_{cylinder} = 1.2 \times 10^3$ Pa, $P_{cube} = 9.8 \times 10^2$ Pa

ATMOSPHERIC PRESSURE

The atmosphere is very important to life on Earth. This blanket of gas surrounding the Earth keeps the planet warm, protects us from radiation and meteoroids, and provides us with oxygen to breathe. In a sense, we can view ourselves as living at the bottom of a great "ocean" of gas. There is pressure associated with our atmosphere. Many times we don't consciously think about this pressure because we have become acclimated to it. However, we sure notice it when the pressure changes such as when we go to higher altitudes. Air pressure results from gravity pulling on the atmosphere, and the resulting force is spread over Earth's surface. Literally, we feel the weight of the air above us. A similar feeling can be experienced when swimming near the bottom of a deep pool.

The Earth is *approximately* a sphere with a radius of 3950 miles. If the air pressure is around 14.7 psi, we can estimate the total weight of the atmosphere. First, let's change the radius from miles to inches:

$$3950 \text{ mi}\left(\frac{5280 \text{ ft}}{1 \text{ mi}}\right)\left(\frac{12 \text{ in}}{1 \text{ ft}}\right) = 2.50 \times 10^8 \text{ in}$$

Now, let's calculate the approximate surface area of Earth, using the surface area of a sphere equaling $4\pi r^2$:

$$\text{Area} = 4 \cdot \pi \cdot (2.50 \times 10^8 \text{ in})^2 = 7.87 \times 10^{17} \text{ in}^2$$

Now, plugging in the pressure of 14.7 psi applied over an area of 787 quadrillion square inches gives us a total weight of the atmosphere of 11.6 quintillion pounds, or 5.79 quadrillion tons.

$$F = \text{Area} \cdot \text{Pressure} = (7.87 \times 10^{17} \text{ in}^2)\left(\frac{14.7 \text{ lb}}{\text{in}^2}\right) = 1.16 \times 10^{19} \text{ lb}$$

MEASURING PRESSURE

Barometer

Pressure can be measured with many different types of devices. A common instrument used to measure atmospheric pressure is called the *barometer*. A simple mercury barometer consists of a tube closed at one end and open at the other. The tube is filled with mercury and inverted in a larger reservoir of mercury that is open to the atmosphere. This geometry will result in a vacuum at the top of the glass tube as the mercury inside the tube runs downward, out of the tube and into the reservoir of mercury (Figure 3.2).

The pressure of the atmosphere will push down on the surface of the mercury reservoir. There is essentially no pressure at the top of the inverted tube, and the pressure of the atmosphere will support a column of mercury to a height *h*. Thus, we have two opposing forces in balance with each other. There is the force of the weight of the mercury in the column and the force due to the air pushing on the surface of the mercury reservoir. As air pressure increases, there is more force on the mercury reservoir, and that pushes the mercury

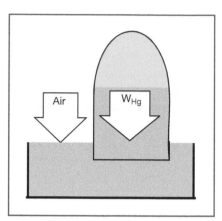

Figure 3.2 A Barometer

higher into the column. The height to which the column rises is dictated by the following equation:

$$P_{atmosphere} = \rho g h$$

where $P_{atmosphere}$ is the atmospheric pressure in Pa, ρ is the density of the fluid (mercury in this case), g is the acceleration due to gravity, and h is the height of the column in meters.

We can easily derive this equation from the definition of pressure. In order to support this column of mercury, the air must exert a pressure equal to the pressure exerted by the mercury. Therefore, $P_{atmosphere} = P_{mercury}$. So, let's calculate the pressure exerted by a column of mercury, and this value will be equal to the atmospheric pressure. First, let's start by imagining a column of mercury of height h. Let us also imagine that this column of mercury is a regular cylinder and the circular base of the cylinder has an area A. The volume of the mercury is, therefore, equal to the height of the cylinder times the area of the circular base. Thus, the volume of a regular cylinder is given by $V = \pi r^2 h$. We can use the density of mercury to convert the volume of the mercury into its mass:

$$\text{Mass} = \text{density} \times \text{volume} = \text{density} \times \text{area} \times \text{height}$$

Or, symbolically

$$m = \rho \times A \times h$$

We can use Newton's second law to calculate the weight of the mercury column. Remember that weight is a force.

$$W = mg = (\rho \times A \times h) \times g$$

Now, let's plug this expression for the force into the definition of pressure:

$$P = \frac{\text{Force}}{\text{Area}} = \frac{(\rho \times A \times h) \times g}{A}$$

Notice the area terms cancel out, leaving

$$P = \rho g h$$

If the pressure is 1.00 atm, how high will a column of mercury be? To answer this question we need to rearrange the pressure equation above by solving for h to get:

$$h = \frac{P_{atmosphere}}{\rho g}$$

We can now substitute in the various quantities using SI units (the density of Hg is 13.69/cm³ which is equal to 13,600 kg/m³ or "g"?):

$$h = \frac{101{,}325 \text{ Pa}}{\left(13{,}600 \frac{\text{kg}}{\text{m}^3}\right) \cdot \left(9.8 \frac{\text{m}}{\text{s}^2}\right)} = 0.760 \text{ m} = 760 \text{ mm}$$

This example illustrates the fact that 1 atm of pressure will support a column of mercury 760 mm tall.

Example: How tall of a column of water would 1.00 atm of pressure support? The density of water is 1000 kg/m³.

Answer: 10.3 m

Manometer

Pressure differences can be measured with U-shaped tubes filled with a fluid of known density such as mercury. These devices are called *manometers* (Figure 3.3).

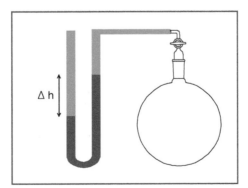

Figure 3.3 A Manometer

To use such a device, one has to connect one end of the manometer to the system whose pressure you want to measure and leave the other end open to the atmosphere. The pressure difference between the system and the atmosphere can be calculated by measuring the height difference, Δh, of the fluid in the U-shaped tube and substituting it into the following equation:

$$\Delta P = \rho g \Delta h$$

The pressure of the system can subsequently be determined from:

$$\Delta P = P_{\text{system}} - P_{\text{atmosphere}}$$

Notice that a positive ΔP means the system has a pressure higher than atmospheric pressure, whereas a negative ΔP means the object has a pressure lower than atmospheric pressure. Manometers become inconvenient for measuring gauge pressures of greater than about 1 atm because of the large column of mercury that must be contained. Less dense liquids, such as water, can be used in a manometer to measure smaller pressure differences.

Aneroid Bellows Gauge

Aneroid gauges are devices that do not require the presence of a liquid such as mercury to operate. In fact, the word "aneroid" has its origins in a Greek word literally meaning without liquid. These types of gauges rely on the expansion or contraction of bellows as the pressure changes. There are two basic types of aneroid bellows gauges, both of which are illustrated in Figure 3.4.

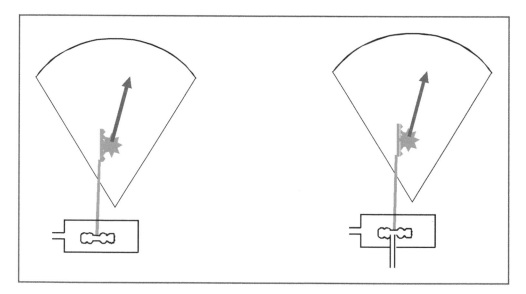

Figure 3.4 Aneroid Barometers

The gauge shown on the left is sealed with respect to the atmosphere and is used to measure changes in absolute pressure. The gauge on the right is open to the atmosphere and sample, and therefore it is used to measure gauge pressure.

Bourdon Gauge

Bourdon gauges are used on gas cylinders and are also considered a type of aneroid gauge. These devices have a coiled tube (shown in Figure 3.5) and are used to measure the pressure difference between the pressure exerted by the gas in a cylinder and the atmospheric pressure. The coiled tube is mechanically coupled to a pointer (shown in red). As a gas at a pressure above atmospheric pressure enters the coiled tube, it causes it to slightly uncoil, kind of like those New Year's Eve paper noisemakers. This causes the pointer to move over a numerical scale, thereby indicating the gauge pressure in the tank.

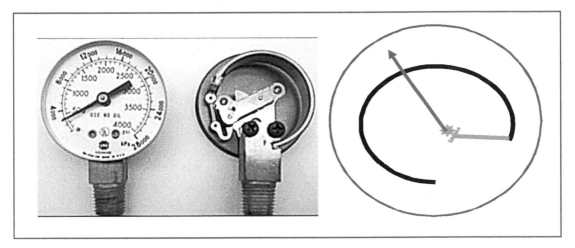

Figure 3.5 Bourdon Gauge

Gauge Pressure and Total Pressure

You must be cautious when dealing with pressures. It is vital that you specify what type of pressure is being utilized and how that pressure was determined. Barometers measure the actual pressure or the absolute pressure. Manometers and Bourdon gauges measure gauge pressure. Because these devices work against atmospheric pressure, their readings are affected by atmospheric pressure. Gauge pressure is the pressure of a system above (or below) atmospheric pressure. The total pressure includes the atmospheric pressure plus the gauge pressure. Thus, we can write:

$$P_{total} = P_{gauge} + P_{atmosphere}$$

Suppose a tank of oxygen gas has a gauge that reads 45 psi. If the atmospheric pressure at that time is 14 psi, what is the total pressure in the tank?

Answer: 59 psi

Suppose you have a cylinder of oxygen gas, and the Bourdon gauge indicates a pressure of 0 psi. If the atmospheric pressure is 14 psi, what is the pressure exerted by the oxygen gas in the cylinder? What would the gauge read if the cylinder was taken up into the mountains where the atmospheric pressure was only 10 psi?

The pressure of the oxygen inside the "empty" cylinder is equal to the atmospheric pressure, 14 psi. If the cylinder were taken into the mountains, the pressure gauge would register 4 psi.

OSCILLOMETRY

Automated, noninvasive blood pressure (NIBP) measurement devices have essentially become the norm at many hospitals, clinics, and medical offices. Automated NIBP instruments are especially attractive when regularly repeated measurements are needed. Many automated NIBP measurement devices are based on oscillometry, which is a method that relies on the measurement of blood pressure oscillations. One particularly effective technique used to measure the rapid variations in pressure is based on the piezoelectric transducer. Generally speaking, a transducer is a device that interconverts a signal between the nonelectrical and electrical domains. In this case, the nonelectrical signal is a pressure change. The piezoelectric transducer subsequently gets distorted by the pressure change, resulting in an electrical signal (a voltage) that is proportional to the pressure change. Systolic and diastolic blood pressures are calculated based on the oscillatory pressure readings using computerized algorithms. Modern quartz watches also rely on the piezoelectric effect, which allows them to keep very accurate time.

Summary

- Newton's First Law: An object at rest or moving at constant speed in a straight line will continue in that state until a net external force acts upon it.
- Newton's Second Law:

$$\vec{F} = m\vec{a}$$

- Newton's Third Law: For every action, there is an equal and opposite reaction.
- Mass is the amount of matter contained in an object.
- Average velocity:

$$\vec{v}_{ave} = \frac{\Delta \vec{r}}{\Delta t}$$

- Average speed:

$$v_{ave} = \frac{\Delta x}{\Delta t}$$

- Vectors are quantities that have magnitude and direction.
- Average acceleration:

$$\vec{a}_{ave} = \frac{\Delta \vec{v}}{\Delta t}$$

- A force is a push or a pull.
- The *magnitude* of the force of gravity:

$$F = G\frac{m_1 m_2}{r^2}$$

- The *magnitude* of an object's weight:

$$W = mg$$

- The SI unit of mass is the kilogram, kg, while the SI unit of force (and hence weight) is the newton, N. The British unit of mass is the slug, while the British unit of force (and hence weight) is the pound, lb.
- Pressure:

$$\text{Pressure} = \frac{\text{Force}}{\text{Area}}$$

- Common units of pressure:

$$1 \text{ atm} = 760 \text{ torr} = 760 \text{ mm Hg} = 101{,}325 \text{ Pa}$$
$$1 \text{ atm} = 1.013 \text{ bar}$$
$$1 \text{ bar} = 100{,}000 \text{ Pa} = 10^5 \text{ Pa} = 100 \text{ kPa}$$
$$\frac{1 \text{ lb}}{\text{in}^2} = 6895 \text{ Pa}$$

■ A barometer is a device used to measure atmospheric pressure. The height of a column of fluid, typically mercury, is a direct measure of atmospheric pressure:

$$P_{atmosphere} = \rho g h$$

■ A manometer is a U-shaped tube, filled with a fluid of known density, used to measure pressure differences. The height difference of the fluid in the two arms of the manometer indicates the pressure difference:

$$\Delta P = \rho g \Delta h$$

■ Aneroid gauges measure pressure without the use of a fluid.
■ The relationship between total pressure, gauge pressure, and atmospheric pressure is:

$$P_{total} = P_{gauge} + P_{atmosphere}$$

Review Questions for Physics (Part 1)

1. What is the difference between velocity and acceleration?

2. What is a force?

3. State Newton's Second Law and what it means.

4. What is the difference between mass and weight?

5. What are the metric units of weight and mass?

6. What is the weight in newtons of a person having a weight of $1\overline{5}0$ lb?

7. You apply a force of $3\overline{0}$ N to a syringe. What is the pressure (in units of N/cm^2) inside the syringe if the diameter is 0.75 cm?

8. A patient has a systolic blood pressure of $10\overline{0}$ mmHg. Express this pressure in units of atm, torr, and kPa.

9. What is the pressure exerted on a table by a block of lead measuring 3.00 cm on each edge? The density of lead is 11.2 g/cm^3.

10. Describe how a barometer works.

11. Describe how a Bourdon gauge works.

12. What is the difference between gauge pressure and absolute pressure?

13. Calculate the area (in m^2) of a circle (circle 1) that has a diameter of 10.0 cm. Calculate the area (in m^2) of a circle (circle 2) that has a diameter of 5.00 cm. Now, calculate the ratio (area of circle 1)/(area of circle 2), and compare it to the ratio of (radius of circle 1)/(radius of circle 2). What effect does cutting the diameter by a factor of 2 have on the area?

14. For a 7.0 N force, calculate the pressure if the force is distributed over the area of a circle with a diameter of 10 cm. For a 7.0 N force, calculate the pressure if the force is distributed over the area of a circle with a diameter of 5.0 cm. What is the ratio between the pressures you just calculated?

15. Calculate the total pressure (P_{total}) if the gauge pressure (P_{gauge}) on an oxygen cylinder reads 10 psi and atmospheric pressure ($P_{atmosphere}$) is 14.7 psi.

16. Convert $20\overline{0},000$ Pa to atm and mmHg.

17. Convert 32 psi to Pa.

18. Using Newton's Second Law, calculate the acceleration of a 2 kg mass acted on by a 6 N force. Specify the proper SI units.

19. Solve Newton's Second Law for m, explicitly showing that the unit for m will be kg.

20. Is a pressure of 29 in of mercury above or below 1 atm?

21. Indicate whether these combinations of units indicate velocity, acceleration, work, pressure, force, or power.
    a. foot · pounds
    b. inches/day^2
    c. kg·m^2/s^3
    d. liter · atmosphere

22. Newton's First Law is most closely associated with:
    a. conservation of energy
    b. conservation of momentum
    c. equivalence between force and mass times acceleration
    d. inertia

23. Newton's Second Law states $\vec{F} = m\vec{a}$. If an object A has a greater mass than object B, then
    a. Object A weighs more than object B, under all conditions.
    b. Identical forces acting on both objects cause object A to undergo a greater acceleration than object B.
    c. Object A always has a greater inertia than object B.
    d. All of these statements are correct.

24. You are riding the roller coaster at the state fair. During which of these are you NOT accelerating?
    a. When the roller coaster speeds up going down a hill
    b. When the roller coaster slows down going up a hill
    c. When the roller coaster makes a hairpin turn
    d. None of the above

25. A gas cylinder has a weight of 100 lb. What is the unit of its metric weight?
    a. pounds
    b. kilograms
    c. newtons
    d. joules

26. What is the metric weight (*not* mass) of the cylinder in the previous question?
    a. 980
    b. 45
    c. 445
    d. 100

27. A gas cylinder has a weight of 300 N. This gas cylinder has a diameter of 10 in, or an area of 0.05 m^2. How much pressure does this cylinder exert on the floor in units of N/m^2?
    a. 300 N/m^2
    b. 6000 N/m^2
    c. 15 N/m^2
    d. 612 N/m^2

28. The metric unit of pressure defined by N/m² is the:
    a. pascal
    b. bar
    c. torr
    d. atmosphere

29. An ACME operating table has four legs. Each leg is square, closed on the end, and measures 0.10 m × 0.10 m. If the table weighs 1200 N and the patient on the table weighs 800 N, what is the total pressure exerted on the operating room floor?
    a. $2\bar{0}0{,}000$ N/m²
    b. $1\bar{0}0{,}000$ N/m²
    c. $5\bar{0}{,}000$ N/m²
    d. $2\bar{0}00$ N/m²

30. What is the weight of the patient in the previous question in pounds?
    a. 363 lb
    b. 81 lb
    c. 180 lb

31. The cylinder pressure on an oxygen tank reads $3\bar{0}0$ kPa. What is the pressure in psi?
    a. 43.5
    b. 445
    c. 101
    d. 14.7

32. Which of these has an operating principle that most resembles a Bourdon gauge?
    a. Keeping soda pop in a straw by holding your finger on the end
    b. Holding a champagne bottle by the neck so the contents spurt out when you open it. (N.B.: If it's good champagne, keep your hot little hands off the neck.)
    c. That New Year's Eve "roll out" party favor
    d. The low-pressure center in the middle of the "swirly thing" in a draining bathtub

33. Which type of pressure-measuring device uses a column of mercury?
    a. Bourdon gauge
    b. Aneroid barometer
    c. Torricelli barometer
    d. None of these

34. When is the gas pressure inside a cylinder equal to zero?
    a. When the pressure gauge reads zero
    b. When no more gas comes out of the open valve
    c. When the cylinder is evacuated with a vacuum pump

35. Which of the following plugged syringes will have the greatest internal pressure?
    a. Applied force of F and radius r
    b. Applied force of $2F$ and radius r
    c. Applied force of F and radius r/2
    d. Applied force of $2F$ and radius 2r

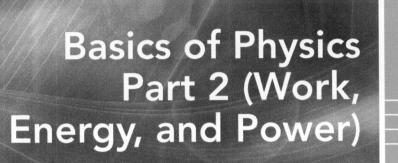

Basics of Physics Part 2 (Work, Energy, and Power)

In the previous chapter we began to explore the most basic concepts in physics, a discussion that led to the concept of pressure, which is, of course, central to the regulation and administration of gases. In this chapter we continue our survey of fundamental physics concepts. Specifically, we will discuss three closely related concepts: work, energy, and power. In addition, we will discuss the three laws of thermodynamics and heat capacity. These concepts undergird a myriad of physiological processes, from passive activities such as osmosis and evaporation of anesthetic medications, to those activities upon which we actively expend our energy resources, such as breathing.

WORK

Our everyday usage of the term *work* is probably significantly different from the scientific definition. We have already studied forces, and forces can be used to do work. In physics, a force does work when it acts on an object and displaces the object in the direction of the force. Mathematically, we can define work as

$$W = F \cdot d \cdot \cos(\theta)$$

where W is the work done, F is the *magnitude* of the force, d is the *magnitude* of the displacement, and θ is the angle between the line of the force and the direction of the displacement. For our purposes, we will assume that the direction of the force and displacement are the same; therefore $\theta = 0°$ and $\cos(0°) = 1$. Thus, our equation for work simplifies to:

$$W = F \cdot d$$

According to this definition of work, if there is no force, no work is done. Likewise, if the magnitude of the displacement is zero, no work is done. We will also assume, for simplicity's sake, that the magnitude of the force is constant over the displacement. In the real world, forces can be highly variable, requiring the use of calculus to calculate the amount of work involved.

The metric system (SI) unit of work is the joule (J), which is a combination of two of the basic SI quantities, the newton and the meter. One joule of work is expended when a force of 1 N acts over a displacement of 1 m.

$$1 \text{ J} = 1 \text{ N} \cdot \text{m} = 1\left(\frac{\text{kg} \cdot \text{m}}{\text{s}^2}\right) \cdot \text{m} = 1 \frac{\text{kg} \cdot \text{m}^2}{\text{s}^2}$$

You might occasionally see the centimeter–gram–second (cgs) system units instead of the SI units. In the SI system, the unit of force is the newton, while in the cgs system, the unit of force is the dyne. In this system, the unit of work is called the erg and is equal to a dyne-centimeter. Applying a force of 1 dyne over a displacement of 1 cm expends 1 erg of work.

$$1 \text{ erg} = 1 \text{ dyn} \cdot \text{cm} = 1\left(\frac{\text{g} \cdot \text{cm}}{\text{s}^2}\right) \text{cm} = 1 \frac{\text{g} \cdot \text{cm}^2}{\text{s}^2}$$

In the British system, the unit of work is called the foot-pound (ft·lb), because the pound is a unit of force and the foot is a unit of displacement. These units of work are all related by simple conversion factors.

$$1 \text{ J} = 0.738 \text{ ft} \cdot \text{lb}$$
$$1 \text{ erg} = 10^{-7} \text{ J} = 7.38 \times 10^{-8} \text{ ft} \cdot \text{lb}$$

This brings us to some example problems. Before tackling these problems, some discussion on problem-solving strategy is in order. As with any word problem, the first step in solving the problem is to completely read through it once. Don't stop in the middle somewhere just because you don't understand a term or get hung up on a fuzzy concept. *Completely* read through the problem! Next, identify the concepts that are involved, identify the key words, and finally list what information is provided. Many times, students become uneasy when they are provided with too much information. Students then tend to force unnecessary information into the answer. Remember, just use what is needed.

Let's apply this strategy to an example. How much work is done lifting a 100-kg patient up a distance of 0.020 m to slide him onto the operating table?

This problem involves the concept of work, so a good equation to start with is $W = F \cdot d$. To apply this equation, we must obviously know the magnitude of some force and the magnitude of a displacement. The distance is 0.020 m, but what is the force? The key phrase in this particular problem is "100-kg patient." Since the patient has mass, the patient has a weight, which is a force. Using $W = m \cdot g$, we can readily calculate the patient's weight:

$$\text{Weight} = 100 \text{ kg} \cdot 9.8 \frac{\text{m}}{\text{s}^2} = 980 \frac{\text{kg} \cdot \text{m}}{\text{s}^2} = 980 \text{ N}$$

The work can now be calculated since we have both the force and a distance.

$$\text{Work} = F \cdot d = 980 \text{ N} \cdot 0.020 \text{ m} = 19.60 \text{ N} \cdot \text{m} = 2.0 \times 10^1 \text{ J}$$

Let's try another one. How much work is done by pushing on a 100-kg patient on a 75-kg operating table and you can't get the darned thing to move?

This problem also involves the concept of work, so a good equation to start with is still $W = F \cdot d$. Note that we must consider the mass of both the patient and the operating table. A complete reading of the problem will yield some very valuable information. The object to which the force is being applied, the patient on the table, does not move. Therefore, $d = 0$ m and the work done is 0 J.

These kinds of problems, however, can get very complicated. As an example, consider the following. How much work is done rolling a 100-kg patient on a 75-kg operating table 10 m across an operating room? Be careful! This is a much harder problem than it first appears to be. There are many factors that must be considered in this problem. Are the patient and bed initially at rest, or are they already rolling along? Are they moving (or to be moved) at a constant velocity or a changing velocity? Are we expected to stop the patient and bed when it reaches its destination? Some of these activities result in work, and some do not. For example, gravity operates vertically, in the downward direction. You will move the patient in the horizontal direction. No component of the gravity acceleration vector operates in the horizontal direction. So, the mass of the patient and the operating table has *no force* due to gravity in the horizontal direction. So, if there is no friction and the patient is rolling at a constant velocity (i.e., the acceleration is zero), then no force is needed to keep the patient rolling. If no force is applied, then no work is done. Many physics problems specify there is no friction, or at least, friction is to be ignored. If we consider friction, and we realistically cannot ignore it, then the work is done to overcome friction. Starting or stopping the patient and bed requires the application of a force, in accordance with Newton's first law, and therefore cause work if the patient and bed move at all during the application of the force. The lesson from this example is to recognize that work is the product of the magnitudes of force and displacement, but you need to carefully consider all forces that operate on the system, as well as the direction in which the forces operate.

Kinetic Energy and Work

Work and energy are very closely related in physics. We've already defined work. Let's now focus on *kinetic* energy. Kinetic energy is also called energy of motion because one way to define it is one-half the product of mass times the square of the speed.

$$\text{Kinetic energy} = KE = \tfrac{1}{2}mv^2$$

Kinetic energy is the energy a mass has by virtue of being in motion. Focusing on the units for KE, we find them to be equivalent to the units of work:

$$KE = \tfrac{1}{2}\,\text{kg} \cdot (\tfrac{\text{m}}{\text{s}})^2 = \frac{\text{kg} \cdot \text{m}^2}{\text{s}^2} = \text{J}$$

We thus see that work and kinetic energy both have units of joules. To understand the relationship between work and kinetic energy, we simply turn to our everyday experiences. You see a patient lying on a table in the wrong room. You take it upon yourself to push the patient–table system into the correct room. You have thus performed work on the patient–table system. In performing this work, you changed the kinetic energy of the system. Initially, the system had no kinetic energy because the system's speed was 0 m/s. However, once the system started moving, it acquired kinetic energy because you expended work.

The relationship between work and kinetic energy is given by the Work–Energy or Work–Kinetic Energy Theorem, which states that *the total work done on a system is equal to the system's change in kinetic energy.* Mathematically, the Work–Energy Theorem is given by:

$$W_{total} = \Delta KE = KE_{final} - KE_{initial} = \frac{1}{2}mv^2_{final} - \frac{1}{2}mv^2_{initial}$$

Gases and Work

Gases that move can also do work, and breathing is moving gases. Therefore, breathing accomplishes work. Let's start with a simple model by imagining a container equipped with a frictionless piston that contains 2.5 L of air at a pressure of 600 Pa above atmospheric pressure (Figure 4.1). The gas will expand until the pressures are equalized.

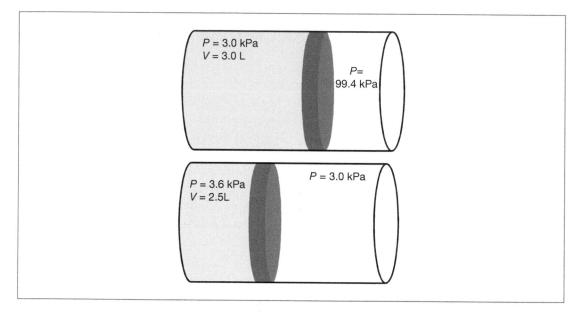

Figure 4.1 Expanding Gases and Work

This is very much like your lungs, by the way. As your diaphragm contracts, it reduces the pressure in the thoracic cavity to about 600 Pa below atmospheric pressure, and about 500 mL of air is drawn into your lungs. Note that we have used an idealized object (a frictionless piston) in this example. While such an object does not exist, it is frequently useful to incorporate such idealizations into working models to help us understand more complicated systems. Now, back to our frictionless piston with 2.5 L of air. How much work is done when the piston slides to a new volume of 3.0 L, assuming the pressure remains constant?

To answer this question we need to go back to our definition of work. From before we know:

$$W = F \cdot d$$

Now, let's assume we have a right circular cylinder with height h and cross-sectional area A. Of course the area is the product of π times the square of the radius ($A = \pi r^2$). Note that we do not necessarily have to have a right circular cylinder for this argument to be valid, but we'll use it anyway to make things simple. The volume of the cylinder is the product of the area and the height.

$$V = A \cdot h$$

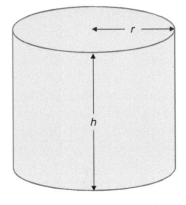

Notice that h can be interpreted as the distance our piston moved. So, let's substitute d for h and then solve the volume equation for d:

$$d = \frac{V}{A}$$

Finally, let's substitute this expression into our equation for work.

$$W = F \cdot d = F \cdot \frac{V}{A} = \frac{F}{A} \cdot V$$

In the above equation, we can recognize the fact that F/A is the pressure. Making this substitution, we get:

$$W = P \cdot V$$

This is a deceptively simple equation that needs differential calculus to solve exactly in the general case, because both pressure and volume can change. If the pressure for a given process is held constant, the above expression becomes a regular algebraic equation which we can solve exactly without resorting to calculus.

$$W = P \cdot \Delta V$$

Let's now apply our new work equation to an example. How much work is done when a volume of a gas increases from 2.5 to 3.0 L against a constant pressure of 600 Pa?

This problem involves the concept of work done on or by a gas under constant pressure conditions; thus we can use $W = P \cdot \Delta V$. In order to end up with SI units, we need to convert everything to SI units. Examining the data, we see that "liters" are not official SI units and must be converted to the official SI unit of volume, m³. There are 1000 L in one cubic meter.

$$2.5 \text{ L} \times \frac{1 \text{ m}^3}{1000 \text{ L}} = 2.50 \times 10^{-3} \text{ m}^3$$

$$3.0 \text{ L} \times \frac{1 \text{ m}^3}{1000 \text{ L}} = 3.00 \times 10^{-3} \text{ m}^3$$

We can now substitute this information into the work equation to get:

$$W = P \cdot \Delta V = P \cdot (V_{final} - V_{initial}) = 600 \text{ Pa} \ (3.00 \times 10^{-3} \text{ m}^3 - 2.50 \times 10^{-3} \text{ m}^3)$$

$$W = 600 \text{ Pa} \cdot 5.0 \times 10^{-4} \text{ m}^3 = 0.300 \text{ Pa} \cdot \text{m}^3 = 0.30 \text{ J}$$

Since 1 J is equal to 1 Pa $\cdot$ m³, the work done is 0.30 J.

STATE FUNCTIONS

State functions are mathematical functions that describe the "state" of a system. The state of a given system can be described using one or more variables such as volume, pressure, or temperature. State functions are unique in the fact that they are independent of the path by which a system gets to a particular state. Internal energy, U, is a state function, for example. You know the value of internal energy by specifying the state parameters, and you know the change in internal energy whenever you specify the change in the state parameters. For a state function, the path in going from the initial state to the final state is irrelevant. All you need to know are the initial and final conditions. Imagine you begin with a sample of gas at a certain temperature, pressure, and volume. These parameters determine the initial internal energy of the system.

Now, let's play. *First* let's double the pressure, and *then* cut the volume in half. The system will have a new internal energy. Suppose, though, that we *first* halved the volume and *then* doubled the pressure. Since internal energy is a state function, and the system winds up at the same final state, the change in internal energy will be the same, even though we changed the parameters in a different order.

Work, on the other hand, is not a state function, and therefore depends on the path one takes in getting from the starting point to the ending point. As an example, consider pushing a box of supplies on the floor from point A, around an interior hallway that leads back to point A. You did work pushing the box around this closed circuit, even though your starting and ending points are exactly the same. The amount of work you have to do depends critically on the choice of your pathway, or how you execute the task. For example, what if you put the box on a cart and then wheeled it around the path and then set it back on the floor? Have you done the same amount of work? Of course not. Work is not a state function.

When considering gas systems, the amount of work done by the gas is, of course, also path dependent. In other words, it depends on how you change the state parameters such as the volume, pressure, and/or temperature.

Let's consider three different paths from point A to point B. Path 1 is a two-step process involving an isochoric (constant volume) decrease in pressure followed by an isobaric (constant pressure) volume expansion of the system, as shown in Figure 4.2.

The first step requires no work, because there is no change in volume. Since $W = P \cdot \Delta V$, when ΔV is zero, work is zero. The second step increases the volume against a constant pressure, so the work can be calculated as $W = P \cdot \Delta V$. You can also determine the work graphically. In this case, the magnitude of the amount of work done for path 1 is simply the area under the horizontal arrow highlighted in green.

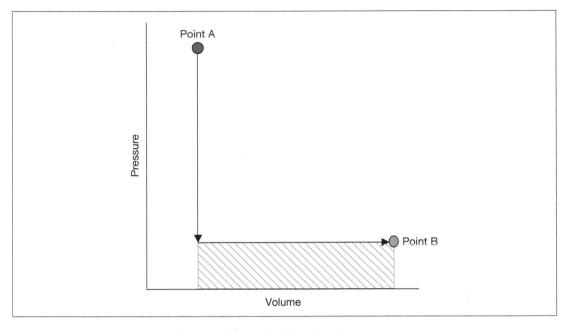

Figure 4.2 Path 1 for Gas Expansion

Path 2 is another two-step process going from point A to point B that involves an isobaric volume expansion of the system followed by an isochoric pressure reduction, as shown in Figure 4.3.

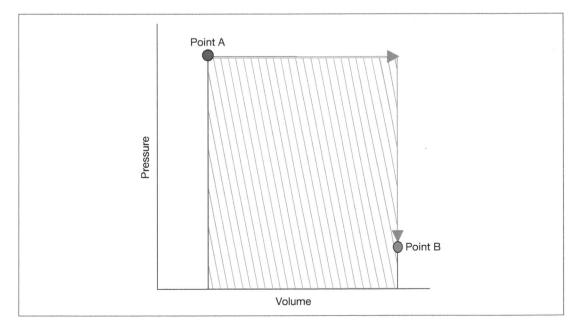

Figure 4.3 Path 2 for Gas Expansion

In this example, the first change results in work, because volume is changing against a pressure. The work is equal to $P \cdot \Delta V$. The second step does not involve a change in volume, so the work is zero. Again, the magnitude of the amount of work done for path 2 is the purple-highlighted area. Note that this purple area is much larger than the green area under path 1, indicating path 2 does a much larger amount of work than path 1 (Figure 4.2). This is because our path choice in the second path had the gas expanding against a greater pressure than in the first path.

Path 3 is a single-step process in which both the volume and pressure are simultaneously changed. This process is shown in Figure 4.4.

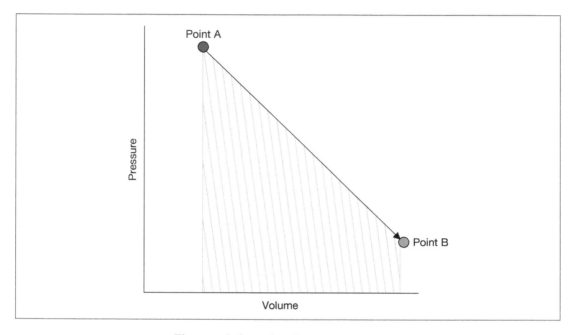

Figure 4.4 Path 3 for Gas Expansion

As can be clearly seen in this figure, the gray area represents the work done in this specific process. It is also much different from paths 1 and 2. Clearly, the amount of work in a given process is highly dependent on how the pressure and volume change. For changing volume and/or pressure of a gas, the work is equal to the area under the curve describing the process.

The Real World

Reality is always more complicated than idealized models used by scientists and engineers. For example, the relationship $W = P \cdot \Delta V$ is valid only under reversible conditions, that is, when ΔV is accomplished in an infinite number of steps and the pressure is held constant for all of the steps. That is, the expansion would take forever. Under these conditions this is known as reversible work. Obviously, we never obtain reversible work. However, the slower the change in volume, the closer the approximation.

ENERGY

The concept of energy is very abstract, although in today's world the topic of energy and the growing energy demands of our global population have become front-page issues. In chemistry and physics, energy is defined as the capacity to do work. Throughout the ages, scientists have observed certain relationships in nature known as laws. These laws can typically be described by simple mathematical relationships or statements. For example, we have already discussed Newton's three laws. With respect to energy there is a very important law known as the Law of Energy Conservation. This law can be stated in many different ways. For instance, the amount of energy in the universe is constant and is constantly being converted from one form to another. Alternatively, we can say, energy is neither created nor destroyed, but only converted to other kinds of energy.

Units of Energy: Joules

The SI unit of energy is the joule (J), exactly the same unit as work. As an example, assume you have a 1-kg mass that you want to accelerate at 1 m/s^2 over a distance of 1 m. The amount of energy you would use is:

$$W = F \cdot d = m \cdot a \cdot d$$
$$1 \text{ J} = (1 \text{ kg}) \cdot \left(1 \tfrac{m}{s^2}\right) \cdot (1 \text{ m}) = 1 \frac{kg \cdot m^2}{s^2}$$

In the first equation, note that Newton's second law ($F = m \cdot a$) was used. Giving this 1 kg mass 1 J of energy also requires that you expend 1 J of work.

Units of Energy: "Big C" Calories and "Little C" Calories

An older unit of energy that is still common in physical chemistry circles is the calorie. One calorie was defined to be the amount of energy necessary to raise the temperature of 1 g of water from 14.5 to 15.5°C. Today, one calorie (thermochemical calorie) is defined to be *exactly* 4.184 J. The ubiquitous food Calorie (note the capitalization) is a kilocalorie or 1000 cal.

The Calorie is an immense amount of energy. For example, most of us probably consume about 2000 kcal each day. Let's convert this into joules, and then compare it to the kinetic energy of a moving truck. First, let's use the relationship that 1 kcal = 4184 J.

$$2000 \text{ kcal}\left(\frac{4184 \text{ J}}{1 \text{ kcal}}\right) = 8.37 \times 10^6 \text{ J}$$

Kinetic Energy

As we said earlier, kinetic energy is the energy a mass has by virtue of being in motion. Let's imagine a truck going down the highway. If the mass of the truck is 1500 kg, how fast must the truck be going in order to have the same amount of energy as is contained in 2000 kcal? Well, from the definition of kinetic energy, we can write:

$$KE = \tfrac{1}{2} mv^2$$

We know the energy and mass terms, so let's solve the equation for speed and do the arithmetic.

$$v = \sqrt{\frac{2KE}{m}} = \sqrt{\frac{2\,(8.37 \times 10^6 \text{ J})}{1500 \text{ kg}}} = 106 \frac{\text{m}}{\text{s}}$$

So, to have a kinetic energy equivalent to the food energy we consume each day, a truck would need to be traveling at 106 m/s. Just in case that doesn't impress you, let's convert the speed into miles per hour, where we find a truck with a mass of 1500 kg needs to be going 236 mph in order to attain a kinetic energy equivalent to 2000 kcal!

$$106 \frac{\text{m}}{\text{s}} \left(\frac{1 \text{ mi}}{1609 \text{ m}} \right) \left(\frac{3600 \text{ s}}{\text{hr}} \right) = 236 \frac{\text{mi}}{\text{hr}}$$

Potential Energy

Energy is constantly being interconverted from one form to another. Energy that is stored by virtue of position is called potential energy. Only conservative forces, which are forces that perform no work around a closed path, can change the potential energy of a system. Potential energy can be recovered and used later. So potential energy is sometimes called stored energy. Some examples of potential energy are an airplane up in the air, a battery, and the chemical energy stored in foods.

Let's examine the gravitational potential energy of a ball which is thrown straight up into the air. The ball starts at rest in your hand, then goes up, slows down, stops, and then falls back to you. The instant you release the ball, it has a maximum amount of kinetic energy. As it travels upwards, though, the velocity decreases and, therefore, the kinetic energy decreases. The kinetic energy is not destroyed, however, but converted into potential energy. At the top of the flight, the kinetic energy is zero, because the ball is not moving, and the potential energy is at a maximum. Then, as the ball begins to fall, the potential energy is converted back into kinetic energy.

We can say that the kinetic energy of the ball is constantly changing because an isolated object such as a ball can have kinetic energy. Potential energy, on the other hand, is a property of a system and not of an isolated object. So for the ball example, we must say that the potential energy of the ball–ground system is constantly changing. The potential energy of the system is large when the ball is high in the air, and it is low when the ball is near the ground. Unlike kinetic energy which is equal to $\frac{1}{2}mv^2$, there is no single formula for potential energy. The specific formula depends upon the situation at hand. For instance, gravitational potential energy is given by:

$$\text{Potential energy} = PE = mgh$$

In the above formula m is the mass of the object, g is the acceleration due to gravity, and h is the height above some arbitrarily chosen zero height (normally the ground). Notice this equation is very similar to that of work, $W = Fd$. This should make sense, because if we want to raise an object some distance above the ground, thereby giving it gravitational potential energy, we have to do work.

The potential energy of a mass connected to an ideal spring is given by:

$$PE = \frac{1}{2} kx^2$$

In this formula k is called the spring constant and depends on the specific spring being used, while x is the distance away from the equilibrium position (the position when the spring is neither stretched nor compressed).

Other forms of potential energy include sugar or fat, thermal energy, and light. Potential energy, like all forms of energy, has official SI units of joules. In addition, all energy quantities are scalar and have no inherent direction.

Internal Energy

The internal energy of a system is the sum of all the kinetic and potential energies of the particles comprising the system. Internal energy is not a macroscopic quantity, but rather a quantity that results from the state of the particles at a molecular level. Usually, internal energy is represented by the letter U. Let's consider an idealized system of a gas in which molecules are considered point particles that neither attract nor repel one another. In addition, we will assume that the individual gas molecules move about freely inside a container, occasionally colliding with one another and the walls of the container. Let us also assume that our container of gas is at a pressure higher than that of the atmosphere. In other words, we are dealing with a compressed gas cylinder containing an ideal gas. Finally, let's put the cylinder in an airplane flying at 30,000 feet. What kind of energy does this cylinder of gas have? The cylinder itself is moving along with the airplane, so it has kinetic energy. The cylinder–Earth system has potential energy since the cylinder is at 30,000 feet with respect to the ground. But the gas molecules are moving about inside the cylinder, so they also have kinetic energy separate and distinct from the kinetic energy of the cylinder. Compressed gases are kind of like springs, so the molecules have potential energy since they are at a higher pressure than the surrounding atmosphere.

THERMODYNAMICS

Thermodynamics can loosely be defined as the study of energy, how it is interconverted from one form to another, and how it flows into and out of thermodynamic systems. The system is the part of the universe that is under consideration. It might be a beaker on a lab bench, or it might be a patient. Everything that is nearby the system is called the surroundings. Almost always, you, as the observer, are part of the surroundings, not the system. This is really important, because all of our upcoming conversations will be from the point of view of the system. It's easy to get confused, so remember, you're part of the surroundings, not the system.

There are four fundamental laws that sum up many years of human observation and experimentation in this field.

The Zeroth Law of Thermodynamics

Two objects, A and B, are found to be in thermal equilibrium. That is, they are at the same temperature. In addition, objects B and C are separately found to be in thermal equilibrium.

Therefore, objects A and C will be in thermal equilibrium if placed in thermal contact with each other, and no heat will flow between A and C.

The First Law of Thermodynamics

The change in the internal energy of a system is equal to the sum of the heat processes that cause energy to flow into/out of the system and the work done by/on the system:

$$\Delta U = Q + W$$

The signs of the heat and work are very important and are defined to be:

$Q > 0 \rightarrow$ endothermic process: energy flows into the system
$Q < 0 \rightarrow$ exothermic process: energy flows out of the system
$W < 0 \rightarrow$ work done by the system on the surroundings—expansions
$W > 0 \rightarrow$ work done on the system by the surroundings—compressions

The Second Law of Thermodynamics

Heat spontaneously flows from a hot body to a cold body when the two bodies are brought into thermal contact. You have doubtless seen applications of this law innumerable times. For example, when a patient gets the chills as she comes out from under anesthesia, and you put a warm blanket on her, you fully expect the heat to flow from the blanket to that patient, and not the other way around.

The Second Law is sometimes stated as the Entropy Law. Entropy is a measure of randomness or disorder in a system. Systems that are more randomized, chaotic, or evenly mixed have more entropy. The Second Law states the entropy of the universe is constantly increasing. One clear implication of the Second Law is that the universe *never*, and a system almost never, spontaneously becomes more organized. So, "hot" molecules will not spontaneously separate themselves from "cold" molecules. Mixtures of oxygen and nitrous oxide will not spontaneously separate and send the oxygen to the patient separately from the nitrous oxide. IV fluids will mix evenly throughout the circulatory system and not congregate in just the left arm.

The Third Law of Thermodynamics

It is not possible to lower the temperature of an object to absolute zero.

These four laws form the foundation of thermodynamics and have never been known to be violated. These laws govern the physical behaviors that we observe every day. These laws can also be seen as goals toward which we must strive in order to make our technological world more thermodynamically efficient and more sustainable for future generations.

Let's take a look at some thermodynamic systems. Let's assume that we have some volume of a gas contained in a cylinder that is fitted with a frictionless piston. We then allow the gas to expand against a constant pressure (600 Pa) to a new volume that is 500 mL greater than the original volume. In the process a quantity of 3 J of energy is absorbed by the system. What are the work, heat, and internal energy changes for this system? Since the pressure is constant, this is called an isobaric process. Let's start by calculating the work that is involved. In order to do this, we must convert the values involved to the proper SI units. The pressure is already in pascals (Pa), but the change in volume is in milliliters (mL). We thus need to convert the volume to cubic meters (m^3).

$$500 \text{ mL} \times \frac{1 \text{ L}}{1000 \text{ mL}} \times \frac{1 \text{ m}^3}{1000 \text{ L}} = 5.00 \times 10^{-4} \text{ m}^3$$

This value is subsequently substituted into the work equation to get the magnitude of the work involved:

$$W = P \cdot \Delta V = (600 \text{ Pa}) \cdot (5.00 \times 10^{-4} \text{ m}^3) = 0.300 \text{ J}$$

Now we must determine the sign of the work. In this case, the system is doing work because there is an expansion; hence $W = -0.300$ J. Since heat is flowing into the system, $Q = +3$ J. The next thing we can calculate is ΔU, which can be found from the First Law of Thermodynamics.

$$\Delta U = Q + W = +3 \text{ J} - 0.3 \text{ J} = +2.7 \text{ J}$$

The internal energy of our system has increased by $+2.7$ J, which would manifest as a temperature increase.

We will now examine the case in which a ventilator is doing the work for the body. This is really the same example we just considered. Let's assume that the ventilator is a constant-pressure device operating at 600 Pa and is used to increase the volume of air in the lungs by 500 mL (5.00×10^{-4} m³). The work performed by the ventilator is therefore:

$$W = P \cdot \Delta V = (600 \text{ Pa}) \cdot (5.00 \times 10^{-4} \text{ m}^3) = 0.300 \text{ J}$$

Heat Versus Temperature

Heat and temperature are not the same thing. At first glance, temperature can be viewed as a rather arbitrary numerical ranking of "hotness" or "coldness." In contrast, heat is energy that is transferred as a result of a temperature difference. In one sense it can be viewed as an energy flow. In another sense it can be viewed as a process. When two bodies of unequal temperature are brought into contact, a flow of heat occurs from the hot object to the cold object until the two temperatures become equal.

The temperature of an ideal gas is proportional to the average kinetic energy of the particles in a sample of the gas. The relationship between kinetic energy and temperature is derived from the Kinetic Molecular Theory of Gases and is given by:

$$KE = \frac{3}{2} kT$$

where T is the absolute temperature in kelvin. The Boltzmann constant (k) is equal to:

$$k = \frac{\text{Ideal Gas Law Constant}}{\text{Avogadro's Number}} = \frac{R}{N_A} = 1.38 \times 10^{-23} \frac{\text{J}}{\text{K}}$$

For a given sample of a single type of gas molecules, not all of the molecules will have the same speed or kinetic energy, but rather the sample will exhibit a distribution of speeds and kinetic energies. The distribution is highly temperature dependent and is called the Maxwell speed distribution, named after a Scottish physicist, James Clerk Maxwell, who originally discovered the distribution. Example distributions for O_2 at $T = 300$ K and $T = 1100$ K are shown in Figure 4.5. Note how the most probable speed, which is represented by the tallest point on

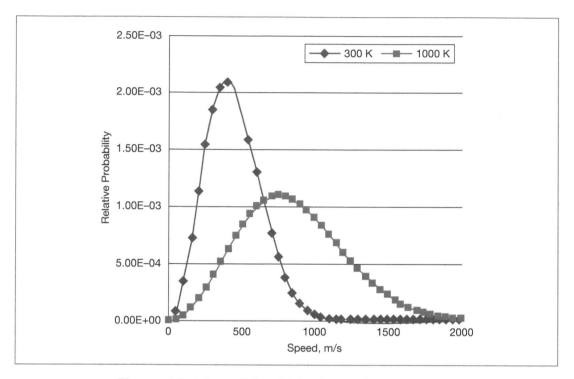

Figure 4.5 Maxwell Speed Distribution for Oxygen Gas

each curve, shifts to higher speeds as the temperature of the system is increased. As the average speed increases, the average kinetic energy increases. So we see that we can relate the temperature of the sample to the average kinetic energy of the particles in the sample. In addition, note how the peak becomes wider at higher temperatures.

The internal energy of an ideal gas is directly related to the random, disordered motion of the gaseous molecules or atoms. If we plot relative probability (that a given molecule has a given kinetic energy) *versus KE*, a plot similar to the one in Figure 4.5 is obtained. By integrating this new function, we can calculate the internal energy of an ideal gas at a given temperature. In case you have not had calculus, "integrate" means to find the area under the curve. So, the temperature of the sample depends on the average kinetic energy, which can be calculated from the average speed, or where the top of the curve is located. The "heat" or internal energy is the area under the curve. With solids and liquids, these calculations are much more complicated because of the extensive interactions between the molecules.

SPECIFIC HEAT

We've all had the experience in the kitchen where we have heated some water in a metal pot on top of the stove. You've no doubt noticed that the metal pot gets hot very quickly while the water seems to take much longer to warm up. Or why do we many times burn our palate even after we've blown on that piece of cheese pizza and the crust doesn't feel that hot anymore? Both of these are related to a concept known as heat capacity. Heat capacity is defined to be the ratio between the amount of heat added to or taken away from an object and the change in temperature of the object.

$$\text{Heat capacity} = C = \frac{Q}{\Delta T}$$

The SI units of heat capacity are J/K, as can be easily seen from the definition of C. Heat capacity is an extensive property of matter. That is, larger samples have greater heat capacities.

The heat capacity of a given object depends not only on the mass of the object but also on the type of material contained in the object. Thus, a more convenient quantity is typically used in thermodynamics. This quantity is called the specific heat, c. The specific heat is defined to be:

$$\text{Specific heat} = c = \frac{Q}{m \cdot \Delta T}$$

Specific heat and heat capacity are almost always positive quantities. The official SI units for specific heat are thus J/(kg · K). Whenever a quantity is divided by a mass, the term *specific* is usually placed in front of the name of the quantity. This is the origin of the "specific" in specific heat. Occasionally texts use units of J/(kg · C°) or J/(kg · °C). Technically, from a unit perspective this is not correct; however, since specific heat involves a temperature interval and the size of a Kelvin is identical to a Celsius degree, the values are numerically the same. You may also see books where specific heat has been defined per gram of material instead of per kilogram. In these texts, the units for c will be J/(g · K), J/(g · C°), or J/(g · °C).

Going back to the water on the stove, water has a very large specific heat, whereas typical metals do not. Thus, much energy must flow into water before its temperature will significantly change. Conversely, not much energy is required to change the temperature of the metal pot. Materials with large specific heats are good thermal insulators, because large amounts of heat cause only small changes in temperatures. Good thermal conductors have low specific heats.

The definition of specific heat can be rearranged to get:

$$Q = m \cdot c \cdot \Delta T$$

We are now equipped with the necessary tools to solve problems involving specific heat and heat flow. Imagine a 10.0 g sample of copper that undergoes an increase in temperature from 25.0 to 100.0°C when 289 J of heat is added. What is the specific heat of copper?

We are asked to calculate the specific heat of copper. We can easily do this using the definition of specific heat, ensuring that we use the proper SI units.

$$c = \frac{Q}{m \cdot \Delta T} = \frac{+289 \text{ J}}{(0.010 \text{ kg}) \cdot (75.0 \text{ K})} = 385 \, \frac{\text{J}}{\text{kg} \cdot \text{K}}$$

How much heat is needed to raise the temperature of $45\overline{0}$ g of water from 25.0 to 85.0°C? The specific heat of water is 4.18 J/(g · K).

Answer: 1.13×10^5 J must be added.

POWER

So far, we have talked about work and energy, but we haven't mentioned anything about the time in which work is done or energy gets expended. Power is the rate of doing work or, conversely, the rate of expending energy. The average power delivered during a time t when an amount of work W is done is given by:

$$\text{Average power} = P = \frac{W}{t}$$

The official SI unit of power is the watt (W), which is defined as joules per second:

$$1\ \text{W} = 1\ \frac{\text{J}}{\text{s}} = 1\ \frac{\frac{\text{kg} \cdot \text{m}^2}{\text{s}^2}}{\text{s}} = 1\ \frac{\text{kg} \cdot \text{m}^2}{\text{s}^3}$$

Be careful not to confuse power with work when examining equations. In this text, the unit of power, the watt, is identified by a nonitalicized uppercase "W", while the symbol for work is an italicized upper case "W".

Alternatively, power can be represented in a modified form which may be useful in certain situations. Starting with the definition of power and rearranging the equation, we get:

$$P = \frac{W}{t} = \frac{F \cdot d}{t} = F \cdot \left(\frac{d}{t}\right)$$

Finally, we recognize the fact that $v = \frac{d}{t}$ and we subsequently make this simple substitution:

$$P = F \cdot \left(\frac{d}{t}\right) = F \cdot v$$

This last equation shows us that the average power is equal to the force involved times the speed.

Let's now calculate the power expenditure in breathing. Assume the patient takes 15 breaths per minute and, based on our previous calculations, the work required for one breath is 0.300 J. We know that $P = W/t$; therefore we need to calculate the time required for one breath:

$$\text{Time for one breath} = t = \frac{60\ \text{s}}{15\ \text{breaths}} = 4.0\ \text{s}$$

The time can now be substituted into the power equation to yield the correct answer:

$$P = \frac{W}{t} = \frac{0.300\ \text{J}}{4.0\ \text{s}} = 0.075\ \text{W}$$

What is the average power consumption of a person who uses the 2000 kcal of food energy each day? Well, it is an easy conversion to find the energy used is 8.37×10^6 J. One day contains 8.64×10^4 s. Therefore, our person has a power usage of about a typical incandescent light bulb.

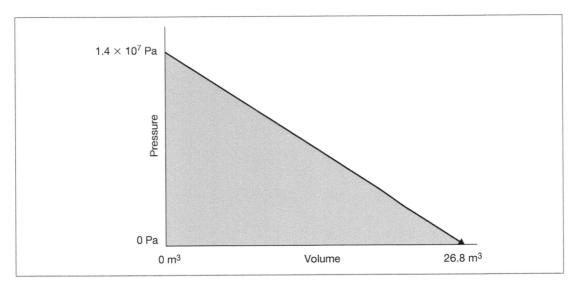

Figure 4.6 Work Done by the Rupture of a Gas Cylinder

$$P = \frac{8.37 \times 10^6 \text{ J}}{8.64 \times 10^4 \text{ s}} = 97 \text{ W}$$

Let's try another one. You know gas cylinders are very dangerous. How much work can 200 L of gas at an initial pressure of 2000 psi accomplish, if the cylinder ruptures and the gas expands to a new volume of 27,000 L?

Well, the change in volume is easy enough: 27,000 L – 200 L = 26,800 L or 26.8 m³. The initial pressure in the cylinder is 2000 psi, which is equal to 1.38×10^7 Pa. Unfortunately, the gas is not expanding against a constant pressure, so we can't use our handy work equation. The gas is expanding against a pressure difference that begins at 2000 psi above atmospheric pressure and decreases to 0 psi above atmospheric pressure. However, we can approximate a solution to this problem graphically. But first a comment. This is by no stretch of the imagination a reversible process. It is far too fast, too rapid, and once the neck is broken off the gas cylinder, there is nothing you can do to get that genie back into the bottle. What we'll find represents the maximum amount of work that can be accomplished, but don't let this lull you into a false sense that gas cylinders are not dangerous.

To help with the math, let's imagine the gas expands in a linear fashion. The area under the curve is a triangle, and we can calculate the area of a triangle as 1/2(base) · (height). The base of our triangle is 26.8 m³. The height is 1.38×10^7 Pa. So the area under the curve is about 185 million joules, or 185,000 kJ!

$$\text{Area} = \text{Work} = \tfrac{1}{2}(1.38 \times 10^7 \text{ Pa})(26.8 \text{ m}^3) = 1.85 \times 10^8 \text{ J}$$

If the release occurs over a period of, say 10 s, the power production is 18.5 megawatts (MW), comparable to a small nuclear reactor.

Summary

- In the world of physics, work is defined as:

$$W = F \cdot d \cdot \cos(\theta)$$

- Kinetic energy:

$$KE = \frac{1}{2} mv^2$$

- Work–Kinetic Energy Theorem:

$$W = \Delta KE = \frac{1}{2} mv^2_{\text{final}} - \frac{1}{2} mv^2_{\text{initial}}$$

- Work done by a gas under constant external pressure:

$$W = P \cdot \Delta V$$

- State functions describe a system and are independent of the path by which a system gets to a particular state. Examples of state functions are volume, pressure, and temperature.
- Energy is the capacity to do work. Energy of motion is called kinetic energy, while the stored energy of a system is called potential energy.
- Gravitational potential energy:

$$PE = mgh$$

- Potential energy of a mass-spring system:

$$PE = \frac{1}{2} kx^2$$

- The internal energy of a system is the sum of all the kinetic and potential energies of the particles comprising the system.
- Thermodynamics is the study of energy and how it is interconverted from one form to another and how energy flows into and out of thermodynamic systems.
- The Zeroth Law of Thermodynamics: Two objects, A and B, are found to be in thermal equilibrium. Objects B and C are separately found to be in thermal equilibrium. Therefore, objects A and C will be in thermal equilibrium if placed in thermal contact with each other.
- The First Law of Thermodynamics: The change in internal energy of a system is equal to the sum of the heat processes that cause energy to flow into/out of the system and the work done by/on the system.

$$\Delta U = Q + W$$

- The Second Law of Thermodynamics: Heat spontaneously flows from a hot body to a cold body when the two bodies are brought into thermal contact.
- An alternate form of the Second Law of Thermodynamics: The entropy of the universe is constantly increasing.
- The Third Law of Thermodynamics: It is not possible to lower the temperature of an object to absolute zero.
- Heat is the energy that is transferred as a result of a temperature difference while temperature is an arbitrary numerical ranking of "hotness" or "coldness."
- The average kinetic energy of an ideal gas:

$$KE = \frac{3}{2} kT$$

- Heat capacity:

$$C = \frac{Q}{\Delta T}$$

- Specific heat:

$$c = \frac{Q}{m \cdot \Delta T}$$

- Average power:

$$P = \frac{W}{t}$$

Review Questions for Physics (Part 2)

1. Describe the difference and the relationship between work and energy.

2. What is the equation that defines mechanical work? How do you know that work is being done?

3. What are the metric units of work?

4. How much work is done on a hydraulic lift when a car having a mass of 1500 kg is lifted 3.0 m straight up?

5. How much work is released when 50.0 L of oxygen expands (slowly and reversibly) to a new volume of 2000 L against a constant pressure of 101 kPa?

6. What is kinetic energy? Give an example.

7. What is potential energy? Give some examples.

8. What is heat?

9. How does temperature differ from internal energy?

10. What is the difference between heat and internal energy?

11. State the First Law of Thermodynamics and describe every term.

12. What is an endothermic process? What is exothermic?

13. Explain why the temperature in an endothermic process drops.

14. What is a state function?

15. What is specific heat?

16. How much heat is required to raise the temperature of 200 g of water by 25.0°C? Assume no phase changes occur.

17. What is power? What are the units of power?

18. A potato chip has about 10 Cal. If the energy in a potato chip were released in your body at an "explosive rate," say over a period of 1.0 ms, how much power is released?

19. Which of the following statement(s) are true?
    a. No work is done by a gas expanding against a constant pressure
    b. No work is done when the pressure of a gas is increasing while the volume remains constant
    c. Both of these statements involve work

20. What are the metric units for work?
    a. pascals
    b. joules
    c. newtons
    d. watts

21. Which of these combinations of base units and/or derived units express work?
    a. (liters) × (pascals)
    b. (kilograms) × (meters)^2/(second)^2
    c. (newtons) × (meters)
    d. All of these represent work

22. A ventilator designed by the well-known surgeon W. E. Kyotay (of ACME Hospital and Mortuary Supply Company) has two settings. This first introduces 50 L of air against a steadily increasing pressure (from 0 psi to 15 psi). The second setting blows the same volume of air into the patient at a constant pressure of 15 psi. Which setting does more work per cycle?
    a. The first setting
    b. The second setting
    c. Both settings accomplish identical work

23. The cylinder pressure on an oxygen tank reads $30\overline{0}$ kPa. What is the pressure in psi?
    a. 43.5
    b. 445
    c. 101
    d. 14.7
    e. None of these statements is correct.

24. The average kinetic energy of a sample of gas molecules is most closely related to:
    a. The heat content of the sample
    b. The potential energy of the sample
    c. The temperature of the sample
    d. The internal energy of the sample

25. The unit of power is:
    a. ohm
    b. joule
    c. watt
    d. what

26. Let's suppose you expend about 2500 kcals per day. How many kilojoules of energy is this?
    a. 2500
    b. 10,460
    c. 596
    d. 4.184

27. Over the course of 24 hours (or 86,400 s), is your average power consumption more or less than the 25-W light bulb in your refrigerator (assuming that it does, in fact, stay on all the time)? Don't forget that the answer in the previous question is in units of kilojoules.
    a. You require more power (on average)
    b. You require less power (on average)
    c. You require the same power (on average)
    d. More information is needed

28. Suppose pumping gas into a gas cylinder requires the expenditure of 3800 J of work. You notice the *cylinder* gets warm. What can you say about the internal energy of *gas* inside the cylinder?
    a. The internal energy has increased by 3800 J
    b. The internal energy has increased by more than 3800 J
    c. The internal energy has increased by less than 3800 J

29. Which statement is true?
    a. Kinetic energy is the energy an object has by virtue of being in motion
    b. Potential energy is stored energy
    c. A change in internal energy can be calculated from heat and work processes
    d. All of these are true

30. Select the true statement about temperature:
    a. The absolute temperature of a gas sample can be calculated from the average kinetic energy of gaseous particles in the sample.
    b. The temperature can be calculated from the sum of kinetic and potential energies of the atoms in the sample.
    c. The temperature always increases when heat is added to a sample.

31. The specific heat of a good thermal insulator should be:
    a. Very large
    b. Very small
    c. Equal to 1

32. Match each of the concepts with the correct word or phrase. Each answer can be used only once.

    1. This describes the change in velocity with time.
    2. This equals the product of an applied force times distance.
    3. These kinds of quantities have both magnitude and direction.
    4. An object at rest remains at rest until a force acts upon it.
    5. This is the ability to do work.
    6. For every action there is an equal and opposite reaction.
    7. This results from the acceleration of gravity acting upon an object's mass.
    8. This describes an object's innate tendency to resist acceleration.
    9. Force divided by area gives ____.
    10. Newton's second law tells us that in order to accelerate a mass, one must apply one of these.
    11. Objects in motion have this kind of energy.
    12. What kind of energy is "stored energy"?
    13. The change in position with time is the ____.
    14. Kilograms are units of ____.
    15. The formula to calculate the weight of an object is derived from this law.
    16. The rate of using energy is ____.

    a. Potential energy
    b. Acceleration
    c. Inertia
    d. Newton's First Law
    e. Newton's Second Law
    f. Newton's Third Law
    g. Force
    h. Power
    i. Weight
    j. Vector
    k. Work
    l. Energy
    m. Kinetic energy
    n. Velocity
    o. Pressure
    p. Mass

33. You arrive for work at the brand-new six-story hospital. Pretty much everything in this facility is state of the art, even the fancy, mahogany-paneled elevators. It's too bad, though, that the elevators are all out of order. Unfortunately, your unit is on the sixth floor. Calculate the power required to move your weight up six floors assuming your mass is 55.0 kg, the sixth floor is 22.0 m above the ground level, and it takes you 95.0 s to walk up all six flights of stairs.

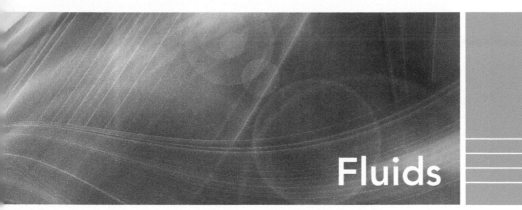

Fluids

The world of fluids is complicated, yet very fascinating. Fluids are literally all around us in the form of the Earth's atmosphere. Fluids make many things possible including air travel and life itself. At work, the nurse anesthetist can readily see the consequences of fluid behavior by observing water flowing from a tap; through the delivery of a metered-dose of QVAR® from an inhaler; from the infusion of IV fluids through a catheter; and via the administration of inhaled anesthetics using an anesthesia workstation. The behavior of fluids is so important to the field of nurse anesthesia that we have devoted an entire chapter to the subject. It is vital that nurse anesthetists have a fundamental understanding of what fluids are and how they behave under the influence of external forces.

FLUIDS: A DEFINITION

Even though many of us are familiar with the term *fluid*, we probably don't know the true scientific definition. You may think that a fluid is a liquid like a normal saline solution. While this is true, a fluid does not have to be a liquid. In the scientific usage of the word, a fluid is any material that has the ability to flow. Thus both liquids and gases are considered fluids. Basic forces, like those that result from gravity or pressure differences, cause fluids to flow. When fluids are placed in a container, they assume the shape of the container, unlike solids that keep their shape.

HYDROSTATICS

The study of fluids is divided into two major areas, hydrostatics and hydrodynamics. Hydrostatics is the study of fluids that are not moving, whereas hydrodynamics is the study of fluids in motion. For fluids that are static, two useful and important properties are density and pressure. We've already discussed density, so we'll move directly to pressure.

Pressure

You are probably well aware that the pressures at great depths under the ocean are extremely large, and the deeper you go, the greater the pressure that is exerted. Let's start small and look at a beaker filled with water. The beaker has a diameter of 10.0 cm, and the height of the water is 20.0 cm. How much pressure does the water exert on the bottom of the beaker? To answer this question, we simply need to recall that $P = F/A$. The force involved is the weight of the water contained in the beaker. The weight of the water can be calculated from the mass of the water, and the mass of the water can be calculated from the volume, using density as a conversion factor.

$$W = mg = \rho V g = \rho A h g$$

Note in this expression we used the fact that $m = \rho V$ and $V = Ah$. Remember that m is the mass of the fluid, ρ is its density, V is the fluid volume, A is the cross-sectional area of the fluid, and h is the fluid height. If we persevere in manipulating the equations *before* filling in any numbers, we can save ourselves some effort. We first apply the definition of pressure to get:

$$P = \frac{F}{A} = \frac{W}{A} = \frac{\rho A h g}{A} = \rho h g = \rho g h$$

Note in the last step we simply rearranged the symbols in order to write the expression in a manner typically used in textbooks. This expression $P = \rho g h$ allows us to calculate the pressure at any depth in any fluid, as long as we know the density of the fluid. We are now ready to substitute the appropriate values into the pressure equation to get:

$$P = \rho h g = \left(1000\ \frac{\text{kg}}{\text{m}^3}\right)\left(9.8\ \frac{\text{m}}{\text{s}^2}\right)(0.200\ \text{m}) = 1.96 \times 10^3\ \text{Pa}$$

How much pressure will the water exert at a depth of only 5.00 cm? Using the pressure equation we calculate:

$$P = \rho h g = \left(1000\ \frac{\text{kg}}{\text{m}^3}\right)\left(9.8\ \frac{\text{m}}{\text{s}^2}\right)(0.0500\ \text{m}) = 4.90 \times 10^2\ \text{Pa}$$

If we take the expression:

$$P = \rho h g$$

and simply omit the acceleration due to gravity, we get the nonstandard but fairly common units of "pressure" expressed in terms of mass per area. An example would be g/cm² or (g · wt)/cm². Notice in the second expression the abbreviation "wt" is added. It should be emphasized that these types of units are not true units of pressure and should be avoided.

In the two previous examples involving the water in a beaker, we have not taken into account the weight of the atmosphere pressing down on the top of our water–beaker system. From our personal experience, we know that the total pressure at the top of the water will be less than the total pressure at the bottom. To derive the general expression for pressure as a function of depth in a fluid, we need to look at our water–beaker system from a perspective of force, as shown in Figure 5.1.

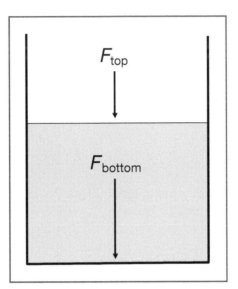

Figure 5.1 Pressure Under a Fluid

Only the weight of atmosphere is pushing down on the top surface of the water. Both the weight of the atmosphere and the weight of the water are pushing down on the bottom of the beaker. Obviously the force pushing down on the bottom of our beaker is larger than the force pushing down on the top of the water. In fact, the difference between F_{bottom} and F_{top} is simply the weight of the water in the beaker. We'll assume our beaker is open at the top to the atmosphere. Since $P = F/A$, we can say:

$$P_{top} = P_{atmosphere} = \frac{F_{top}}{A}$$

This expression can be solved for F_{top} to get:

$$F_{top} = \left(P_{atmosphere}\right)A$$

Likewise, at the bottom of the container, F_{bottom} is equal to the force at the top of the water plus the weight of the water (W).

$$F_{bottom} = F_{top} + W$$

We know from before that the weight of the water is given by:

$$W = \rho Ahg$$

This can be substituted into the F_{bottom} equation to get:

$$F_{\text{bottom}} = F_{\text{top}} + W = F_{\text{top}} + \rho Ahg$$

Hang on, we're almost finished. We can finally obtain the pressure by dividing the equation for F_{bottom} by A to give:

$$\frac{F_{\text{bottom}}}{A} = \frac{F_{\text{top}}}{A} + \frac{\rho Ahg}{A}$$

Simplifying, we end up with:

$$P_{\text{bottom}} = P_{\text{top}} + \rho gh$$

This equation can be written in a more general form to indicate the pressure at any two points in a fluid, not just the top and the bottom. This leaves us with:

$$P_2 = P_1 + \rho gh$$

This is our general expression for pressure as a function of depth in a fluid. Note in the expression that point 1 is above point 2 by a height h. Let's now apply what we've learned.

Let's calculate the pressure at a depth of $10\bar{0}0$ m below the surface of the ocean, assuming that ocean water has a density of 1.03×10^3 kg/m³. At sea level the atmospheric pressure is 1 atm or 101,325 Pa.

We need to apply our pressure as a function of depth equation to solve this problem. P_1, at the surface of the ocean, will be equal to atmospheric pressure. Substituting the appropriate values into our equation, we get

$$P_2 = P_1 + \rho gh = 101{,}325 \text{ Pa} + \left(1.03 \times 10^3 \tfrac{\text{kg}}{\text{m}^3}\right)\left(9.8 \tfrac{\text{m}}{\text{s}^2}\right)(10\bar{0}0 \text{ m})$$
$$= 1.02 \times 10^7 \text{ Pa}$$

Note at this depth the pressure is approximately 100 times what it is at the surface of the ocean.

Pressure at the Same Depth

Assume we have a point particle suspended in a fluid with density ρ. Since this is a *point* particle, it occupies no space or volume. This idealized particle is represented as the medium blue dot in Figure 5.2. At this particular depth in the fluid, the fluid exerts the same pressure in all directions on the particle. No matter where we put our point particle, as long as it is in the fluid, the fluid will act the same; that is, it will exert the same pressure in all directions.

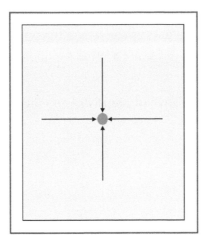

Figure 5.2 Pressure in a Fluid

Pressure Versus Container Shape

As illustrated in Figure 5.3, the pressure is independent of the container shape. You will notice that the height of the fluid is the same in the various arms of the apparatus, indicating the same pressure.

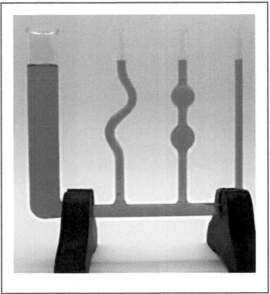

Figure 5.3 Pressure Independent of Shape

Pascal's Principle

Suppose we have a confined fluid at a given pressure and decide to increase the pressure of the fluid. What happens? Pascal's Principle gives us the answer:

When an external pressure is applied to a confined fluid, it is transmitted unchanged to every point within the fluid.

Pascal's Principle can easily be understood by examining the pressure versus depth equation. Let's assume we have a confined fluid system where the pressure on top of the fluid is atmospheric pressure. Assuming point P_1 is at the top of the fluid, the pressure at any point P_2 will initially be:

$$P_2 = P_1 + \rho gh = P_{atmosphere} + \rho gh$$

We now decide to increase the pressure on top of the fluid by 5 Pa. The pressure at any point P_2 will now be:

$$P_2 = P_1 + \rho gh = (P_{atmosphere} + 5 \text{ Pa}) + \rho gh$$

Notice how the pressure at point P_2 is simply the original pressure plus 5 Pa.

Let's apply some of this knowledge to a real situation. You're applying a pressure of 3 psi on the plunger of a syringe with a plugged needle held in a horizontal position. How does the pressure in the barrel compare to the pressure in the needle? Some of you may have a gut instinct that the pressure must be greater in the needle since it has a smaller cross-sectional area. Don't be fooled! We see that this is simply an application of Pascal's Principle since we are increasing the pressure on an enclosed fluid. If we increase the pressure by 3 psi on the plunger, the pressure will increase everywhere within the fluid by the same amount. However, if the needle were not plugged, we would have an entirely different situation. Then the fluid would be moving, and we would employ hydrodynamics, not hydrostatic equations.

Buoyancy

Why do certain things sink while others float? For instance, many types of wood float, yet a steel bar sinks very rapidly. However, most ships are made of steel, and they float. A simple experiment can provide us with some important clues. Imagine picking a friend up and throwing her into the air. You might say that you cannot do this. Now imagine picking the same friend up and throwing her into the air while you are both in a swimming pool. Can you do it now? Clearly the water is providing some kind of force that makes this feat much easier. We call this force the buoyant force. All fluids exert a buoyant force on objects immersed in them. Archimedes' Principle provides a solution to this conundrum.

Archimedes' Principle

Imagine a cube of some material completely immersed in a liquid of density ρ as shown in Figure 5.4.

Note that only one of the cube's six faces is shown. Since we are dealing with a cube, the length of each side is the same; therefore we'll represent this length as L. As can be seen in the figure, the fluid is exerting forces on all the faces of the cube. It should be noted that these forces are perpendicular to the faces, so they are called *normal forces*. This is a general term used to describe forces that are perpendicular to a surface. Forces F_3 and F_4 obviously cancel out and have no net effect on the cube. The same cannot be said for forces F_1 and F_2. Clearly F_2 is larger than F_1 because of the difference in depth L between the top surface of the cube and the bottom surface.

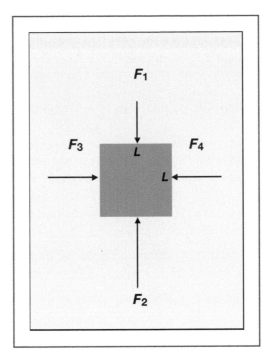

Figure 5.4 Buoyancy

Remembering the fact that $P = F/A$, we can readily represent the force on the top of the cube as:

$$F_1 = P_1A = P_1(L \cdot L) = P_1L^2$$

On the bottom of the cube, the force is:

$$F_2 = P_2A = P_2L^2$$

From before we know that the pressure at any point P_2 relative to P_1 is:

$$P_2 = P_1 + \rho gh$$

Thus, for our present situation, we'll represent the pressure on top of the cube as P_1 and the pressure on the bottom of the cube as P_2. The pressure P_2 is thus given by:

$$P_2 = P_1 + \rho gL$$

In this last equation, we recognized the fact that h is simply the length of the cube, which is L. We can now substitute this pressure expression into the equation for F_2 to get:

$$F_2 = P_2A = P_2L^2 = (P_1 + \rho gL) \cdot L^2 = P_1L^2 + \rho gL^3 = P_1L^2 + \rho gV$$

Notice in the last step L^3 was replaced by V since L^3 is the volume of a cube. Finally, the buoyant force is just the difference between the upward directed F_2 and the downward directed F_1. Since F_2 is bigger than F_1, there will be a net upward force. This buoyant force is what makes things lighter in a swimming pool. In equation form, the buoyant force F_b is:

$$F_b = F_2 - F_1 = (P_1 L^2 + \rho g V) - P_1 L^2 = \rho g V$$

Looking at the units on $\rho g V$, we see that they must be equal to newtons, since the left-hand side of the above equation is a force. Thus, $\rho g V$ represents the weight (which is a force) of the displaced fluid. In our example the object was *completely* immersed in a fluid; however, the object can only be partially immersed and the above derivation will still be valid. In the case of partial immersion, V would represent the volume of the object that is immersed.

This finally brings us to Archimedes' Principle:

An object immersed either totally or partially in a fluid feels a buoyant force equal to the weight of fluid displaced.

If we represent the true weight of an object as W_{true}, the apparent weight $W_{apparent}$ of an object due to total or partial immersion is thus:

$$W_{apparent} = W_{true} - F_b$$

Let's now apply our newly found knowledge to a practical example. How much will a $70\overline{0}$ N person weigh under water, if he has a volume of 65 L?

We are being asked to calculate the apparent weight $W_{apparent}$ of a person who is totally immersed in water. To do this, we need to know the buoyant force and the true weight. The true weight is provided in the problem. In order to calculate F_b, we will need the density of water, which we will assume to be 1000 kg/m³. In addition, we will need the volume in cubic meters (m³) instead of liters. Thus,

$$V = 65 \text{ L}\left(\frac{1 \text{ m}^3}{1000 \text{ L}}\right) = 6.50 \times 10^{-2} \text{ m}^3$$

Substituting this into the buoyant force expression, we get:

$$F_b = \rho g V = \left(1000 \text{ } \frac{\text{kg}}{\text{m}^3}\right)\left(9.8 \text{ } \frac{\text{m}}{\text{s}^2}\right)(6.5 \times 10^{-2} \text{ m}^3) = 637 \text{ N}$$

Finally, we calculate the apparent weight:

$$W_{apparent} = W_{true} - F_b = 70\overline{0} \text{ N} - 637 \text{ N} = 63 \text{ N}$$

Let's now examine possible scenarios involving an object with density ρ_{object} and a fluid with density ρ_{fluid}. If the density of the object is greater than the fluid ($\rho_{object} > \rho_{fluid}$), the object will sink. Based upon this observation, we can conclude that the buoyant force is less than the true weight of the object. If the density of the object is less than the density of the fluid ($\rho_{object} < \rho_{fluid}$), the object will float. In this case we can conclude that the buoyant force is greater than the weight of the object. If the buoyant force equals the weight of the object, we say that the object is neutrally buoyant and thus $\rho_{object} = \rho_{fluid}$, and the object will remain stationary in the fluid.

Hydrometers

A hydrometer is a very simple device used to measure the specific gravity of liquids such as urine or milk. A typical hydrometer has a weighted end to keep it upright in the liquid of interest. These devices are also usually calibrated. When placed in a liquid, a hydrometer will sink until it displaces an amount of fluid exactly equal to its weight. If the fluid is dense, it will displace only a small amount of fluid and thus not sink very deep. If the density of the fluid is not very high, the hydrometer will sink deeper. The user simply reads the specific gravity of the liquid from the calibrated scale in the neck of the hydrometer. Two examples are shown in Figure 5.5. The fluid on the left has a greater specific gravity than the fluid on the right.

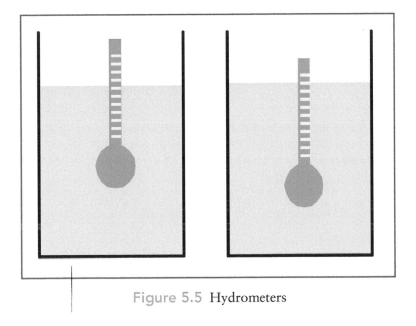

Figure 5.5 Hydrometers

HYDRODYNAMICS: MOVING FLUIDS

We will now change our focus to hydrodynamics and consider moving fluids. Here, in addition to the effects of gravity (weight and pressure), we need to consider several unique properties of moving fluids. In general, there are two types of flow. Smooth flow is known as *laminar flow*, while flow that is not smooth is called *turbulent flow*. Laminar flow is characterized by an unchanging flow pattern where adjacent layers of fluid smoothly slide past each other. Turbulent flow has a continuously varying pattern of flow. In a nutshell, laminar flow is smooth and orderly, whereas turbulent flow is chaotic and abruptly changing. The highly complex topic of turbulent flow is outside the scope of this text. Not surprisingly, laminar flow is much easier to study and much more difficult to obtain. We will thus only consider laminar flow.

Flow Rate

Let's examine a fluid flowing through a straight horizontal piece of pipe as shown in Figure 5.6. The flow of the fluid is represented by the arrow, and the volume of fluid is represented by the

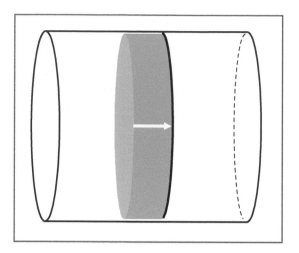

Figure 5.6 Flow Rate

blue section. Flow rate is the volume of fluid passing a particular point (e.g., the black line) per unit time. Thus flow rate will have units of volume divided by time such as gallons per minute or liters per hour. However, the official SI units of flow rate are cubic meters per second (m^3/s). So, if the volume of the blue section is 1 m^3 and it takes 1 s to flow past the black line, the flow rate is 1 m^3/s.

Speed and Diameter

Let's now change the diameter of the pipe by making it smaller in one discrete step, as shown in Figure 5.7. This is the same as what happens when you put your thumb over the end of a garden hose or when blood moves from a large vessel into a smaller one.

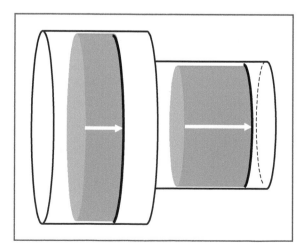

Figure 5.7 Flow Rate and Diameter

What happens to the speed of the water when you do this? Of course it increases tremendously, and you're able to squirt someone who is a great distance from where you are standing, someone you couldn't have reached had you not placed your thumb over the end of the hose. We thus see as the diameter of the tube decreases, the speed of the fluid increases. If there are no leaks in the tube, the flow rate has to be the same throughout the system. As seen in Figure 5.7, the volumes of the two blue sections remain the same while their shapes change. We are thus assuming the density remains constant. We call a fluid with such behavior incompressible. In addition, each of the blue sections takes the same amount of time to move past the black marker points. Since the length of the blue section in the narrower tube is longer, that section travels a greater distance per unit time. Therefore, the fluid flows at a greater speed through the narrow part of the tube.

Equation of Continuity

Let's now develop a more quantitative expression for the relationship between the cross-sectional area A of a tube and the speed of the incompressible fluid v flowing through it. Figure 5.8 shows the same pipe as illustrated before. In this case, however, note that the left-hand side of the pipe has a cross-sectional area equal to A_{left}, while the right-hand side has a cross-sectional area equal to A_{right}.

Video 5.1

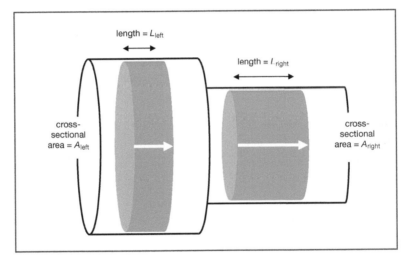

Figure 5.8 Equation of Continuity

If, again, there are no leaks in the system, the flow rate must be the same everywhere throughout the pipe; therefore we can write:

$$\text{Flow rate}_{\text{left}} = \text{Flow rate}_{\text{right}}$$

The flow rate on the left side of the pipe is equal to the volume of fluid passing through it per unit time; therefore:

$$\text{Flow rate}_{\text{left}} = \frac{\text{volume}_{\text{left}}}{\text{time}} = \frac{V_{\text{left}}}{t} = \frac{A_{\text{left}} \cdot L_{\text{left}}}{t} = A_{\text{left}} \cdot \left(\frac{L_{\text{left}}}{t} \right) = A_{\text{left}} \cdot v_{\text{left}}$$

Note in this last equation we used the fact that the volume on the left, V_{left}, is equal to the cross-sectional area of the left side of the pipe times the length of the left side of the pipe. In addition, we used the fact that L_{left}/t is simply the speed of the fluid, v_{left}, in the left side of the pipe, since the magnitude of velocity is equal to distance divided by time.

Now, focusing on the right side of the pipe, the flow rate is equal to:

$$\text{Flow rate}_{right} = \frac{\text{volume}_{right}}{\text{time}} = \frac{V_{right}}{t} = \frac{A_{right} \cdot L_{right}}{t} = A_{right} \cdot \left(\frac{L_{right}}{t}\right) = A_{right} \cdot v_{right}$$

Equating the left- and right-side flow rates, we get the equation of continuity for an incompressible fluid:

$$A_{left}v_{left} = A_{right}v_{right}$$

Many texts will use numbered subscripts instead of left and right; therefore the equation of continuity can also be written as:

$$A_1v_1 = A_2v_2$$

There we have it, a quantitative expression that relates the cross-sectional area A of a tube and the speed of the fluid v flowing through it. Many times pipes will have a circular cross-sectional area, so the flow rate can also be written as:

$$\text{Flow rate} = Av = \pi r^2 v$$

For example, blood is moving at 15 cm³/s through the aorta. If the diameter of the aorta is 1.0 cm, what is the speed of the blood?

We are asked to calculate the speed of a fluid through the aorta given the diameter and the flow rate. We can rearrange the flow-rate expression and solve it for speed to get:

$$v = \frac{\text{flow rate}}{A}$$

We will assume a circular cross-sectional area for the aorta, remembering the fact that the radius is half of the diameter. Substituting this information into the above expression, we get:

$$v = \frac{\text{flow rate}}{A} = \frac{15 \frac{cm^3}{s}}{\pi(0.5 \text{ cm})^2} = \frac{15 \frac{cm^3}{s}}{0.7854 \text{ cm}^2} = 19.0986 = 19 \frac{cm}{s}$$

Notice in this calculation we used cgs units instead of SI units.

Let's now do a more complicated example involving a syringe. The barrel of a syringe has a diameter of 1.0 cm, while the diameter of the needle is 0.020 cm. If you apply a pressure on the plunger so that the medicine moves at 1.0 cm/s through the barrel, how fast does it move through the needle?

We will use the equation of continuity, assigning the subscript 1 to the barrel of the syringe and the subscript 2 to the needle. We are asked to calculate the speed of the medicine through the needle, which is represented by v_2. Our first step will be to solve the equation of continuity for v_2:

$$v_2 = \frac{A_1 v_1}{A_2}$$

Next, we simply fill in the other information, once again remembering the fact that a radius is one half of a corresponding diameter.

$$v_2 = \frac{A_1 v_1}{A_2} = \frac{(\pi r_1^2) v_1}{(\pi r_2^2)} = \frac{(\pi (0.50 \text{ cm})^2)(1.0 \frac{\text{cm}}{\text{s}})}{(\pi (0.010 \text{ cm})^2)} = \frac{0.7854 \frac{\text{cm}^3}{\text{s}}}{3.14159 \times 10^{-4} \text{ cm}^2} = 2.5 \times 10^3 \frac{\text{cm}}{\text{s}}$$

The Bernoulli Equation

We have seen the effect of changing the diameter of a pipe on the speed of the fluid flowing through the pipe. It turns out that as the speed of a fluid increases, the pressure exerted by the fluid decreases. This phenomenon is called the *Bernoulli effect*. The Bernoulli effect provides the lift for airplanes and is why shower curtains get sucked toward you when you first turn on the shower. It also provides the basis for a Venturi flowmeter.

To explain this, we need to apply the Work–Energy Theorem to this situation. Remember, the Work–Energy Theorem is given by:

$$W_{\text{total}} = \Delta KE = KE_{\text{final}} - KE_{\text{initial}}$$

As before, let's assume we have a pipe in which the diameter changes in one discrete step. We will also assume that the fluid is flowing from left to right as shown in the Figure 5.9, where the positive x direction is to the right. The blue area on the far left represents a fluid element

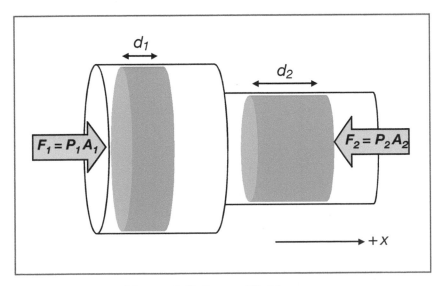

Figure 5.9 Bernoulli's Theorem

with thickness d_1 that is being pushed to the right by some pressure P_1 from the left. Note that this pressure exerts a force F_1 on the volume element given by $F_1 = P_1A_1$. Since we are assuming that the fluid is incompressible, the volume element changes shape but does not change volume when it enters the portion of the pipe having a narrower diameter.

Note its thickness is now d_2. In the narrower portion of the pipe, the volume element will experience a force F_2 due to pressure P_2 which is opposed to force F_1. Since work is force times distance and the positive x direction is to the right, F_1 will do an amount of work on the volume element equal to F_1d_1 while F_2 will do an amount of work equal to $-F_2d_2$. The total amount of work is therefore:

$$W_{total} = W_1 + W_2 = F_1d_1 - F_2d_2$$

Since $F_1 = P_1A_1$ and $F_2 = P_2A_2$, substitution yields:

$$W_{total} = F_1d_1 - F_2d_2 = P_1A_1d_1 - P_2A_2d_2$$

We can now substitute the Work–Energy Theorem (assuming the mass remains constant in the volume element):

$$W_{total} = \Delta KE = KE_{final} - KE_{initial} = \tfrac{1}{2}mv_2^2 - \tfrac{1}{2}mv_1^2$$

into the left-hand side of the previous equation to yield:

$$\tfrac{1}{2}mv_2^2 - \tfrac{1}{2}mv_1^2 = P_1A_1d_1 - P_2A_2d_2$$

We know that $V_1 = A_1d_1$ and $V_2 = A_2d_2$. Therefore:

$$\tfrac{1}{2}mv_2^2 - \tfrac{1}{2}mv_1^2 = P_1V_1 - P_2V_2$$

From the definition of density, we know that $m = \rho V$; hence:

$$\tfrac{1}{2}\rho_2 V_2v_2^2 - \tfrac{1}{2}\rho_1 V_1v_1^2 = P_1V_1 - P_2V_2$$

In addition, since the fluid is incompressible, $V_1 = V_2 = V$ and $\rho_1 = \rho_2 = \rho$; therefore we can simplify the equation to yield:

$$\tfrac{1}{2}\rho Vv_2^2 - \tfrac{1}{2}\rho Vv_1^2 = P_1V - P_2V$$

Since every term has a V in it, we can divide both sides of the equation by V to get one form of the Bernoulli equation:

$$\tfrac{1}{2}\rho v_2^2 - \tfrac{1}{2}\rho v_1^2 = P_1 - P_2$$

Thus, we see that the change in pressure exerted by a fluid depends on the change in speed and the density of the fluid.

An alternate yet equivalent form of the Bernoulli equation can be obtained by collecting all the terms with the same subscripts onto the same side:

$$P_1 + \frac{1}{2}\rho v_1^2 = P_2 + \frac{1}{2}\rho v_2^2$$

Be very cautious when using the Bernoulli equation. Pay special attention to the subscripts and the units. When in doubt with what units to use, always revert to SI units, where the pressure will be given in pascals, the speed in meters per second, and the density in kilograms per cubic meter.

Okay, let's give this equation a test drive. What is the pressure differential across a flat roof having an area of 240 m² when the wind blows at 25.0 m/s? The density of air is 1.29 kg/m³.

To solve this problem we're going to have to make some assumptions. First, we'll assume that the speed of air inside the house is 0.00 m/s, which is fairly reasonable. The physical situation is illustrated in Figure 5.10. Starting with the Bernoulli equation and making the necessary substitutions, we get:

$$P_1 - P_2 = \frac{1}{2}\rho v_2^2 - \frac{1}{2}\rho v_1^2$$
$$= \frac{1}{2}\left(1.29 \frac{kg}{m^3}\right)\left(25.0 \frac{m}{s}\right)^2 - \frac{1}{2}\left(1.29 \frac{kg}{m^3}\right)\left(0.00\frac{m}{s}\right)^2 = 403 \text{ Pa}$$

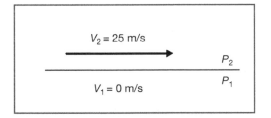

Figure 5.10 Wind Blowing Over a Roof

Note that we do not know anything about the absolute pressure, only the difference in pressure between the inside and outside of the house. This is why wind storms can blow roofs off of houses. If we multiply the pressure times the area, we find the force on the roof from the air inside the house amounts to 96,700 N, or 10.9 tons.

$$(403 \text{ Pa})(240 \text{ m}^2) = 96,700 \text{ N}$$

Venturi Tube Flowmeter

Venturi tube flowmeters are devices used to measure fluid speeds in pipes. Figure 5.11 illustrates the general construction of a Venturi tube. These are very simple devices that consist of a middle section with a small diameter connected on both ends to larger diameter sections via smooth transitions in order to prevent turbulence. A U-tube containing a fluid of known

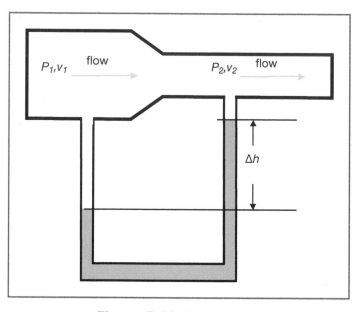

Figure 5.11 Venturi Tube

density connects the large and small diameter tubes. The U-tube is a manometer, and we can use it to measure differences in pressure.

Notice that one end of the U-tube is connected to the smaller diameter section of the Venturi tube, whereas the other end is connected to one of the larger diameter sections. If the fluid is flowing from left to right, according to the Bernoulli equation, $v_2 > v_1$ and $P_2 < P_1$. The fact that the pressure is lower in the narrow part of the tube is the primary scientific basis for the operation of an aspirator. The fluid levels in the manometer will reflect the pressure difference between P_1 and P_2. A measure of the height difference Δh and a knowledge of the density of the fluid in the manometer will yield the pressure difference between P_1 and P_2 using:

$$P_1 - P_2 = \rho g \Delta h$$

Finally, the speed of the fluid v_1 can be calculated using the Bernoulli equation in combination with the equation of continuity. Specifically, starting with the Bernoulli equation,

$$P_1 - P_2 = \tfrac{1}{2}\rho v_2^2 - \tfrac{1}{2}\rho v_1^2 = \tfrac{1}{2}\rho(v_2^2 - v_1^2)$$

We can replace v_2 in the above equation with the rearranged continuity equation

$$v_2 = \frac{A_1 v_1}{A_2}$$

to get:

$$P_1 - P_2 = \tfrac{1}{2}\rho(v_2^2 - v_1^2) = \tfrac{1}{2}\rho\left(\left|\frac{A_1 v_1}{A_2}\right|^2 - v_1^2\right)$$

We have now eliminated v_2 from the equation. Applying some (more) algebra, we obtain

$$P_1 - P_2 = \frac{1}{2}\rho\left(\left(\frac{A_1}{A_2}\right)^2 v_1^2 - v_1^2\right) = \frac{1}{2}\rho v_1^2\left(\left(\frac{A_1}{A_2}\right)^2 - 1\right)$$

We'll call this last equation the *Venturi tube flowmeter equation*, a highly descriptive, if not generally recognized, name. This is the equation that can be used to determine v_1, the speed of a fluid in a pipe.

Let's try an example. Cyclopropane has a density of 0.001816 g/cm^3. What pressure difference is generated by a flow rate of 50 cm^3/s through a Venturi tube having radii of 0.50 and 0.030 cm?

In order to apply the Venturi tube flowmeter equation, we will need to know the cross-sectional areas A_1 and A_2 of the two different parts of the Venturi tube. Let's be careful and work in SI units.

$$A_1 = \pi r_1^2 = \pi(0.50 \text{ cm})^2 = 0.7854 \text{ cm}^2 \times \left(\frac{1 \text{ m}}{100 \text{ cm}}\right)^2 = 7.854 \times 10^{-5} \text{ m}^2$$

$$A_2 = \pi r_2^2 = \pi(0.030 \text{ cm})^2 = 2.827 \times 10^{-3} \text{ cm}^2 \times \left(\frac{1 \text{ m}}{100 \text{ cm}}\right)^2$$
$$= 2.827 \times 10^{-7} \text{ m}^2$$

In addition, we were given the flow rate but we need speed. We therefore need to calculate the speed associated with a flow rate of 50 cm^3/s. This is easily accomplished using:

$$\text{Flow rate} = A_1 v_1$$

Rearrangement of this expression followed by substitution of the appropriate values yields:

$$v_1 = \frac{\text{Flow rate}}{A_1} = \frac{50 \frac{\text{cm}^3}{\text{s}}}{0.7854 \text{ cm}^2} = 63.662 \frac{\text{cm}}{\text{s}} \times \frac{1\text{m}}{100 \text{ cm}} = 0.63662 \frac{\text{m}}{\text{s}}$$

We need to do one more preliminary calculation, that is, convert the density of cyclopropane to SI units.

$$\rho = 0.001816 \frac{\text{g}}{\text{cm}^3} \times \frac{1 \text{ kg}}{1000 \text{ g}} \times \left(\frac{100 \text{ cm}}{1 \text{ m}}\right)^3 = 1.816 \frac{\text{kg}}{\text{m}^3}$$

Substituting all of the necessary information into the Venturi tube flowmeter equation, we get:

$$P_1 - P_2 = \frac{1}{2}\rho v_1^2\left(\left(\frac{A_1}{A_2}\right)^2 - 1\right) = \frac{1}{2}\left(1.816 \frac{\text{kg}}{\text{m}^3}\right)\left(0.63662 \frac{\text{m}}{\text{s}}\right)^2\left(\left(\frac{7.854 \times 10^{-5} \text{ m}^2}{2.827 \times 10^{-7} \text{ m}^2}\right)^2 - 1\right)$$

Simplification yields:

$$P_1 - P_2 = \frac{1}{2}\left(1.816 \frac{\text{kg}}{\text{m}^3}\right)\left(0.63662 \frac{\text{m}}{\text{s}}\right)^2(7.71845 \times 10^4 - 1) = 2.8 \times 10^4 \text{ Pa}$$

Let's try another one. A Venturi tube is attached to an Entonox line. The manometer in the Venturi tube is filled with water and shows a pressure difference of 20 mmH$_2$O. If the radii of the Venturi tube are 1.0 and 0.10 cm, what is the flow rate of the anesthetic gas? Assume the density of the Entonox is 1.9 kg/m^3.

First, let's convert the pressure difference from units of mmH$_2$O into pascals. Mercury has a density 13.6 times greater than the density of water, so a column of mercury that exerts the same pressure as a column of water is 13.6 times shorter. So, the pressure is 1.47 torr. Now, using the usual conversion factors, we find the pressure difference is 196 Pa.

Now inspection of the Venturi tube equation reveals that we know the pressure difference, the density, and the areas. Thus, we can solve for the speed in the larger tube. Remember the product of speed and area gives the flow rate.

$$P_1 - P_2 = \frac{1}{2}\rho v_1^2 \left(\left| \frac{A_1}{A_2} \right|^2 - 1 \right)$$

Let's start by solving for v_1.

$$v_1 = \sqrt{\frac{2(P_1 - P_2)}{\rho\left[\left|\frac{A_1}{A_2}\right|^2 - 1\right]}} = \sqrt{\frac{2(196Pa)}{1.9\,\frac{kg}{m^3}\left[\left|\frac{2\pi(0.01\ m)^2}{2\pi(0.001\ m)^2}\right|^2 - 1\right]}}$$

Ouch. A bit of careful arithmetic shows that the speed in the larger tube is 0.144 m/s (we'll keep one additional significant figure in this intermediate answer to prevent a rounding error). If the radius of the large tube is 0.010 m, the area is 3.14 × 10^{-4} m^2. Therefore, the flow rate is:

$$\text{Flow rate} = (3.14 \times 10^{-4}\ m^2)(0.144\ \tfrac{m}{s}) = 4.52 \times 10^{-5}\ \tfrac{m^3}{s}$$

This is equivalent to 45 mL/s or 2.7 L/min.

VISCOSITY

Ideal fluids are those in which there is no loss of energy due to friction, there are no interactions between the molecules that make up the fluid, and there are no interactions between the fluid molecules and the pipe, tubing, or container. In everyday life we deal with real fluids, not ideal fluids. Viscosity is a measure of a fluid's resistance to flow. Fluids with high viscosity such as honey do not flow very readily, while fluids with low viscosity such as water flow more easily. Let's take a closer look at a real fluid flowing through a horizontal pipe of length L and cross-sectional area A, as shown in Figure 5.12.

Since the molecules that make up the fluid can interact with each other and the pipe, the resulting flow pattern may be surprising. The speeds of the molecules next to the pipe walls are essentially zero, while those molecules near the center of the pipe are moving the fastest. The closer a molecule is to a wall, the slower it moves. We thus see that adjacent regions of the fluid will have different speeds and the faster regions will flow past the slower ones—hence the shape of the flow pattern.

A force must be introduced in order to maintain the flow of a real fluid. For example, you must turn the water spigot on to cause water to flow from a garden hose. Thus, the force

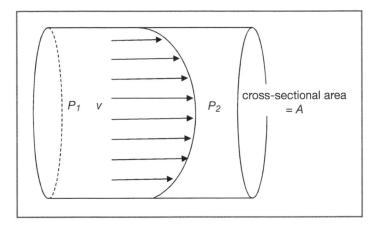

Figure 5.12 Viscosity Forces

required to maintain the flow of a real fluid is due to a pressure difference over the length L of pipe. Empirically (that is, by experiment), it can be shown that the required difference in pressure to maintain flow is proportional to the pipe length L and the average speed v of the real fluid, and inversely proportional to the cross-sectional area A of the pipe. The coefficient of viscosity η of a fluid (also called the viscosity of the fluid) is defined such that the following relationship is true:

$$P_1 - P_2 = 8\pi\eta \frac{vL}{A}$$

We'll call this equation the *viscosity equation.* Solving the viscosity equation for η will help us to understand the SI units of this quantity:

$$\eta = \frac{(P_1 - P_2)A}{8\pi v L} = \frac{\text{Pa} \cdot \text{m}^2}{\frac{\text{m}}{\text{s}} \cdot \text{m}} = \frac{\text{Pa} \cdot \text{m}^2}{\frac{\text{m}^2}{\text{s}}} = \text{Pa} \cdot \text{s}$$

In the older literature you will find the viscosity unit of poise, defined to be:

$$1 \text{ poise} = 1 \frac{\text{dyne} \cdot \text{s}}{\text{cm}^2} = \frac{0.1 \text{ N} \cdot \text{s}}{\text{m}^2} = 0.1 \text{ Pa} \cdot \text{s}$$

In other words,

$$1 \text{ Pa} \cdot \text{s} = 10 \text{ poise}$$

The viscosity equation can be solved for v to get:

$$v = \frac{(P_1 - P_2)A}{8\pi\eta L}$$

However, we know that flow rate is given by:

$$\text{Flow rate} = Av = \frac{V}{t}$$

Thus, we can multiply both sides of the rearranged viscosity equation by A to obtain flow rate:

$$\text{Flow rate} = \frac{V}{t} = Av = \frac{(P_1 - P_2)A^2}{8\pi\eta L}$$

If our pipe has a circular cross section, then $A = \pi r^2$; therefore $A^2 = \pi^2 r^4$. This can be substituted into the flow-rate equation to obtain:

$$\text{Flow rate} = \frac{V}{t} = Av = \frac{(P_1 - P_2)A^2}{8\pi\eta L} = \frac{(P_1 - P_2)\pi^2 r^4}{8\pi\eta L}$$

Final simplification yields Poiseuille's equation:

$$\text{Flow rate} = \frac{V}{t} = \frac{(P_1 - P_2)\pi r^4}{8\eta L}$$

Poiseuille's equation applies to laminar flow only. Let's now try some examples.

Blood has a viscosity of 0.0015 Pa·s. If a pressure of $1\overline{0}0$ mmHg (or 13,000 Pa) is applied to the aorta ($r = 0.010$ m; $L = 1.0$ m), what flow results?

This is a fairly straightforward plug-and-chug Poiseuille's equation problem. Using SI values only, we obtain:

$$\text{Flow rate} = \frac{(1.3 \times 10^4 \text{ Pa})\pi(0.010 \text{ m})^4}{8(1.5 \times 10^{-3} \text{ Pa·s})(1.0 \text{ m})} = 3.4 \times 10^{-2} \frac{\text{m}^3}{\text{s}}$$

Now, for one more. An IV bottle hangs 2.0 m above a patient. The tube leads to a needle having a length of 0.040 m and a radius of 4.0×10^{-4} m. This configuration results in a pressure of $1\overline{0}00$ Pa above the patient's blood pressure. If the viscosity of the liquid is 0.0015 Pa·s, what flow rate results?

This is another application of Poiseuille's equation. Notice in this case $P_1 - P_2 = 1000$ Pa.

$$\text{Flow rate} = \frac{(1.0 \times 10^3 \text{ Pa})\pi(4.0 \times 10^{-4} \text{ m})^4}{8(1.5 \times 10^{-3} \text{ Pa·s})(0.04 \text{ m})} = 1.7 \times 10^{-7} \frac{\text{m}^3}{\text{s}}$$

or 0.17 mL/s, which equals $1\overline{0}$ mL/min.

Video 5.2

There are certain situations in patient care when it is necessary to increase the fluid flow rate through a catheter or needle. Careful inspection of Poiseuille's equation yields three possible ways to accomplish this. First and foremost, increasing the pressure differential ($P_1 - P_2$) across the catheter or needle can increase the flow rate. Using a larger diameter catheter or needle will also allow for a greater flow rate. Finally, using a shorter catheter or needle will result in a greater fluid flow rate. In many cases, due to time constraints, it is impractical or not possible to use a larger diameter and/or shorter catheter or needle. Thus, increasing the pressure differential becomes the only realistic option.

The type of fluid flow that will actually be present in a given situation can be difficult to model and predict. Flow type depends upon the physical arrangement of the particular location and other factors such as the fluid's speed, density, and viscosity. The Reynolds number, N_R, is a dimensionless quantity that is used to characterize fluid flow. Specifically, for the case of a fluid flowing through a straight, smooth pipe, the Reynolds number is defined to be:

$$N_R = \frac{2r\rho v}{\eta} = \frac{d\rho v}{\eta}$$

where r is the pipe's radius in meters; ρ is the fluid's density in kg/m³; v is the average speed of the fluid; η is the viscosity of the fluid in Pa·s; and d is the diameter of the pipe. Note in the second form of the equation the relationship $d = 2r$ was used.

Situations in which the calculated Reynolds number is less than approximately 2000 will typically have laminar flow, while Reynolds numbers greater than 3000 will result in turbulent flow. When the Reynolds number is between these two values, the fluid flow will be unstable, and the flow can transition between laminar and turbulent flow.

Summary

- A fluid is any material that has the ability to flow.
- Hydrostatics is the study of fluids at rest.
- Hydrodynamics is the study of fluids in motion.
- Pressure:

$$P = \frac{F}{A} = \rho g h$$

- Pressure at any two points in a fluid:

$$P_2 = P_1 + \rho g h$$

- At a given point in a fluid, the fluid exerts the same pressure in all directions.
- The pressure at a given depth is independent of the container shape.
- Pascal's Principle: An external pressure applied to a confined fluid is transmitted unchanged to every point within the fluid.
- Archimedes' Principle: An object immersed either totally or partially in a fluid feels a buoyant force equal to the weight of the fluid displaced.

$$F_b = \rho g v$$

- Flow rate is the volume of fluid that flows per unit time.
- Equation of continuity for an incompressible fluid:

$$A_1 v_1 = A_2 v_2$$

- The Bernoulli equation:

$$P_1 + \frac{1}{2}\rho v_1^2 = P_2 + \frac{1}{2}\rho v_2^2$$

- Venturi tube flowmeter equation:

$$P_1 - P_2 = \frac{1}{2}\rho v_1^2 \left(\left(\frac{A_1}{A_2}\right)^2 - 1 \right)$$

- Coefficient of viscosity:

$$P_1 - P_2 = 8\pi\eta\frac{vL}{A}$$

■ Poiseuille's equation:

$$\text{Flow rate} = \frac{V}{t} = \frac{(P_1 - P_2)\pi r^4}{8\eta L}$$

■ The Reynolds number, which is useful in characterizing fluid flow, is defined to be:

$$N_R = \frac{2r\rho v}{\eta} = \frac{d\rho v}{\eta}$$

Review Questions for Fluids

1. Describe fluids, hydrostatics, and hydrodynamics.

2. What is the total pressure on a diver in pascals at a depth 45 m under the sea? Assume the atmospheric pressure is 101 kPa. Sea water has a density of 1.025 g/cm^3.

3. Describe Pascal's Principle.

4. Describe buoyancy.

5. Describe the Archimedes' Principle.

6. A spherical diving bell has a diameter of 6.0 m. What is the buoyant force in newtons when it is fully submerged? The specific gravity of sea water is 1.025 g/cm^3.

7. Describe laminar and turbulent flow.

8. Describe flow rate of a fluid.

9. If water is moving at a speed of 0.15 m/s through a tube with a cross-sectional area of $0.01\overline{0}$ m^2, what is the flow rate?

10. How does the equation of continuity relate flow rate of a fluid to the size of the tube through which it flows?

11. A sample of helium gas flows at a rate of $10\overline{0}$ cm^3/s through a tube having a radius of $0.50\overline{0}$ cm. What is the speed of the He?

12. If the He from the previous question enters a tube having a radius of $0.050\overline{0}$ cm, what is the speed of the He?

13. Describe the Bernoulli effect.

14. Cyclopropane moves with a speed of 1.0 m/s through a tube of diameter 1.0 cm and enters a tube with a diameter of $0.01\overline{0}$ cm. By how much does the speed increase, and what is the pressure drop? The density of cyclopropane is 1.8 kg/m^3.

15. Describe how a Venturi tube works.

16. Describe viscosity and how it is measured.

17. Describe how Poiseuille's theorem relates fluid flow to viscosity and tube size.

18. What flow rate will a pressure of $2\overline{0}$ Pa drive nitrous oxide through a $1\overline{0}$ m tube of radius $0.001\overline{0}$ m if the density of N_2O is 2.0 kg/m^3 and the viscosity is 1.8×10^{-5} Pa·s?

19. Explain why moving an IV bag higher above the patient results in an increased flow rate of the IV fluid.

20. A rapid infusion system has a 16-gauge catheter with an internal radius of 0.705 mm and length of 55 mm. Calculate the flow rate for blood ($\eta = 0.0015$ Pa · s) when the blood is delivered at a pressure of 2450 Pa and 4900 Pa above the patient's blood pressure. Report both answers in m^3/s and mL/min.

The Gas Laws

THE EMPIRICAL GAS LAWS

The empirical gas laws date back to as early as the seventeenth century and describe the inter-relationship between four quantities that describe the state of a gas, namely, pressure, temperature, volume, and the number of moles. Each law is named after the scientist who discovered or first reported the relationship. However, besides the historical interest, there is no particular necessity to learn the names of the gas laws. We believe it is more useful for you to develop an intuitive understanding about how changing one of these quantities affects the others. So, let's imagine some fixed volume of gas. How does changing one parameter at a time affect the volume? If we increase the pressure on the sample, doesn't it follow that the sample will get squished smaller; that is, the volume will decrease? Therefore, volume and pressure are inversely proportional. Most things expand when heated, and gases are no exception. Therefore, volume and temperature are directly proportional. Finally, it should stand to reason that a larger number of gas molecules will occupy a larger volume, assuming pressure and temperature remain constant. Therefore, the volume of a gas is directly proportional to the number of moles. That's all there is to the gas laws, except for a few mathematical details. The gas laws, coupled with the Kinetic Molecular Theory of Gases, make the gaseous state the best understood of the three common states of matter. Now, let's take a closer look at some of the details.

The Volume–Pressure Relationship: Boyle's Law

Robert Boyle reported that the volume of a fixed sample of gas is inversely related to pressure, as long as the temperature is constant. As the pressure increases, the volume decreases, and vice versa. This is the basis of breathing. During inspiration, the diaphragm muscle contracts and pulls downward, thereby increasing the volume of

Video 6.1

the thoracic cavity. The gas pressure in the lungs decreases to less than the ambient atmospheric pressure, and this pressure gradient causes air to move into the lungs. You can rationalize a gas moving in response to a difference in pressure as a diffusion process. At a higher pressure the gas is more concentrated, and at a lower pressure the gas is less concentrated. Diffusion spontaneously occurs from the area of higher concentration to an area of lower concentration. During expiration, the process is reversed. The diaphragm muscle relaxes, and this causes the volume of the thoracic cavity to decrease. The decrease in volume causes an increase in pressure in the lungs, relative to the ambient atmospheric pressure, and the air in the lungs is expelled.

The mathematical statement of Boyle's Law is usually formulated as:

$$P_1 V_1 = P_2 V_2$$

where P is the pressure and V is the volume.

This relationship has four quantities. To exactly solve a problem, you need three of the four variables. Boyle's Law is not sensitive to units of pressure or volume. You can measure pressure in units of pascals, psi, atm, or whatever. Likewise, you can measure volume in units of cubic meters (m^3), milliliters, or gallons. However, once you have selected a set of units for one side of the equation, you have to use the same units on the other side.

For example, the B size portable oxygen cylinder has a radius of 4.1 cm and a height of 29.5 cm. If the oxygen in a full cylinder is at a pressure of $20\overline{0}0$ psi, what volume will the oxygen occupy when it is released at a pressure of $2\overline{0}$ psi? Notice we said that the total pressure on the oxygen is $20\overline{0}0$ psi, not that the gauge on the cylinder reads $20\overline{0}0$ psi. In this problem we are dealing with absolute pressures in both the initial state and the final state, so using absolute pressure is fine. Just don't mix gauge pressures and absolute pressures with these kinds of problems. If the patient is using oxygen at a rate of 2.0 L/min, how many hours will this cylinder last?

Step 1: Let's calculate the volume of the oxygen in the cylinder, which, of course, will be equal to the volume of the cylinder. The volume of a regular cylinder is given by $V_{cylinder} = \pi r^2 h$, where r is the radius of the cylinder's base and h is the height of the cylinder. Okay, a gas cylinder isn't quite a regular cylinder, but it is close enough for this example.

$$V_{cylinder} = \pi(4.1 \text{ cm})^2(29.5 \text{ cm}) = 1600 \text{ cm}^3 \text{ (to two significant figures)}$$

Step 2: We know that $P_1 = 2000$ psi, and $V_1 = 1600$ cm³. The final pressure, P_2 is stated to be 20 psi. Therefore, we can solve Boyle's Law for V_2 and substitute in these values:

$$V_2 = \frac{P_1 V_1}{P_2} = \frac{(2000 \text{ psi})(1600 \text{ cm}^3)}{2\overline{0} \text{ psi}} = 1.6 \times 10^5 \text{ cm}^3$$

Step 3. The oxygen will occupy a volume of 160,000 cm³, or 160 L at the lower pressure. We can now use the flow rate as a conversion factor to calculate that this cylinder will last for about 1 hr and 20 min.

$$160 \text{ L}\left(\frac{1 \text{ min}}{2.0 \text{ L}}\right) = 80 \text{ min}$$

The Volume–Temperature Relationship: Charles's Law

Jacques Charles was an 18th-century hot-air balloonist who is credited with discovering the relationship between the volume and temperature of a gas. Charles's Law

Video 6.2

states that the volume of a gas is proportional to its absolute temperature, as long as the pressure and amount of gas are held constant. In algebraic terms, Charles's Law is given as:

$$\frac{V_1}{T_1} = \frac{V_2}{T_2}$$

where V is the volume in any unit and T is the Kelvin temperature.

Notice there is a key difference between Charles's Law and Boyle's Law. While the volume and pressure measurements in Boyle's Law can be measured in any units, as can the volume term in Charles's Law, we do not have this freedom with the temperature. As the relationship is formulated, the temperature must be expressed in kelvins.[1] Recall the Kelvin scale starts at absolute zero, the coldest possible temperature. One kelvin is equal to 1°C, and the relationship between the Kelvin scale and the Celsius scale is:

$$°C = K - 273.15$$

Charles's Law provides an important argument for the existence of some minimum temperature. Let's imagine filling a syringe with air to a volume of 10.00 mL when the temperature is 25°C. The plunger of the syringe is free to move, so the pressure on the gas will remain constant and equal to the atmospheric pressure. When we place the syringe in a hot water bath whose temperature is 100°C, the volume of the gas expands, pushing the plunger out to a volume of 12.52 mL. Now, let's watch the volume of the gas change as the water bath cools. We could even add some ice or ice and salt to drive the temperature even lower. We could easily collect data down to –20°C or so. If we were to graph our volume–temperature data, we would get a graph that looks as in Figure 6.1.

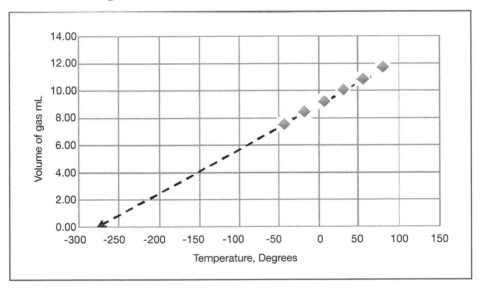

Figure 6.1 Volume–Temperature Relationship of a Gas (Celsius Scale)

---

[1] Actually, there are other "absolute" temperature scales, but the Kelvin scale is the official SI measure of temperature.

We see a linear relationship between temperature and volume, just as Charles's Law predicts. Can the volume decrease indefinitely as we lower the temperature? Take a close look at the dotted arrow and what happens at really low temperatures. Eventually, we get to a temperature where the volume equals zero. If we lowered the temperature any further, the volume would become negative. Since a negative volume is not physically possible, the temperature corresponding to $V = 0$ must be the coldest possible temperature. This temperature is called absolute zero.

If we change the temperature scale to kelvins, we still obtain a straight line, but this line goes through the origin (see Figure 6.2). Since the value of the y-intercept is zero, the equation of this line is:

$$V = mT$$

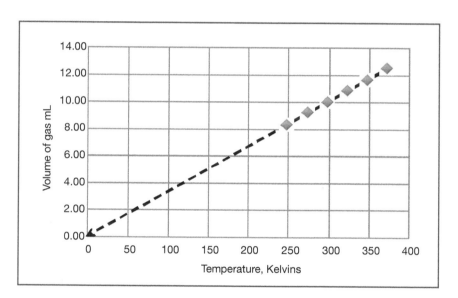

Figure 6.2 Volume–Temperature Relationship of a Gas (Kelvin Scale)

where V is volume, T is temperature, and m is the slope of the line.

Of course, $V = mT$ is another way to write Charles's Law. The equation of the line using the Celsius data has the form $V = mT + b$. That is, there is a nonzero y-intercept. If the line does not go through the origin, then Charles's Law would have to include a term to correct for the nonzero value of the y-intercept (b). Of course, this is doable, but it is not as convenient. To avoid this hassle, we always use the Kelvin temperature.

Now for an example. The temperature of 34.5 mL of neon is increased from −35.0°C to 75.0°C. What is the new volume? The initial temperature (−35.0°C) corresponds to 238 K, and the final temperature corresponds to 348 K. If we solve Charles's Law for V_2 and substitute in the values for V_1, T_1, and T_2, we find the volume of the neon will increase to 50.4 mL.

$$V_2 = \frac{V_1}{T_1}T_2 = \frac{34.5 \text{ mL}}{238 \text{ K}} 348 \text{ K} = 50.4 \text{ mL}$$

The Volume–Mole Relationship: Avogadro's Law

While most beginning chemistry students associate the name of Avogadro with moles and Avogadro's number, Amedeo Avogadro's principal contribution to the chemical sciences is the hypothesis that, at equal temperatures and pressures, equal volumes of gas contain equal numbers of particles. This work made the determination of several atomic and molecular masses possible. In other words, the volume of a gas is directly proportional to the number of gas molecules, as long as temperature and pressure are held constant.

$$\frac{V_1}{n_1} = \frac{V_2}{n_2}$$

For example, nitric oxide (NO) is produced in internal combustion engines by oxidizing nitrogen in the air. Nitric oxide further reacts with oxygen in the air to give nitrogen dioxide (NO_2). Imagine a perfectly flexible balloon filled with 0.040 moles of nitric oxide. The temperature and pressure are such that the volume of the balloon is 1.0 L. A second balloon is filled with 0.020 moles of oxygen at the same temperature and pressure as the first balloon, and this balloon has a volume of 0.50 L. The two balloons are connected by a valve. Now imagine opening the valve and allowing the gases to mix and react according to the following equation:

$$2\,NO + O_2 \rightarrow 2\,NO_2$$

Let's also imagine that our balloons are really tough and don't burst from the vigorous reaction that has just occurred. Let's wait until the temperature returns to the initial value. Since our imaginary balloons are perfectly flexible, the pressure also remains constant. What total volume will the gaseous product occupy?

Well, first of all, does it make sense that the volume of the oxygen-filled balloon is half that of the nitric oxide–filled balloon? Of course. The ratio of volume to moles is equal, which satisfies Avogadro's Law.

$$\frac{1.0\ L\ NO}{0.040\ mol\ NO} = \frac{0.50\ L\ O_2}{0.020\ mol\ O_2} = 25$$

Now, let's look at the system after the reaction. The initial volume of NO and O_2 together is 1.50 L. There are 0.060 moles of NO and O_2 in total. When the reaction is over, all of the NO and O_2 will be gone, having combined to give 0.040 moles of NO_2. Let's solve Avogadro's Law for V_2 and plug in the known quantities.

$$V_2 = \frac{1.5\ L}{0.060\ mol} \cdot 0.040\ mol = 1.0\ L$$

Historically, the observation that volumes of reacting gases were always simplified to ratios of small, whole numbers is called Gay-Lussac's Law of Combining Volumes. In the preceding example, the volumes of NO to O_2 to NO_2 fit the pattern of 2:1:2. This observation further strengthened Dalton's argument for an atomic theory of matter.

Combined Gas Law

The three preceding laws can be formulated together as a single combined gas law that has the form:

$$\frac{P_1 V_1}{n_1 T_1} = \frac{P_2 V_2}{n_2 T_2}$$

This is really the only gas law you need to know to calculate the change of a gas's state which results from changing one or more of the state parameters (volume, pressure, temperature, or moles). If one of the parameters does not change, just drop it from the equation.

Let's have some fun playing with the gas laws. A gas cylinder contains a solution of 5% (v/v) carbon dioxide in oxygen.[2] The volume of the cylinder is 11 ft³, the *pressure gauge* on the cylinder reads 2000 psi, and the barometric pressure of the atmosphere is 14.5 psi.

What is the mole percent of the carbon dioxide in the cylinder? Well, let's take a look at the combined gas law. In this problem, temperature and pressure are constant, so let's focus on the volume and mole terms. A little rearranging, as well as eliminating the terms that don't change, gives:

$$\frac{V_1}{V_2} = \frac{n_1}{n_2}$$

We see that the ratio of the volumes is equal to the ratio of the moles. So, if a sample contains 5% (v/v) of carbon dioxide, that means there is 5 L of CO_2 in 100 L of the mixture. It also means that there are 5 moles of CO_2 in 100 moles of the mixture. For gases, and only for gases, the volume percent of each component is always equal to the mole percent.

Now, let's imagine the gas is delivered to a patient through a gas scrubber filled with Ascarite to remove the carbon dioxide. Ascarite is sodium hydroxide on asbestos. Ascarite II uses silica instead of asbestos. As the gas flows over this high-surface-area medium, carbon dioxide in the sample reacts with the sodium hydroxide to give the nonvolatile product sodium bicarbonate.

$$CO2 \ (g) + NaOH \ (s) \rightarrow NaHCO3 \ (s)$$

The sodium bicarbonate can further decompose by reacting with another NaOH to give sodium carbonate and water. The sodium carbonate is nonvolatile, and water is trapped on the sodium hydroxide as a nonstoichiometric hydrate.

$$NaHCO_3 \ (s) + NaOH \ (s) \rightarrow Na_2CO_3 \ (s) + H_2O \ (l)$$

What volume of pure oxygen gas can be delivered at an absolute pressure of 14.5 psi? First, we need to decide what's changing and what's constant. The initial volume is $V_1 = 11$ ft³, and we are looking for V_2. The *gauge pressure* of the gas in the cylinder is 2000 psi, and the atmospheric pressure is 14.5 psi. We are looking for the new volume at an absolute pressure of 14.5 psi, so we need the initial pressure expressed as an absolute pressure, not a gauge pressure. So the initial pressure $P_1 = 2014.5$ psi. The temperature isn't mentioned, so we will assume that

---

[2] Percent volume to volume is acceptable with gases, because the volumes of gases are additive.

temperature is constant and drop the T terms out of the equation. Now, the Ascarite scrubber removes the carbon dioxide; therefore the moles will change, decreasing by 5%. So, if we initially have n_1 moles of gas in the container, n_2 will be $0.95n_1$. Let's plug these quantities into the combined gas law:

$$\frac{(2014.5 \text{ psi})(11 \text{ ft}^3)}{n_1} = \frac{(14.5 \text{ psi})V_2}{0.95n_1}$$

You might notice that we don't have a number for n_1. Although in the next section we will learn how to calculate the number of moles in a sample under a given set of conditions, in this case we don't need a number for n_1, because our problem-solving strategy results in this term dividing out from both sides. Now, let's solve the expression for V_2 and do the arithmetic:

$$V_2 = \frac{(2014.5 \text{ psi})(11 \text{ ft}^3)(0.95)}{(14.5 \text{ psi})} = 1.45 \times 10^3 \text{ ft}^3$$

Let's try another one. A scuba diver encounters a shark at a depth of 35 ft under the sea, where the pressure is 2.07 atm and the temperature is 18.0°C. Our startled diver emits a bubble of gas having a volume of 75.0 mL. What will the new volume be at the surface, where $P = 0.967$ atm and $T = 23.0$°C?

Answer: 163 mL

THE IDEAL GAS LAW

The work of the empirical gas laws made it possible to formulate a state function to completely describe the state of a gas under a given set of conditions. This is called the ideal gas law, which has the form:

$$PV = nRT$$

where P is pressure, V is volume, n is moles, T is absolute temperature, and R is a constant.

As the name implies, the ideal gas law exactly describes the behavior of an ideal gas under all conditions. Unfortunately, we encounter only real gases. Ideal gases don't exist. Fortunately, though, as long as the pressure on the gas isn't too high and the temperature isn't too low, real gases *approximate* ideal behavior.

Standard Conditions and the Universal Gas Constant

The ideal gas law can be derived from the combined gas law. One interpretation of the combined gas law is that volume is inversely proportional to pressure and directly proportional to absolute temperature and to moles. That is, the product of pressure and volume, divided by the product of absolute temperature and moles will equal a constant value. If we state the combined gas law for a single set of conditions, we can obtain the ideal gas law:

$$\frac{PV}{nT} = \text{constant}$$

For no particularly obvious reason, this constant is symbolized as R. If we use R as the constant, and with a little rearrangement, we arrive at the normal statement of the ideal gas law:

$$PV = nRT$$

A value for R can be determined if we specify a set of conditions. One commonly cited set of conditions is called *standard temperature and pressure*, or *STP*. The standard temperature is defined to be exactly 0°C or 273.15 K. The standard pressure is 1 bar or 100 kPa. If you took chemistry a few years ago, you probably learned that standard pressure was 1 atm. The definition of standard pressure was changed to bring it in accordance with SI units. Given the following relationships (i.e., conversion factors), you are easily able to make necessary conversions.

$$1 \text{ atm} = 101.325 \text{ kPa} = 760 \text{ torr}$$

The standard molar volume of a gas is the volume that exactly 1 mole of an ideal gas occupies under STP conditions. The standard molar volume of a gas is 22.71 L.

$$\text{Standard molar volume} = 22.71 \text{ L}$$

Now we can calculate a value for R. The numerical value of R will depend on our choice of units for STP. Chemists frequently use a value of 0.08205 L · atm/mol/K, because these are units that are easily measured in the laboratory. If we convert 100 kPa into 0.9869 atm and use a standard molar volume of 22.71 L, we get:

$$R = \frac{(0.9869 \text{ atm})(22.71 \text{ L})}{(1 \text{ mol})(273.15 \text{ K})} = 0.08205 \, \frac{\text{L} \cdot \text{atm}}{\text{mol} \cdot \text{K}}$$

The value of R in SI units is 8.314 J/mol/K. We can calculate this from the standard molar volume in cubic meters and the standard pressure in pascals.

$$R = \frac{(100{,}000 \text{ Pa})(2.271 \times 10^{-2} \text{ m}^3)}{(1 \text{ mol})(273.15 \text{ K})} = 8.314 \, \frac{\text{Pa} \cdot \text{m}^3}{\text{mol} \cdot \text{K}}$$

These units of R can be simplified. Recall that a pascal is equivalent to newtons divided by square meters. If we simplify the meter terms in the numerator, we get newtons times meters, which is joules.

$$8.314 \, \frac{\text{Pa} \cdot \text{m}^3}{\text{mol} \cdot \text{K}} = 8.314 \, \frac{\frac{\text{N}}{\text{m}^2} \cdot \text{m}^3}{\text{mol} \cdot \text{K}} = 8.314 \, \frac{\text{N} \cdot \text{m}}{\text{mol} \cdot \text{K}} = 8.314 \, \frac{\text{J}}{\text{mol} \cdot \text{K}}$$

R is called the universal gas constant because it appears in several seemingly unrelated physical relationships. The units of R are significant. *Joules per mole per kelvin* can be interpreted as dividing energy among the molecules in a sample at a specified temperature. For example, R

is a factor in describing the relationship between temperature and kinetic energy of a collection of molecules, in distributing energies and velocities among the molecules in a sample, and in describing the heat capacity ratio of a gas.

Recall that when we used some variation of the combined gas law, the problem always specified some change in one or more of the conditions. The ideal gas law is most useful when the problem specifies a single set of conditions.

For example, let's calculate the moles and weight of oxygen in a full cylinder, if the volume of the cylinder is $30\overline{0}$ L, the pressure of the gas is 136 atm, and the temperature is 298 K. The only tricky item is which value of R do we use? We either have to use a value of R that has units that match our pressure and volume units, or convert the pressure and volume units to match the units in R. Since the pressure is in atmospheres and the volume is in liters, we can use 0.08205 L · atm/mol/K. We simply substitute these values into the ideal gas law and solve for the number of moles:

$$n = \frac{PV}{RT} = \frac{(136 \text{ atm})(300 \text{ L})}{\left(0.08205 \frac{\text{L} \cdot \text{atm}}{\text{mol} \cdot \text{K}}\right)(298 \text{ K})} = 1670 \text{ mol}$$

The molecular mass of O_2 is 32.0 g/mol:

$$1670 \text{ mol}\left(\frac{32.0 \text{ g}}{\text{mol}}\right) = 5.34 \times 10^4 \text{ g}$$

Finally, we can calculate the weight of the oxygen to be 523 N or 118 lb.

Gas Density

Unlike solids and liquids, the density of a gas depends *very* strongly on the temperature and pressure. Also, unlike solids and liquids, we can easily calculate the density of a gas if we know the temperature and pressure. For example, what is the density of air under normal conditions (25°C and 750 torr)? Recall that air is approximately 79% nitrogen and 21% oxygen (by volume). If we want to calculate the density of a sample, we need to know its mass and volume. Since density is an intrinsic physical property, we can take any sample size we want, so let's take a sample volume of 1.0 L.

Since nonreacting gaseous volumes are additive, a 1.0-L sample of air would contain 0.79 L of nitrogen and 0.21 L of oxygen. Now, we can use the ideal gas law to find the moles of each component.

$$n_{N_2} = \frac{\left(\frac{750}{760} \text{ atm}\right)(0.79 \text{ L})}{\left(0.08205 \frac{\text{L} \cdot \text{atm}}{\text{mol} \cdot \text{K}}\right)(298 \text{ K})} = 0.0319 \text{ mol N}_2$$

$$n_{O_2} = \frac{\left(\frac{750}{760} \text{ atm}\right)(0.21 \text{ L})}{\left(0.08205 \frac{\text{L} \cdot \text{atm}}{\text{mol} \cdot \text{K}}\right)(298 \text{ K})} = 0.00848 \text{ mol O}_2$$

The molecular masses of nitrogen and oxygen are 28 g/mol and 32 g/mol, respectively, so we can find the mass of each component:

$$0.0319 \text{ mol N}_2 \frac{28 \text{ g}}{\text{mol}} = 0.893 \text{ g}$$

$$0.00848 \text{ mol O}_2 \frac{32 \text{ g}}{\text{mol}} = 0.271 \text{ g}$$

Now the mass of the air sample will be the sum of the mass of the nitrogen and the mass of the oxygen. Therefore, the density of air under these conditions is:

$$d_{\text{air}} = \frac{0.893 \text{ g} + 0.271 \text{ g}}{1.0 \text{ L}} = 1.2 \frac{\text{g}}{\text{L}}$$

DALTON'S LAW OF PARTIAL PRESSURES

Dalton's Law states the total pressure of a gaseous mixture is the sum of the partial pressures of each of the component gases. The partial pressure of a gas in a mixture is the pressure the gas would exert by itself under identical conditions of temperature and volume. Let's imagine three steel containers each having an identical volume and insulated to maintain a constant temperature. Each container is absolutely empty; that is, the pressure inside each is 0 atm. Now, let's put enough nitrogen gas into the first container so that the pressure rises to 0.80 atm. Let's put enough oxygen gas into the second container so that the pressure rises to 0.20 atm. Now, if we were to transfer all of the nitrogen from the first container and all of the oxygen from the second container into the third container, what would the pressure be? Well, we put in enough nitrogen to exert a pressure of 0.80 atm and enough oxygen to exert a pressure of 0.20 atm. The mixture of gases will exert a total pressure of 1.00 atm.

In symbols, Dalton's Law is given by:

$$P_{\text{total}} = \Sigma P_i = P_1 + P_2 + \cdots + P_n$$

where P_{total} is the total pressure in the gas mixture and P_i is the partial pressure of each component gas. The sigma (Σ) is a shorthand way of saying "add these things up."

The partial pressure of each component is proportional to the moles of that component in the sample. So, if half of the molecules in the sample were oxygen, the partial pressure of oxygen would be equal to half the total pressure. In more general terms, the partial pressure of each component in a gaseous mixture is equal to the *mole fraction* of the component times the total pressure.

$$P_i = \chi_i P_{\text{total}}$$

The mole fraction of a component (χ_i) in a mixture is equal to the moles of the component divided by the number of moles of every component.

$$\chi_i = \frac{\text{mol}_i}{\text{mol}_{\text{total}}}$$

This is an easy relationship to derive. Let's start with the ratio of the partial pressure of one of the components divided by the total pressure. Since the ideal gas law states $PV = nRT$, we can substitute nRT/V for each of the pressure terms, and we get this expression:

$$\frac{P_i}{P_{total}} = \frac{\frac{n_i RT}{V}}{\frac{n_{total} RT}{V}}$$

Since all of the components are in the same container, the volumes are the same, as are the temperatures. Of course, the gas constant is the same, so V, R, and T drop out of the right side, which gives us:

$$\frac{P_i}{P_{total}} = \frac{n_i}{n_{total}} = \chi_i$$

Since the ratio of n_i to n_{total} is the mole fraction, we obtain, after rearranging:

$$P_i = \chi_i P_{total}$$

Suppose we have a 10.0-L sample that contains 0.350 moles N_2 and 0.0700 moles O_2 at a temperature of $30\overline{0}$ K. What is the total pressure in the sample, and what is the partial pressure of each component?

Step 1: Let's find the mole fraction of each component.

$$\chi_{O_2} = \frac{0.0700 \text{ mol } O_2}{0.350 \text{ mol } N_2 + 0.0700 \text{ mol } O_2} = 0.167$$

$$\chi_{N_2} = \frac{0.350 \text{ mol } N_2}{0.350 \text{ mol } N_2 + 0.0700 \text{ mol } O_2} = 0.833$$

Notice the sum of the mole fractions is 1.00, or 100% of the sample.

Step 2: Let's find the total pressure of this mixture using the ideal gas law. There are a total of 0.42 moles of gas in this sample.

$$P_{total} = \frac{(0.420 \text{ mol})(0.08205 \text{ L} \bullet \text{atm/mol/K})(30\overline{0}\text{K})}{(10.0 \text{ L})} = 1.03 \text{ atm}$$

Step 3: Dalton's Law gives the partial pressure of each gas.

$$P_{O_2} = (0.167)(1.03 \text{ atm}) = 0.172 \text{ atm}$$
$$P_{N_2} = (0.833)(1.03 \text{ atm}) = 0.858 \text{ atm}$$

Notice that the sum of the two partial pressures is equal to the total pressure in the sample.

$$0.172 \text{ atm} + 0.858 \text{ atm} = 1.03 \text{ atm}$$

Relative Humidity

Relative humidity measures the saturation of water in air. One way to calculate relative humidity is to divide the amount of water in the air by the solubility of water in the air.

$$\text{Relative Humidity} = \frac{\text{concentration of water in the air sample}}{\text{solubility of water in air}} 100\%$$

The solubility of water in air is the maximum amount of water that a given volume of air can accommodate, and the solubility of water in air increases with increasing temperature. So the relative humidity is also a function of temperature. For example, air at 25°C will dissolve 0.023 g of water per liter of air. At 30°C, the solubility of water in air is 0.030 g/L. So if an air sample at 30°C containing 0.010 g of water per liter is cooled to a new temperature of 25°C, the relative humidity increases from 33% to 43%.

$$\text{Relative Humidity}_{30 \text{ degrees}} = \frac{0.010 \frac{\text{g}}{\text{L}}}{0.030 \frac{\text{g}}{\text{L}}} 100\% = 33\%$$

If the sample is cooled to a temperature where the actual concentration of water exceeds the solubility of water in air, dew will form. This temperature is, obviously enough, called the *dew point temperature.*

Another way to express the relative humidity of water in air is by dividing the partial pressure of water in the air by the vapor pressure of water at that temperature.

$$\text{Relative Humidity} = \frac{\text{partial pressure of water}}{\text{vapor pressure of water}} 100\%$$

So, let's imagine a glass of water at a constant temperature. Let's also put some plastic wrap tightly over the top of the glass, so that the water cannot evaporate out into the surrounding room and escape our system. Some of the water molecules in the liquid sample will vaporize, and some of the water molecules in the air will condense. The processes of evaporation and condensation will continue until the air is saturated with water vapor, and the relative humidity reaches 100%.

Since the glass is sealed by the plastic wrap, the water molecules are confined to our system and zip around with the other gas molecules, bouncing off the sides of the glass, the plastic wrap, and the surface of the water. That is to say, the gaseous water molecules exert a pressure, and this pressure is called the vapor pressure. At higher temperatures, more water vaporizes into the air, and, therefore, exerts a higher pressure. So if the vapor pressure is higher at higher temperatures, then the air has a greater capacity for water at that temperature. So, the maximum solubility of water in air is directly related to the vapor pressure of water.

Let's test this out. We stated earlier that the solubility of water in air at 25°C is 0.023 g/L. Let's use the ideal gas law to calculate the mass of water contained in a 1.0 L sample of air when the relative humidity is 100%. The vapor pressure of water at 25°C is 23.76 torr or 0.03126 atm. Therefore, the partial pressure of water is 0.03126 atm. First, we use the ideal gas law to calculate the moles of water contained in our 1.0-L sample:

$$n_{water} = \frac{(0.03126 \text{ atm})(1.0 \text{ L})}{(0.08205 \text{ L} \cdot \text{atm/mol/K})(298 \text{ K})} = 1.28 \times 10^{-3} \text{ mol}$$

The molecular mass of water is 18 g/mol, so the mass of water in our 1.0-L sample is 0.023 g.

$$g_{water} = 1.28 \times 10^{-3} \text{ mol}\left(\frac{18.0 \text{ g}}{1 \text{ mol}}\right) = 0.023 \text{ g}$$

So, saying the vapor pressure of water is 23.76 torr is equivalent to stating the solubility of water is 0.023 g/L.

Let's try another example. If the relative humidity in air is 35%, and the air temperature is 25°C, what is the partial pressure of water in the air?

Well, at this temperature, the vapor pressure of water is 23.76 torr. A relative humidity of 35% means that the partial pressure of water divided by the vapor pressure of water (times 100%) equals 35%.

$$\text{Relative Humidity} = \frac{\text{partial pressure of water}}{\text{vapor pressure of water}}100\%$$

When we solve this equation for the partial pressure of water, and substitute in the known quantities of vapor pressure and relative humidity, we find the partial pressure of water is 8.3 torr.

$$P_{water} = (0.35)(23.76 \text{ torr}) = 8.3 \text{ torr}$$

Now, let's try a more complex example. What is the partial pressure of oxygen in air when the air temperature is 20°C, the barometric pressure is 740 torr, and the relative humidity is 45%? The vapor pressure of water at this temperature is 17.54 torr, and the mole fraction of oxygen in air is 0.21.

First, we need to recognize the total pressure of air is governed by Dalton's Law. So, we can write:

$$P_{total} = P_{dry \text{ air}} + P_{water}$$

We calculate the partial pressure of water from the relative humidity and the vapor pressure of water.

$$P_{water} = (0.45)(17.54 \text{ torr}) = 7.9 \text{ torr}$$

If we subtract the partial pressure of water from the barometric pressure, we find the partial pressure of dry air.

$$P_{dry\ air} = 740\ torr - 7.9\ torr = 732\ torr$$

Since oxygen is 21% of *dry* air, the partial pressure of oxygen is 21% of the partial pressure of dry air.

$$P_{O2} = (0.21)(732\ torr) = 154\ torr$$

Notice that dry air at 740 torr has a partial pressure of oxygen equal to 154 torr. While this is not a huge difference, it does show that increasing humidity decreases the partial pressure of oxygen available for breathing.

KINETIC MOLECULAR THEORY OF GASES

The Kinetic Molecular Theory of Gases is based on four basic tenets. These tenets exactly describe an *ideal* gas. However, as we shall see, this theory does not describe *real* gases under all conditions.

1. Gases consist of small particles whose volume is negligible compared to the volume of the gas.
2. Gas molecules are in constant, random motion.
3. The molecules in the sample show a range of kinetic energies, but the average kinetic energy depends only on the temperature.
4. There are no attractive or repulsive forces between the gas particles, so all collisions are elastic.

Distribution of Kinetic Energies

In any sample of gas, the molecules are in constant random motion, but that doesn't mean they are all moving at the same speed. Some of the molecules are going faster and some are going slower. The sum of the kinetic and potential energies of the molecules is the internal energy of the sample. If we increase the temperature of the sample, the added energy goes into increasing the average kinetic energy of the molecules. The temperature is directly proportional to the average kinetic energy.

$$(KE)_{ave} = \frac{3}{2}kT$$

where $(KE)_{ave}$ is the average kinetic energy, k is the Boltzmann constant, and T is the temperature in kelvin.

The distribution of kinetic energies among the molecules in a sample is given by the Boltzmann distribution.

$$N = 4\pi\left(\frac{m}{2\pi kT}\right)^{\frac{3}{2}}v^2 e^{-\frac{mv^2}{2kT}}$$

This odious equation states that the fraction of molecules N is a function of the mass m, speed v, the Kelvin temperature T, and a constant modestly called the Boltzmann constant k. It is much more useful to examine this function in graphical form. The graph in Figure 6.3 shows the kinetic energy of the molecules in a gas sample at two different temperatures. The black line represents the sample at a higher temperature than the gray line.

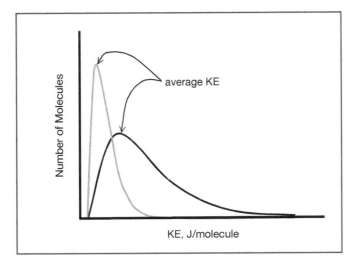

Figure 6.3 Kinetic Energy Distribution in Gas Molecules

We see that no molecules have zero kinetic energy. The number of molecules at a given kinetic energy increases to a maximum and then asymptotically approaches zero once again. The average kinetic energy at a given temperature is just past the maximum of the curve.

We also notice the shape of the curve changes as a function of temperature. At a higher temperature, the curve has a maximum corresponding to a higher kinetic energy. So, we see that a higher temperature means a higher *average* kinetic energy of the molecules in the sample.

Speed of Gas Molecules

The average kinetic energy for a gas depends on the temperature and not the identity of the gas. Therefore, different gases have the same average kinetic energy if their temperatures are the same. In order for a lighter gas, such as helium, to have the same kinetic energy as a heavier gas, such as neon, the lighter gas particles must have a higher speed. We can derive an expression for the average speed starting from the relationship between kinetic energy and temperature.

$$KE = \frac{3}{2}kT$$

Remember that kinetic energy is $1/2mv^2$. The Boltzmann constant k is equal to the ideal gas constant R divided by Avogadro's number. Avogadro's number is the number of molecules in a mole, so the Boltzmann constant treats individual molecules, while the ideal gas constant deals with moles of molecules. So, if we use the molecular mass (M), we need to use the ideal gas constant. Also, we're always safest in physics when we stick to SI units, so let's express the molecular mass in units of kg/mol and R in units of J/mol/K. So, on a molar scale, we can recast the relationship between kinetic energy and temperature as:

$$\tfrac{1}{2}Mv^2 = \tfrac{3}{2}RT$$

If we solve this equation for speed, we arrive at an expression representing the root mean square (rms) speed of a gas. The rms speed is symbolized u.

$$u = \sqrt{\frac{3RT}{M}}$$

Let's calculate the average speed[3] of an oxygen molecule at room temperature (25°C). Since we are dealing with physics here, we need to stick to SI units. Therefore, while the molecular mass of oxygen (O_2) from data off of the periodic table is 32 g/mol, we need to use the SI units of 0.032 kg/mol. Notice that we also use R in the SI units of 8.31 J/mol/K.

$$u = \sqrt{\frac{3(8.31 \text{ J/mol/K})(298 \text{ K})}{0.032 \text{ kg/mol}}} = 482 \text{ m/s}$$

Whoa, is that fast? Well, there are 1609 m in a mile and 3600 s in an hour, so the rms speed is over 1000 mph!

$$\left(\frac{482 \text{ m}}{\text{s}}\right)\left(\frac{1 \text{ mi}}{1609 \text{ m}}\right)\left(\frac{3600 \text{ s}}{\text{hr}}\right) = 1078 \text{ miles per hour}$$

GRAHAM'S LAW OF EFFUSION

Video 6.3

Effusion and diffusion are substantially the same process. Diffusion is movement of a substance from an area of higher concentration to an area of lower concentration. Effusion is the movement of a gas through a small opening. Graham's Law of Effusion states the rate of effusion is inversely proportional to the square root of the molecular mass.

$$\text{Rate of Effusion} \propto \sqrt{\frac{1}{M}}$$

This should make sense. The rate of effusion depends on the speed of the molecules. The average kinetic energy is the same for two different gases at the same temperature. The speed of the gas molecules depends on the kinetic energy $1/2mv^2$.

$$KE = \tfrac{1}{2}mv^2$$

---

[3] Physical chemists refer to the average speed calculated in this method as the rms or root-mean-square speed.

So, the square of the speed is inversely proportional to the mass, if the kinetic energy is constant. Rearranging and taking the square root of both sides of the equation shows that the speed is inversely proportional to the square root of the mass.

$$\sqrt{mv^2} = \sqrt{2 \cdot KE} = \text{constant}$$

When comparing the rates of effusion of two gases, Graham's Law is cast as:

$$\frac{\text{Rate}_1}{\text{Rate}_2} = \sqrt{\frac{M_2}{M_1}}$$

For example, how much faster does helium effuse than methane? The molar mass of helium and that of methane are 4 g/mol and 16 g/mol, respectively. If we insert these values into Graham's Law, we find helium effuses at twice the rate of methane.

$$\frac{\text{Rate}_{He}}{\text{Rate}_{CH_4}} = \sqrt{\frac{16 \frac{g}{mol}}{4 \frac{g}{mol}}} = 2$$

IDEAL GASES AND REAL GASES

Ideal gases obey the ideal gas law at all temperatures and pressures. However, there are no ideal gases, only real gases. Real gases deviate from ideal behavior most strongly at high pressures and/or low temperatures. So, where do the basic tenets of Kinetic Molecular Theory fail?

1. *Gases consist of small particles whose volume is negligible compared to the volume of the gas* is a good approximation, but it is not absolutely correct. The volume occupied by a gas is very much larger than the volume of the individual molecules, but the volume of the gas molecules themselves is not zero. For example, one mole of liquid water occupies a volume of 18 mL at 1 atm and 25°C, but one mole of gaseous water occupies 24,500 mL under the same conditions. So, the water molecules in a gaseous sample of water under these conditions occupies about 0.07% of the volume of the gas sample, and 99.93% of the sample is empty space.

This tenet starts to fail with real gases as the pressures get large, because at higher pressures more gas molecules are crowded into the same sample volume. So, sticking with our example of water, the volume of gaseous water at 25°C and a pressure of 1000 atm occupies a volume of only 24.5 mL. Now the volume of the molecules represents 73% of the sample volume, which clearly violates the first tenet.

2. *Gas molecules are in constant, random motion* is perfectly correct for real gases.

3. *The molecules in the sample show a range of kinetic energies, but the average kinetic energy depends only on the temperature* is an excellent model for the behavior of real gases.

4. *There are no attractive or repulsive forces between the gas particles, so all collisions are elastic* is a good approximation, but it is not absolutely correct. All molecules have attractive forces between them, or we would never be able to coax them into the liquid state. It takes energy to pull these attracting molecules away from each other. This tenet starts to fail for real gases at low temperatures, when the energy needed to overcome the attractive forces between

molecules becomes a greater fraction of the molecular kinetic energy. The observed pressure of the sample will decrease, because the molecules are sticking together.

The Van der Waals Equation

There are many equations that correct for the nonideal behavior of real gases. The Van der Waals equation is one that is most easily understood in the way that it corrects for intermolecular attractions between gaseous molecules and for the finite volume of the gas molecules. The Van der Waals equation is based on the ideal gas law.

$$\left(P + \frac{n^2a}{V^2}\right)(V - nb) = nRT$$

In this equation, P, V, n, and T represent pressure, volume, moles, and temperature, respectively, just as in the ideal gas law. R is still the universal gas constant.

The Van der Waals equation can be solved for P to get:

$$P = \frac{nRT}{(V - nb)} - \frac{n^2a}{V^2}$$

The new terms a and b are constants that are unique to each gas. The a term accounts for intermolecular attractions that reduce the observed pressure in the gas sample. This observed reduction in pressure is increased by a larger number of moles in the sample and decreased by a larger volume. The b term accounts for the volume of the gas molecules, an effect that becomes more important as more moles of gas are introduced into the sample. The trouble with the Van der Waals equation is that it is mathematically complicated. For example, if we try and solve the equation for V, we obtain a cubic equation, and cubic equations cannot be solved exactly.

Summary

■ The gaseous state of matter is characterized by having neither a definite volume nor a definite shape. Gases are compressible and expand to fill the available volume. The volume occupied by a gas is governed by three variables: pressure, temperature, and moles.

■ Boyle's Law states the volume of a gas is inversely proportional to pressure, if temperature and amount of gas are held constant. As pressure increases, the volume decreases, and vice versa. Volume and pressure can be expressed in any units, but both volumes must have the same units and both pressures must have the same units.

$$V_1 P_1 = V_2 P_2$$

■ Charles's Law states the volume of a gas is directly proportional to the Kelvin temperature, if pressure and amount of gas are held constant. As temperature increases, the volume increases, and vice versa. Temperature is expressed in kelvins (never in Celsius). Volume can be in any units, but both volumes must have the same units.

$$\frac{V_1}{T_1} = \frac{V_2}{T_2}$$

■ Avogadro's Law states the volume of a gas is directly proportional to moles of gas, if the temperature and pressure are held constant. As moles of gas increase, the volume increases, and vice versa. The volume can be in any units, but both volumes must have the same units.

$$\frac{V_1}{n_1} = \frac{V_2}{n_2}$$

■ The combined gas law relates the four variables of temperature, pressure, volume, and moles and is a compilation of Boyle's Law, Charles's Law, and Avogadro's Law. The combined gas law is most useful for situations where the value of one or more of the variables is changing. That is, the gas is undergoing a change in volume and/or pressure and/or temperature and/or the number of moles. Any variables that remain constant (or are not mentioned) can be eliminated from the equation.

$$\frac{P_1 V_1}{n_1 T_1} = \frac{P_2 V_2}{n_2 T_2}$$

■ The ideal gas law relates the four variables of temperature, pressure, volume, and moles to each other for a gas sample in a single state. The ideal gas law is most useful for situations where you know three of the four variables, and none of them are changing.

$$PV = nRT$$

- The ideal gas constant or universal gas constant, R, is a fundamental constant of nature that relates energy to temperature for a given amount of material. (Remember that the product of pressure times volume is equal to work or energy.) R commonly has units of 0.08205 L·atm/mol/K or 8.31 J/mol/K.
- Standard conditions, or standard temperature and pressure (STP) are 273K and 1 bar of pressure. Standard temperature was once designated at 1 atmosphere, but has since been redefined as 1 bar or 100 kPa.
- The standard molar volume of an ideal gas is the volume occupied by an ideal gas at STP. The SMV for an ideal gas is 22.71 L/mol.
- Dalton's Law of Partial Pressures states the partial pressure of a gas in a sample is equal to the mole fraction of that gas times the total pressure in the sample. Alternatively, the total pressure in a sample is equal to the sum of the partial pressures of each gaseous component.

$$P_i = \chi_i P_{total}$$

- Alternatively, the total pressure in a sample is equal to the sum of the partial pressures of each gaseous component.

$$P_{total} = P_1 + P_2 + \cdots + P_n = \sum_{i=1}^{n} P_i =$$

- The mole fraction of a gas is equal to the moles of an individual gas in a sample divided by the total number of moles of all gases in a sample.

$$\chi_i = \frac{mol_i}{mol_{total}}$$

- Relative humidity is based on Dalton's Law and relates the partial pressure of water in a sample of gas to the vapor pressure of water at that temperature.

$$\text{Relative Humidity} = \frac{P_{H2O}}{\text{vapor pressure of water}} 100\%$$

- The vapor pressure of water depends on temperature; at higher temperatures, the vapor pressure is higher. Vapor pressure results from molecules in a liquid sample escaping the liquid state into the gas state. The pressure is created by these gaseous molecules colliding with the walls of the container. Vapor pressure is discussed in greater detail in Chapter 7.
- The kinetic molecular theory describes gases as infinitely small, independent particles that experience no forces between each other or with the container. The particles are in constant, random motion, and the average speed of the particles depends only on temperature.
- The average kinetic energy (KE) of the molecules in a gas sample is:

$$(KE)_{ave} = \frac{3}{2}kT$$

where k is the Boltzmann constant and T is the Kelvin temperature.

■ The Boltzmann constant (k) is equal to the gas constant divided by Avogadro's number (N_A):

$$k = \frac{R}{N_A} = \frac{8.31 \frac{J}{mol.K}}{6.02 \times 10^{23} mol^{-1}} = 1.38 \times \frac{10^{-23} J}{K}$$

■ The Boltzmann constant relates energy and temperature for individual molecules, while the universal gas constant, R, relates energy and temperature for moles of molecules.

■ The average speed of a gas molelcule depends on its mass and the temperature.

$$u = \sqrt{\frac{3RT}{M}}$$

where u is the average speed, R is the universal gas constant (8.31 J/mol/K), T is the Kelvin temperature, and M is the molar mass of the gas, in units of kg/mol.

■ Graham's Law of Effusion describes how fast a gas can effuse (diffuse through a pinhole). The rate of effusion is proportional to the average speed of the gas molecules in the sample, which means the rate of effusion is also inversely proportional to the molar mass of the gas.

$$\text{Rate of Effusion} \propto \sqrt{\frac{1}{M}}$$

■ If we compare the rate of effusion between two different gases, we have:

$$\frac{\text{Rate of Effusion}_1}{\text{Rate of Effusion}_2} \propto \sqrt{\frac{M_2}{M_1}}$$

■ The ideal gas law applies precisely to *ideal* (theoretically perfect) gases. For real gases, the ideal gas law introduces noticeable errors at high pressures (where the volume of the individual gas molecules begins to constitute a significant fraction of the total volume of the gas sample) and at low temperatures (where the intermolecular forces among the gas molecules begins to change the kinetic energy of the molecules when they collide).

■ The Van der Waals equation attempts to address the shortcomings of the ideal gas law in describing real gases. The Van der Waals equation introduces two new variables, a (which accounts for the intermolecular forces among real gas molecules) and b (which accounts for the nonzero volume of real gas molecules.) The other variables (P, V, n, and T) represent pressure, volume, moles, and temperature, just as in the ideal gas law.

$$\left(P + \frac{n^2 a}{V^2}\right)(V - nb) = nRT$$

Review Questions for Gases

1. List and describe the empirical gas laws.

2. Describe the ideal gas law. When is it most useful and when does it fail?

3. Under what conditions should you use the ideal gas law, and when should you use the empirical gas laws?

4. List the basic tenets of the Kinetic Molecular Theory of Gases.

5. Do you expect He or SF_6 to most closely obey the ideal gas law, and why?

6. What is R? What are the units of R?

7. What is STP?

8. What is the standard molar volume of an ideal gas?

9. Calculate R in units of mL · bar/mol/K.

10. What is the molar volume of a gas at 25.0°C and $73\bar{0}$ torr?

11. What pressure will 1.0 L of argon at 0.80 atm have if it is compressed to $5\bar{0}0$ mL? Assume no temperature change.

12. What temperature will increase the pressure of 1.00 L of He initially at 25.0°C and 1.00 atm to a new pressure of 1.50 atm? Assume no volume change.

13. If the standard molar volume of an ideal gas is 22.71 L, calculate R in units of Pa · m^3/mol/K. There are 1000 L in a cubic meter. Report your answer to three significant digits.

14. What will happen to the volume of a gas if both the pressure and temperature are increased?

15. Verify the gas laws are equations of state. Imagine you have 2.0 L of oxygen at –45°C and $0.05\bar{0}$ atm. You first increase the temperature to 50°C (holding the pressure constant), and then increase the pressure to 0.75 atm. What is the new volume? What would the new volume be if you increased the pressure first and then the temperature?

16. How many moles of helium at $2\bar{0}$°C are needed to fill a dirigible? Assume the shape is a cylinder $5\bar{0}$ m long with a diameter of $3\bar{0}$ m, and the pressure inside is 1.05 atm.

17. What is the mass in grams of 1 L of helium gas at STP? (Find moles and then convert moles to grams; atomic mass = 4.0 grams/mol) What is the mass in grams of 1 L of air at STP? (Air is 80% N_2 and 20% O_2, so the apparent molecular weight of air is 0.8 · 28 + 0.2·32 = 28.8 g/mol.) Why do helium balloons float?

18. State Dalton's Law in words and symbols.

19. What is the mole fraction of oxygen in an anesthesia mixture that contains 2 moles of oxygen for every 0.5 moles of cyclopropane?

20. If the partial pressure of oxygen in an anesthesia mixture is $70\overline{0}$ mmHg and the total pressure is 15.0 psi, what is the mole fraction oxygen?

21. The vapor pressure of water at room temperature is 22 mmHg. What is the mole percent water in air if the air is saturated with water vapor? Assume the barometric pressure is 760 torr.

22. When do real gases most closely approximate ideal gas behavior?

23. What is the Van der Waals equation, and when is it useful?

24. What is the partial pressure of oxygen in humidified air (RH = 100%) at body temperature, if the barometric pressure is $75\overline{0}$ torr and the vapor pressure of water at 37.0°C is 47.0 torr? Dry air is 20.0% oxygen.

25. Rationalize Boyle's Law in terms of the Kinetic Molecular Theory of Gases.

26. You are scuba diving at a depth of $3\overline{0}$ feet, where the pressure is 2.0 atm. You take a deep breath and fill your lungs with 3.0 L of air. Then you hold your breath and ascend rapidly to the surface, where the air pressure is 1.0 atm. What volume will the air in your lungs now require?

27. What does STP not require?
    a. A temperature of 0°C
    b. Exactly 1 mole of gas
    c. A pressure of 1 bar
    d. All of these are requirements for STP

28. In your laboratory is a steel reaction vessel having a volume of 15.0 L. You charge the vessel with 2.00 moles of H_2 and 1.00 moles of O_2. The temperature of the vessel is $40\overline{0}$ K. What is the partial pressure of the hydrogen? $R = 0.08205$ L atm/mol/K.
    a. 8.80 atm
    b. 4.38 atm
    c. 1.39 atm
    d. 2.78 L

29. Referring again to the previous question, the mixture in the vessel ignites and exactly 2 moles of water vapor is produced. No hydrogen or oxygen is left over. The temperature of the mixture rises to 550 K and the walls of the vessels nearly give way, causing the volume of the container to increase to 15.5 L. Which expression below would give the new pressure inside the container?

    a. $\dfrac{(15.0)(6.56)(550)(2.00)}{(400)(3.00)(15.5)}$

    b. $\dfrac{(15.0)(8.80)(550)(2.00)}{(400)(3.00)(15.5)}$

    c. $\dfrac{(15.5)(6.56)(550)(2.00)}{(400)(3.00)(15.0)}$

    d. $\dfrac{(15.0)(6.56)(550)(3.00)}{(400)(2.00)(15.5)}$

30. Which of these is NOT a tenet of the Kinetic Molecular Theory of Gases?
    a. The volume of the gas molecules is small compared to the volume of the gas sample.
    b. Collisions between molecules are elastic.
    c. The temperature of the gas is a function of the average kinetic energy of the gas molecules.
    d. The velocities of identical atoms are the same, if the temperature of the sample is constant.

31. Two identical balloons are at the same temperature. One is filled with helium and one is filled with argon. When filled, the balloons have identical pressures and volumes. Which statement is true?
    a. The average kinetic energy of the argon atoms is greater than the average kinetic energy of the helium atoms.
    b. The average speed of the argon atoms is greater than the velocity of the helium atoms.
    c. The time it takes for an argon atom to travel across the diameter of the balloon is (on the average) greater than the time it takes for a helium atom (on the average) to travel across the diameter of the balloon.
    d. None of the above statements is correct.

32. Referring again to the previous question, which statement is true?
    a. The helium-filled balloon will weigh less than the argon-filled balloon.
    b. The argon-filled balloon will weigh less than the helium-filled balloon.
    c. Both balloons will have the same weight.
    d. It is not possible to predict the weights, based on the information given.

33. Which temperature scale begins at absolute zero?
    a. The Fahrenheit scale
    b. The Celsius scale
    c. The Kelvin scale
    d. The ideal gas scale

34. Under what conditions are real gases most likely to deviate from ideal gas behavior?
    a. High temperatures and high pressures
    b. High temperatures and low pressures
    c. Low temperatures and low pressures
    d. Low temperatures and high pressures

35. Match each phenomenon with the corresponding physical law or principle. Answers can be used once, more than once, or not at all.

    1. Breathing is based on this law.
    2. When concerned about the nonideal behavior of gases, it is best to use this law.
    3. Uranium is enriched by allowing mixtures of ^{235}U and ^{238}U to race down a long tube.
    4. The composition of a gaseous mixture is measured by the pressure of each gas.
    5. You know your tires are good, but they look flatter in the winter. Assume the pressure has not changed.
    6. This gas law is best to use when dealing with a gas in a single state.

    a. Boyle's Law
    b. Charles's Law
    c. Van der Waals equation
    d. Electronegativity
    e. Dalton's Law
    f. Graham's Law
    g. Ideal gas law

36. Fill in the missing parameters for each of these samples of helium gas. The atomic mass of helium is 4.00 g/mol.

SAMPLE	VOLUME	PRESSURE	AMOUNT	TEMPERATURE
1	? L	0.543 atm	12.6 g	355K
2	56.9 L	? atm	0.589 mol	500K
3	359 mL	355 mmHg	? mol	280K
4	26.5 L	73.8 kPa	1.62 mol	? K
5	? L	1.26 atm	1.26 mol	298K
6	369 m³	? bar	3.52 mol	249°C
7	549 mL	1.28 atm	? g	800K
8	89.2 L	0.594 atm	1.00 mol	? K
9	? L	1.00 atm	1.00 mol	273K
10	? L	760 torr	4.00 g	0.00°C

37. Calculate the final volume for each of the samples of xenon gas undergoing the proposed changes. All pressure measurements are in torr; all volume measurements are in mL; all temperature measurements are in kelvins.

SAMPLE	V_{FINAL}	P_{FINAL} .	T_{FINAL}	$V_{INITIAL}$	$P_{INITIAL}$	$T_{INITIAL}$
1	?	600	300	100	600	350
2	?	600	300	100	800	300
3	?	600	300	100	1200	600
4	?	600	300	100	900	250
5	?	600	300	100	300	200
6	?	600	300	100	400	400

38. A sample of gas contains 1.0 mole of oxygen and 3.0 moles of nitrogen. The temperature is 300 K and the total pressure is 2.00 atm.
    a. What volume would the oxygen occupy separately?
    b. What is the total volume of the sample?
    c. What is the mole fraction of oxygen?
    d. What is the partial pressure of the oxygen in atm?
    e. What is the percent by volume of the oxygen?

39. A sample of anesthesia gas contains 22 grams of nitrous oxide and 16 grams of oxygen gas. The molar mass of oxygen is 32 g/mol, and the molar mass of nitrous oxide is 44 g/mol. The temperature of the mixture is 20.0°C, and the total pressure is 850 torr.
    a. What is the partial pressure of each gas in torr?
    b. What total volume will this mixture of gases occupy?
    c. What is the mole fraction of nitrous oxide?
    d. What is the volume percent of nitrous oxide?

40. What would the "standard" molar volume of a gas be if the standard pressure were to be redefined as 760 torr and the standard temperature were to be redefined as 25°C?

41. Given the standard molar volume of a gas is 22.71 L/mol at a temperature of 273.15K and a pressure of 1 bar, what is the value of the ideal gas constant in units of mL·torr/mol/K?

42. Given the standard molar volume of a gas is 22.71 L/mol at a temperature of 273.15K and a pressure of 1 bar, what is the value of the ideal gas constant in units of L·atm/mol/K?

43. A 1 liter sample of anesthetic contains equal volumes of nitrous oxide and oxygen. If the temperature is 298K and the total pressure is 100 kPa, what is the mole fraction of each gas?

44. The vapor pressure of desflurane is 672 torr at 20°C. Use Dalton's Law to calculate the mole fraction desflurane in a saturated solution in oxygen gas when the total pressure is 800 torr.

45. The vapor pressure of isoflurane is 295 torr at 25.0°C. If oxygen gas at a pressure of $70\overline{0}$ torr is passed over liquid isoflurane at 25.0°C, what is the maximum percent by volume isoflurane that can be obtained? Assume the total pressure of the system is maintained at a pressure of $70\overline{0}$ torr.

46. A 22-L sample of air at 298K contains 0.10 grams of water vapor. What is the relative humidity of the air, if the vapor pressure of water at this temperature is 24.0 torr?

47. A sample of air at a temperature of 298K has a relative humidity of 10.0%. If 1.00 L of this air at a pressure of $76\overline{0}$ torr is placed in a sealed container containing a desiccant that will absorb all of the water from the sample, what will the pressure of the sample be once all of the water is removed? The vapor pressure of water at 298K is 24.0 torr.

48. A sample of air at a temperature of 298K has a relative humidity of 10.0%. If the temperature of 1.00 L of this air is decreased to 273K, what will the relative humidity be at the lower temperature? The vapor pressure of air at 298K is 24.0 torr and 4.5 torr at 273K.

49. A patient complains of having trouble breathing on hot, humid days. Pure air is 21% oxygen and 79% nitrogen. Assume the barometric pressure remains constant at $75\overline{0}$ torr.
    a. What is the partial pressure of oxygen in pure, dry air?
    b. If the relative humidity is 80% and the vapor pressure of water is 32 torr at 30°C, what is the partial pressure of water vapor in the air?
    c. If the total air pressure is $75\overline{0}$ torr and the relative humidity is 80%, what is the combined pressure of the oxygen and nitrogen in the air?
    d. Air is 21% oxygen. What is the partial pressure of the oxygen in the air when the relative humidity is 80% and the temperature is 30°C?

50. You are designing a desiccant unit for an anesthesia machine. The water-removal system must be able to remove all of the water from $10\overline{0}$ L of gas at a temperature of 273K and a pressure of $76\overline{0}$ torr. The relative humidity of the gas is 10%, and the vapor pressure of water at 273K is 4.5 torr.
    a. What is the partial pressure of water in the gas?
    b. How many moles of water are in the gas?
    c. What mass of water must the desiccant be able to absorb? The MW of water is 18 g/mol.

51. What is the average speed (in m/s) of a nitrous oxide molecule when the temperature is $30\overline{0}$K? The molar mass of nitrous oxide is 0.044 kg/mol.

52. What is the average speed of an oxygen molecule when the temperature is 300K? The molar mass of oxygen is 0.032 kg/mol.

53. In a given time interval, oxygen gas can effuse a distance of 1.0 meters. In the same time interval, how far will nitrous oxide effuse?

CHAPTER

7

States of Matter and Changes of State

KINETIC MOLECULAR THEORY OF MATTER

The Kinetic Molecular Theory of Matter attempts to describe all the states of matter and the conversion between the states by considering the structures of molecules comprising matter and how those molecules interact. There are three commonly encountered states of matter: solids, liquids, and gases. There are a few other states of matter, such as plasmas, but these are encountered only under extremely high energy conditions. Therefore, we restrict our conversation to the more mundane states.

Solids are characterized by a definite shape and volume. This is because the molecules, atoms, or ions comprising a solid are held in place with respect to each other by intermolecular forces sufficient to fix the constituents in place with respect to each other. Liquids are characterized by a definite volume, but liquids do not have a definite shape. Instead, liquids conform to the shape of the container. The forces between liquid particles are weaker than in the solid state and allow the particles to flow past each other. However, the forces are strong enough to hold the particles in a liquid sample in a compact mass. In fact, the particles comprising both liquid and solid samples are essentially touching. This is why solids and liquids are not easily compressible, and they are called the *condensed states of matter*. Gases have neither a definite shape nor a definite volume. A gaseous sample will expand to fill the available space. This is because there are essentially no intermolecular forces between the gaseous particles. Because both gases and liquids have the ability to flow, they are called fluids.

In addition to the states of matter, we need some familiarity with the transitions between the states of matter. Melting is the conversion of a solid into a liquid, while the reverse process is called freezing. A liquid vaporizes into a gas, a process also known as vaporization. A gas condenses into a liquid. Conversion of a gas into a solid is called deposition. Sublimation is the conversion of a solid directly into a gas, without forming an intermediate liquid. Each of these processes has a corresponding energy change and a temperature at which it occurs.

Of the three states, the gaseous state is by far the best understood. For this reason, as well as the importance that gases play in anesthesia, gases merit a detailed chapter of their own. In this chapter we explore some of the properties of solids and liquids, with only a cursory examination of gases. We explore the structure and properties of the states of matter, both at the macroscopic level and the molecular level. Finally, we examine the energy changes that accompany changes of state.

Intermolecular Forces

Chemical bonds may hold atoms together into molecules, but intermolecular forces determine how molecules interact with one another. While the intermolecular forces are determined by chemical bonding, it is the intermolecular forces that most directly impact the macroscopic properties of a sample. The state of matter under a given set of conditions, the tendency of a liquid to evaporate, and the solubility of a material in water are just three examples of macroscopic properties that are determined exclusively by intermolecular interactions. Furthermore, key microscopic and chemical processes, such as transcription of DNA, recognition of a substrate by an enzyme, and maintaining proteins in their active, three-dimensional shape are also controlled by intermolecular forces. Not to overstate the case, but if one of the intermolecular forces—hydrogen bonding—were to instantly cease operating, you would explode and then collapse into a puddle of amorphous goo.

All molecules and atoms, as well as oppositely charged ions, have an inherent attraction for each other, as long as they don't get too close together. Then there is an inherent repulsion. These forces are collectively known as *intermolecular forces*, because they occur between different molecules. The intermolecular forces are electrostatic in nature. They arise because of the attraction of opposite charges, according to Coulomb's Law. Since most of the volume of an atom or molecule is an electron cloud, the organization of electrons as chemical bonds between atoms in a molecule, as well as the three-dimensional arrangement of the electron clouds in space, is responsible for the type and strength of the intermolecular forces. To understand intermolecular forces, therefore, we need to first consider how electrons are used to form chemical bonds. Then we need to examine how the tendency to attract or release electrons varies across the periodic table of the elements. Finally, we need to take a look at the shapes of molecules.

CHEMICAL BONDING AND THE OCTET RULE

As discussed in Chapter 2, there are two limiting kinds of chemical bonds: ionic bonds and covalent bonds. The overarching driving force in the formation of chemical bonds of all types is the *octet rule*. Every atom tends to add, remove, or share electrons so as to wind up with eight *valence* electrons. Valence electrons are the electrons in the highest energy, or valence, shell. The noble gases each have a filled valence shell containing eight electrons.[1] While it doesn't sound very scientific, it is very useful to view the *noble gases*, located in group 8A, as the envy of every other atom on the periodic table. That is, every atom strives to be isoelectric to the nearest noble gas. Isoelectric species have the same electron configuration. Elements that are only one or two atomic numbers away from a noble gas form ions. Metals give up an electron or two to achieve a noble-gas-like electron configuration, forming cations. Nonmetals acquire an extra electron or two, forming anions that have electron configurations like a noble

---

[1] This first approximation is sufficiently correct for our purposes here. Interested readers are encouraged to further investigate electronic configurations.

gas. When nonmetal elements combine, neither has the chemical potential to force its partner into forming an ion. So, instead, these atoms pool their electron resources so that each has *access* to as many electrons as the nearest noble gas. Both the ionic and covalent strategies to achieve a noble gas–like electron configuration involve adding, removing, or sharing electrons in order to fill the highest energy electron shell, and that means winding up with eight valence electrons.

Ionic Bonds

Ionic compounds are formed between metals and nonmetals. Metals don't hold onto their electrons very strongly (this is why they are good electrical conductors) and tend to form cations. Nonmetals are much better at attracting electrons and tend to form anions.

Remember that electrons occupy quantum energy shells, and the capacity of each shell is equal to the number of elements in that specific row of the periodic table. The first row has two elements, and the first shell holds two electrons. Helium, located at the end of the first row, has a filled valence shell. Therefore, helium is chemically inert: It does not need to gain or lose any electrons to fill its valence shell. The second row has eight elements, and the second shell has a capacity of eight electrons. Neon, the noble gas situated at the end of the second period, has a filled valence shell. Therefore, neon is also chemically inert.

Representative elements whose atomic numbers only differ by 1 or 2 from the atomic number of a noble gas will gain or lose electrons so as to be isoelectric with the nearest noble gas. For example, a sodium atom has 11 electrons. The first two electrons go into the first energy shell, and this filled shell is like helium. The next eight electrons go into the second energy shell, which brings us to neon. The eleventh and last electron goes into the third energy shell. You see we can describe electrons as belonging to one of two categories. The electrons in filled shells are called *core* or *kernel* electrons. The core electrons are virtually inert, because they represent complete "collections" of electrons, and it is energetically expensive to break up a complete set. Electrons in partially filled shells (or partially empty, depending on your personality type) are called valence electrons. Valence electrons are used in forming chemical bonds. In the case of sodium metal, that last electron is easily lost, leaving an Na^+ cation. A sodium cation has a filled outermost energy shell containing eight electrons, just like neon.

So, sodium has one valence electron, as do Li, K, and so forth. This family of elements (called the *alkali metals*) is in group 1A, and they all form cations with a charge of +1. The next family, Group 2A, contains Be, Mg, and Ca, and each of these elements has two valence electrons and forms cations with a charge of +2.

Going over a few groups, we come to the halogens, or Group 7A, which are headed by fluorine. Fluorine has nine electrons, the first two of which fill the first energy shell (like helium). The last seven electrons are in the partially filled second shell. Therefore, fluorine has seven valence electrons, one short of a filled shell. However, rather than surrender seven electrons, it is easier for fluorine to acquire one additional electron. This gives the fluoride anion a charge of −1, fills the highest energy shell with eight electrons, and gives fluorine an electron configuration just like neon, the nearest noble gas.

Do you see the pattern? For the representative elements, the number of valence electrons is equal to the group number. This is why the authors believe the older AB numbering system for the periodic table is superior to the serial (1–18) numbering system. For the A-group elements, the number of valence electrons is equal to the group number. The B-group elements, or transition elements, are not as easy, and we will skip them for the purposes of our discussion.

An ionic bond results from Coulombic attraction between oppositely charged ions. In the solid state, the oppositely charged ions settle into a highly organized crystalline lattice, in which every cation is attracted to every anion. There are no ionic molecules, just a repeating three-dimensional array of repeating unit cells. Ionic bonds are generally stronger than covalent bonds. This means the ions are very strongly held in place, and for this reason virtually every ionic compound is a solid at room temperature and pressure. Of course, ionic compounds will melt at sufficiently high temperatures. For example, NaCl melts at 800°C. When melted, ionic liquids consist of mobile ions floating around each other. Therefore, ionic melts conduct electricity.

Covalent Bonds

Covalent bonds result from sharing one or more pairs of electrons. While every atom in a molecular species doesn't get eight electrons (or an octet of electrons) all to itself, the sharing strategy gives each atom *access* to eight electrons. But how are electrons shared? In the valence bond approach, sharing electrons results from bringing atoms close enough so that their electron orbitals overlap. Now, electron shells are divided into subshells, and subshells are divided into orbitals. Just think of dividing a house into stories and the stories into individual rooms. An electron orbital is analogous to a room. Electrons occupy rooms, and every room has a corresponding story number, and each story has a corresponding house number. The physical meaning of an orbital is more complicated, though. On the one hand, the shell, subshell, and orbital quantum numbers describe the quantum state, or energy, of an electron in a given orbital. On the other hand, the orbital describes a region in space where you are likely to find the electron.[2]

At the atomic scale, everyday common experience doesn't necessarily apply. For example, most of us started out thinking about electrons as negatively charged particles, but this model doesn't fully explain all the behaviors of electrons. It turns out that electrons also behave as waves. By the way, that doesn't mean the electrons are little particles surfing along on some kind of ethereal wave. Electrons behave as though they were waves, not localized particles.[3] Since they are waves, or at least have wave properties, an electron in a given orbital can be described by a mathematical wave function. The square of this function specifies a probability of finding the electron at some point in space. The thing to remember is this: An orbital is a mathematical function that describes the wave nature of an electron, and, like all mathematical functions, we can add them together.

Valence Bond Theory
A covalent bond results from sharing one or more pairs of electrons between two atoms through the overlap of their valence electron clouds. Let's imagine two hydrogen atoms. Each atom

---

[2] More precisely, it is the square of the orbital function, integrated over 3D space, that gives the probability of finding the electron.
[3] Unless you peek at them, expecting them to be particles. Then they are particles. Quantum mechanics is just weird: It never provides absolute answers, just probabilities. Too bad it's such a successful model for describing what we actually observe.

consists of a single proton that is surrounded by a single electron. The electron is in the first quantum energy level that is described by a wavefunction. The wavefunction of each electron delineates a spherical region in space (in this case, at least) centered in the nucleus where we are likely to find the electron. The electron is bound to the nucleus because the negatively charged electron is attracted to the positively charged nucleus by Coulomb's Law. Now, let's start moving the two atoms closer together. Eventually we will get to a geometry in which the two wave functions share some space in common. That is, the wavefunctions overlap. Mathematically, we achieve this by adding the wavefunctions together. That means the electrons on both atoms are now guided by a compound wavefunction that encompasses both nuclei.

As shown in Figure 7.1, the electrons are now attracted to both nuclei. That means the electron on atom 1 can zip over and bask in the positive glow of the nucleus on atom 2. The same is true for the electron on atom 2. A covalent bond results from this sharing of the negatively charged electrons between the two positively charged nuclei. In addition, the presence of electrons (really electron density) between the two nuclei will shield the nuclei from each other, thereby reducing the magnitude of the Coulombic repulsive force. This stabilizes the molecule by lowering its overall energy.

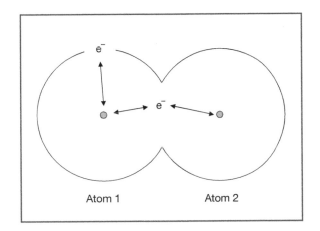

Figure 7.1 Formation of a Covalent Bond

Bond Length and Bond Energy

Figure 7.2 shows how the energy of a molecular system changes as two atoms are brought closer together, forming a covalent bond. At the far right side of the diagram, the atoms are completely separated, and there is no covalent bond between them.

As the atoms come together, and the degree of attraction among the collection of nuclei and electrons increases, the energy of the molecule first begins to decrease, and eventually reaches a minimum value. In the atomic world, having less energy means you are more stable. If the two atoms come too close together, repulsion between the two positively charged nuclei overwhelms the electron–nucleus attractions, and the energy of the molecule rises sharply. Thus, there is an ideal bond length, or distance between the nuclei, at which the energy is at a minimum, and the molecule is at maximum stability. The depth of this "energy well" represents the *bond dissociation energy*, or the amount of energy necessary to break that bond.

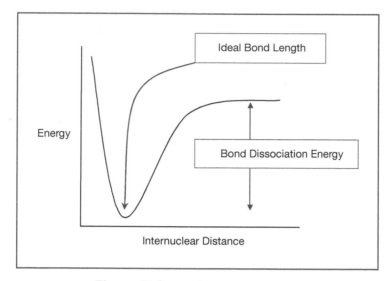

Figure 7.2 Bond Energy Diagram

Some covalent bonds are stronger than others. We see this manifested in some molecules that are more stable than others.

Lewis Structures

Lewis structures are one of the most useful and versatile tools in the chemist's toolbox. G. N. Lewis reported this model for chemical bonding in 1902. Lewis structures are nonmathematical models that allow us to qualitatively describe the chemical bonding in a molecule and then gain insights about the physical and chemical properties we can expect of that molecule. Don't discount the power of Lewis structures just because the underlying mathematics isn't evident. In a Lewis structure, the atoms are represented by their chemical symbol. Lines between atoms represent shared *pairs* of electrons in covalent bonds. Valence electrons that are not used for covalent bonds are lone pairs, and they are represented as pairs of dots on the atom.

Here is a general guide to drawing Lewis structures.

- First, you must determine the atom connectivity: In other words, which atom is bonded to which? For example, in nitrous oxide, is the atom connectivity N–N–O or N–O–N? This is a nontrivial problem for beginners. There are some general guidelines, such as oxygen is rarely a central atom and hydrogens are often bonded to oxygen. The fact is, however, that sometimes more than one arrangement of atoms is possible for a given molecular formula. Compounds that are different but have the same molecular formula are called *isomers*. For now, let's start with the atom connectivity as a given.
- Next we need to count the total number of electron pairs available for bonding. Only the valence electrons are used in forming covalent bonds. Remember that the number of valence electrons for a representative element is equal to the group number. Since electrons have a strong tendency to work in pairs, divide the total number of valence electrons by 2.

- Now, let's connect the atoms with single bonds.
- Then add the remaining valence electron pairs, either as lone pairs or double or triple bonds, so that every atom in the molecule has access to an octet of electrons.

For example, let's consider the water molecule. In water, the atom connectivity is H–O–H, as shown in Figure 7.3.

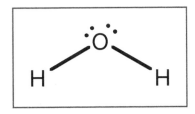

Figure 7.3 Lewis Structure of Water

Each hydrogen has one valence electron and oxygen has six. Thus, there are eight valence electrons, or four electron pairs. Two of the electron pairs are used to give single bonds between the oxygen and each hydrogen. The two remaining electron pairs are placed on oxygen as lone pairs of electrons. Now, let's count electrons. Oxygen has two shared pairs and two lone pairs. Therefore, oxygen has access to eight electrons and is stable. While each hydrogen has access to only two electrons, the closest noble gas to hydrogen is helium, and helium has two electrons. Therefore, the hydrogen atoms are also stable.

Carbon dioxide is a good example of multiple bonding (Figure 7.4). Carbon dioxide, with the atom connectivity of O–C–O, has a total of 16 valence electrons, or eight pairs.

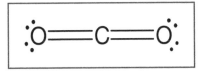

Figure 7.4 Lewis Structure of Carbon Dioxide

After using two of the pairs to establish the carbon–oxygen bonds, we have only six pairs of electrons with which to work. We could place the remaining six electron pairs as three lone pairs of electrons on each oxygen, giving the oxygens an octet. However, the carbon has access to only four electrons, and we have used up all of the valence electrons. When this happens, try multiple bonding. Let's try allowing each of the oxygens to share one of its lone pairs with the central carbon atom. This gives us double bonds from the carbon to each oxygen, and each oxygen still has two lone pairs. Every atom has access to the required octet, and this is now a stable molecule.

Other molecules that are important in anesthesia are nitrogen, carbon monoxide, and oxygen, Their Lewis structures are shown in Figure 7.5.

Some compounds exhibit resonance. That is, there is more than one acceptable Lewis structure for a compound. Another way to say this is that no single Lewis structure fully describes the actual structure. Resonance structures differ only in how the electrons are distributed among

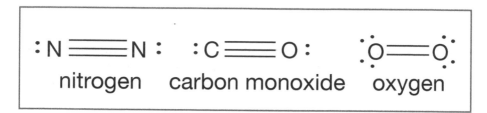

Figure 7.5 Lewis Structures of Some Common Gases

the atoms. If the atom connectivity is different, the two structures are isomers, which can be isolated and observed separately.

Nitrous oxide is a good example of a molecule that shows resonance. The structure on the left in Figure 7.6 has two double bonds, with two lone pairs on each of the distal atoms. The structure on the right has a triple bond. Neither structure fully describes nitrous oxide, nor does either structure actually exist. The real nitrous oxide molecule is a resonance hybrid of the two Lewis structures.

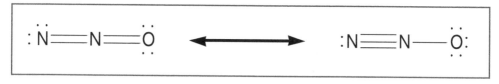

Figure 7.6 Resonance Structures of Nitrous Oxide

VSEPR Theory

Video 7.1

Atoms are bound into molecules by shared pairs of electrons. Electrons dislike each other because like charges repel each other. Therefore, whether they are lone pairs of electrons or bonding pairs of electrons, electron pairs try to get as far apart in space as is geometrically possible. There is a fancy name that summarizes these simple ideas: the VSEPR Theory, which stands for Valence Shell Electron Pair Repulsion Theory. Even though the VSEPR Theory is founded on fundamentally simple ideas, it is a tremendously powerful tool for predicting the shapes of molecules.

Let's stop a moment and recall where this conversation is going. We want to be able to rationalize the physical and chemical properties we observed for matter. To do this, we need to understand the intermolecular forces that operate between the molecules in a sample, all of which are caused by how electrons are arranged in the molecule. To understand the intermolecular forces, therefore, we need to have some ideas about how molecules are formed, what shape they have, and how the negatively charged electrons are distributed around and among the atoms in the molecule. Now, on to molecular geometry.

Molecular geometry describes the spatial arrangement of the atoms in a molecule. Diatomic molecules, such as O_2, have a linear geometry, because two points define a line. What about molecules containing three or more atoms?

Linear Geometry

Linear molecules include the molecular geometry class known as AX_2. The A refers to a central atom, and the X's are stick-on groups bonded to the central atom A. The more formal term for a stick-on group is *ligand*. Carbon dioxide (CO_2) is a good example of an AX_2 molecule, as seen in Figure 7.7. The electrons bonding the X's to the A push the molecule into a linear geometry. The angle between the carbon–oxygen bonds is 180°.

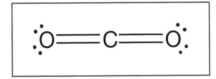

Figure 7.7 Carbon Dioxide, a Linear Molecule

Trigonal Planar Geometry

A trigonal planar species has bond angles of 120°. AX_3 species, such as urea, are trigonal planar (Figure 7.8). Note there are no lone pairs of electrons on the central carbon atom.

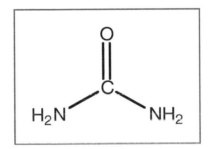

Figure 7.8 Urea, a Trigonal Planar Molecule

Tetrahedral Geometry

Molecules that consist of a central atom with four stick-on groups are denoted as AX_4. You might be tempted first to ascribe a square planar geometry to this class, which would give bond angles of 90° between each of the covalent bonds. However, if we twist the molecule into the third dimension, so each X group points to the corners of a tetrahedron (similar to a pyramid), the bond angles increase to just over 109°, as shown in Figure 7.9.

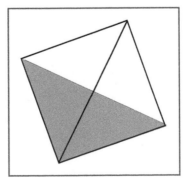

Figure 7.9 A Tetrahedron

Therefore, AX_4 molecules exhibit a tetrahedral geometry. Methane (CH_4, Figure 7.10), the major component in natural gas, and chloroform ($CHCl_3$), which was once used as an anesthetic, both have tetrahedral geometries.

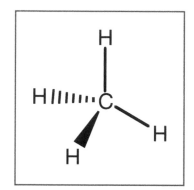

Figure 7.10 Methane, a Tetrahedral Molecule

Bent Geometry

Molecules belonging to the geometry classes AX_2E_2 are bent or nonlinear. In this notation, A is still the central atom, and X is still a stick-on group. The E's represent lone pairs of electrons on the central atom (A). Water is a good example of a bent molecule. Now, all of the electron pairs—the bonding pairs and the lone pairs—are, more or less, tetrahedrally arranged. However, in assigning molecular geometry, we focus on the atoms, not the lone pairs. While lone pairs of electrons are critically important in determining molecular geometry, we ignore them and describe the molecular geometry solely in terms of where the atoms are. So, if we ignore the lone pairs on the oxygen, the H–O–H atoms trace out a bent line, as seen in Figure 7.11. Therefore water is said to have a bent geometry.

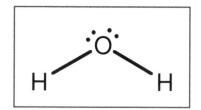

Figure 7.11 Water, a Bent Molecule

Pyramidal

AX_3E species, such as ammonia (NH_3), have a pyramidal geometry (Figure 7.12). The four pairs of electrons (three bonding pairs and one lone pair) approximately point to the vertices of a tetrahedron. As with the bent species, however, only the atoms are considered in describing the molecular geometry.

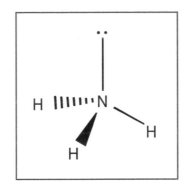

Figure 7.12 Ammonia, a Pyramidal Molecule

Electronegativity

Some atoms have a greater propensity in pulling electrons toward themselves than others, which is a property known as *electronegativity*. This affinity for electrons is a balancing act between the number of protons in the atomic nucleus and the occupancy of the various quantum electron shells. Linus Pauling proposed the first electronegativity scale. Fluorine is the most electronegative element (EN = 4.0), and the closer an atom is to fluorine, the more electronegative it is. For example, oxygen is more electronegative than nitrogen, and chlorine is more electronegative than bromine.

Polar Covalent Bonds

If two atoms of identical electronegativity are bonded together, the bond is nonpolar. That means the shared pair(s) of electrons are evenly distributed between the two nuclei, and there are no "electron hot spots" where excess electron density leads to accumulation of electric charges. Homodiatomic molecules, such as H_2, and O_2, and so on are always nonpolar.

If two atoms of different electronegativity are bonded together, the bond is polar. A greater difference in electronegativity between the two atoms creates a more polar bond. The bonding electrons spend more time around the more electronegative atom. Why? Because it is the nature of an electronegative atom to pull electrons toward itself. The more electronegative the atom, the more time the "shared" electron pair spends on that atom. You probably experienced an analogous situation growing up. Did your parents ever give a special toy to you and your siblings, with the expectation that you all would evenly share it? Well, was the toy symmetrically shared, or did one of you manage to snag more time with it?

This uneven sharing of electrons leads to the formation of *partial charges*. Now, don't confuse partial charges with ionic charges. An ionic charge is a full, integral multiple of the charge on an electron or proton. A partial charge is just what it sounds like: a fraction of a charge. Partial charges are denoted with a lowercase delta: δ+ or δ−. Since the shared pair of electrons spends more time around the more electronegative atom, the electron density increases around that atom. Since electrons have a negative charge, the increase in electron density around the more electronegative atom leads to a buildup of a partial negative charge on that atom. The less electronegative atom, necessarily, has a lower electron density and develops a partial positive charge. As seen in Figure 7.13, the direction in which the bond is polarized is shown with an arrow that points toward the negative end of the bond dipole. A small line is added to the beginning of the arrow. This looks like a plus sign and is a visual clue where the positive end of the dipole is located.

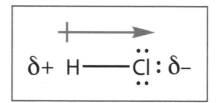

Figure 7.13 Bond Dipole of Hydrogen Chloride

Many beginning chemistry students think HCl is an ionic compound, but hydrogen chloride is a molecular species since both H and Cl are nonmetals. Chlorine is more electronegative than hydrogen because Cl is closer to F on the periodic table. Therefore, the bonding pair of electrons spends more time around chlorine. Thus, the chlorine end of the HCl molecule has a partial negative charge, while the hydrogen end of the molecule has a partial positive charge.

Molecular Polarity

Bond dipoles are vectors. The overall polarity of a molecule depends on the vectoral sum of the bond dipoles. The bond dipole vectors may add together to give a net molecular dipole, or they may cancel out, resulting in a nonpolar molecule. That's right. Just because a molecule contains polar bonds does not ensure the molecule is polar.

Let's take a look at water. A Lewis structure for water is shown Figure 7.14. VSPER Theory predicts water will be a bent molecule. Because oxygen is more electronegative than hydrogen,

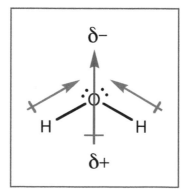

Figure 7.14 Molecular Dipole of Water

the O–H bonds in water are polarized toward the oxygen. We see one O–H bond dipole pointing up and to the left, as indicated with the blue arrow. The other O–H bond dipole is pointing up and to the right. Even without the mathematical details of vector addition, it is obvious that these bond dipoles add to give a net molecular dipole (shown in red) that points straight up. The oxygen side of the water molecule has a partial negative charge, and the hydrogen side of the water molecule has a partial positive charge.

Now, what about carbon dioxide? Well, when we draw a Lewis structure of carbon dioxide, and consider VSEPR Theory, we arrive at a linear molecule, as illustrated in Figure 7.15. Each C–O bond is polarized toward the oxygen. However, the two bond dipoles are pointing in exactly opposite directions and cancel each other out. Therefore, carbon dioxide is a nonpolar molecule.

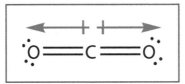

Figure 7.15 The Molecular Dipole in Carbon Dioxide Is Zero

INTERMOLECULAR FORCES

Finally, we have arrived at a point where we can discuss intermolecular forces. There are three principle interactions: dipole–dipole attraction, hydrogen bonding, and London forces. There are also ion–dipole interactions.

Dipole–Dipole Attraction

Also called dipolar attraction, this is the attraction between the opposite (partial) charges of polar molecules. Obviously, dipolar attractions occur only between polar molecules. A dipole–dipole attraction is shown below for a pair of hydrogen chloride molecules. The dipole–dipole attractive force is indicated with a blue dashed line.

Hydrogen Bonding

Hydrogen bonding is a special type of dipolar interaction, and hydrogen bonding is generally stronger than dipole–dipole attractions. Hydrogen bonding is possible only when a hydrogen atom is directly bonded to F, O, or N. Recall that hydrogen is the smallest atom. When a hydrogen atom bonds to one of these highly electronegative atoms, the hydrogen atom is stripped of so much of its electron cloud that it appears as a tiny, focused point of positive charge, which is highly attractive to all species with a partial negative charge. Hydrogen bonding is illustrated in Figure 7.16 by blue dashed lines for a collection of water molecules.

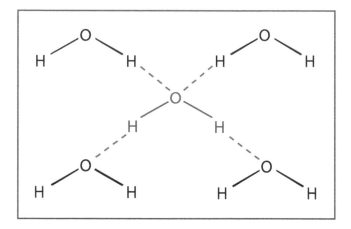

Figure 7.16 Hydrogen Bonding Among Water Molecules

Hydrogen bonding is arguably the most important of the intermolecular forces. Water is a liquid under normal conditions, which is remarkable considering how small the water molecule is. Hydrogen bonding holds the water molecules in the liquid state and prevents the water in our bodies from exploding away into the gas state. Hydrogen bonding holds DNA into a double helix and is the mechanism through which messenger and transfer RNA transcribe and translate the genetic code into polypeptide (protein) chains. Hydrogen bonding is critical in maintaining globular proteins, such as enzymes, in their native and active conformations. Hydrogen bonding gives silk its strength and wool its springiness. Life as we know it would not be possible without hydrogen bonding.

London Forces

If hydrogen bonding is the strongest of the intermolecular forces, London forces are the most ubiquitous. Even though London forces are generally the weakest of the three intermolecular forces, an argument can be made that they are the most important. All molecules and atoms that have electrons exhibit London forces, so all common matter experiences London forces. A London force is the result of an instantaneous dipole that is created whenever electrons in a molecule are unevenly distributed. Larger molecules show more London forces, because large molecules have more electrons.

Let's consider a pair of helium atoms illustrated in Figure 7.17. Each helium atom has two electrons. Let's imagine that at one instant in time both electrons on the first helium atom are over on the left side of the atom. Of course, this arrangement isn't going to last, because the electrons repel each other and because the electrons are moving very fast. However, this

$$^{\delta-}\!:\!\text{He}^{\,\delta+}\text{------}^{\,\delta-}\!:\!\text{He}^{\,\delta+}$$

Figure 7.17 London Forces Between Helium Atoms

arrangement in space is possible, so, for the merest fraction of an instant, that helium atom has a dipole: The left side has a partial negative charge, and the right side has a partial positive charge. This instantaneous dipole affects the neighboring helium atom. The electrons on the other helium atom "see" the nearby partial positive charge and are instantaneously[4] attracted to it. This is a London force.

Of course, this arrangement of electrons in the first helium atom is not stable and disappears in an instant. However, with a large number of atoms or molecules in a sample, the frequency of the London forces is very high. Larger atoms and molecules have more electrons, so the likelihood of an uneven distribution of electrons is greater. It is not necessary for all of the electrons to be clustered on one side of the molecule in order to have a London force. The distribution of electrons just needs to be ever so slightly nonsymmetrical. Therefore, larger molecules show stronger London forces than smaller molecules.

Ion–Dipole Attraction

Ion–dipole attractions occur (surprisingly enough) between an ion and a polar molecule. The strength of this force varies widely and depends on the magnitude of the dipole moment of the polar species and the size of the ion. This attraction allows ionic solids to dissolve in water. Let's imagine a crystal of sodium chloride. The sodium cations and the chloride anions are stacked together, and each ion is attracted to every oppositely charged ion. The strength of these attractions is very large, so why does NaCl dissolve in water? Well, water is a very polar molecule, so if a sodium ion separates from the solid crystal, it is immediately surrounded by an amorphous cluster of water molecules. Each water molecule will be oriented so that the oxygen end, that is the negative end of its molecular dipole, is pointed toward the positively charged sodium cation. All of the water molecules, taken together, compensate the sodium for the loss of its attractive forces to chloride ions. Likewise, the positive ends of the water dipoles cluster around a chloride ion (Figure 7.18).

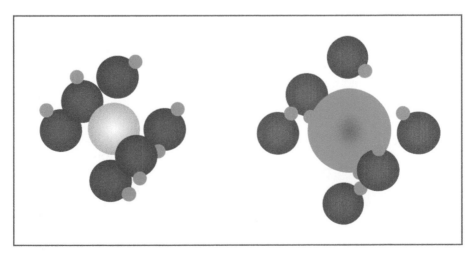

Figure 7.18 Solvated Ions

---

[4] At least within the constraints of Einstein's Theory of Relativity and the speed of light.

INTERMOLECULAR FORCES AND PHYSICAL PROPERTIES

Once we have constructed models of a molecule and have determined which intermolecular forces are operative, we can make some pretty remarkable predictions about the physical and chemical properties of the substance. Let's consider just a few examples.

Solubility

One of the great rules of thumb in chemistry is *like dissolves like*. This means that polar compounds dissolve more easily in other polar compounds, while polar compounds tend to exclude nonpolar compounds. Polar molecules that can participate in dipole–dipole interactions, hydrogen bonding, and/or ion–dipole interactions tend to have much greater degrees of attraction between molecules than nonpolar species that only experience London forces. Thus, the polar molecules will stick together and prevent nonpolar species from entering into their midst. So we expect nonpolar species such as nitrogen, oxygen, and carbon dioxide to be relatively insoluble in a polar solvent like water. In fact, water molecules stick together so much by hydrogen bonding that they *generally* only admit ionic compounds or other compounds that can also hydrogen-bond. Ionic compounds, the ultimate in polar species, are soluble in water, but are generally not soluble in nonpolar organic solvents such as ether. Most organic compounds have backbones consisting of hydrogen and carbon. Since the electronegativity difference between carbon and hydrogen is very small, hydrocarbons are essentially nonpolar. Therefore, the majority of organic compounds are not very soluble in water.

Surface Tension

Substances that have greater intermolecular attractions have greater surface tensions. Molecules in the liquid state are attracted to all neighboring molecules by intermolecular forces. For molecules deep within the sample, these forces operate in all directions, and, therefore, cancel each other out. The forces bind the molecules into a compact state of matter, but there is no net force on any of the interior molecules. This is not the case for the molecules near the surface, however. A molecule near the surface experiences attractive forces to all of the surrounding surface molecules and with the molecules below. However, there are no balancing attractions from above. This collection of unbalanced forces creates a "skin" on the surface of a liquid. The attractive strength of this skin is called the surface tension. Figure 7.19 shows an iron paper clip floating on water. Even though iron is denser than water, and should sink, surface tension prevents the paperclip from sinking.

Surface tension is responsible for many common experiences with water. For example, bugs can walk on water because the force due to the bug's weight is less than the surface tension of water. Water has a high surface tension, and is, therefore, a very cohesive liquid. That means water molecules have a strong attraction for other water molecules. Surface tension and cohesion cause water to readily form spherical drops when falling through air and to coalesce into puddles on a surface. This high cohesion of water is also at least partly responsible for the tendency of the lungs in premature infants to collapse.

For example, water, which can form two hydrogen bonds per molecule, has a surface tension of 0.073 N/m at room temperature against air. Acetone (aka fingernail polish remover) is

Figure 7.19 Surface Tension of Water

a polar molecule, and its dipolar interactions result in a surface tension of 0.024 N/m. Hexane is a nonpolar molecule that only has London forces and has the lowest surface tension of the compounds listed in Table 7.1.

Table 7.1 SURFACE TENSIONS

SUBSTANCE	PRINCIPAL IM FORCE	SURFACE TENSION (N/m)
Water	Hydrogen bonding	0.073
Acetone, $(CH_3)_2C = O$	Dipole–dipole	0.024
Hexane, C_6H_{14}	London forces	0.018

Surface tension is responsible for several interesting phenomena. For example, consider how water beads up on a freshly waxed car. The water is a highly polar substance, and the wax is highly nonpolar. Since the polar water is not attracted to the nonpolar wax, there are few adhesive forces between the water and the waxed surface. Adhesive forces cause different substances to stick to each other. However, the surface tension in water results in significant cohesive forces between the water molecules. Cohesive forces, or cohesion, cause a substance to be more attracted to itself. The height of the bead depends on the wax surface and the surface tension of water.

Figure 7.20 shows a (blue) liquid droplet on a (green) surface, which can be solid, liquid, or gas. While the droplet is based on an approximately spherical geometry, the amount of curvature and the height of the droplet depend on the intermolecular forces between the liquid and the surface, as well as on the surface tension in the liquid. The contact angle is the angle between the surface and the line tangent to the point where the edge of the droplet meets the

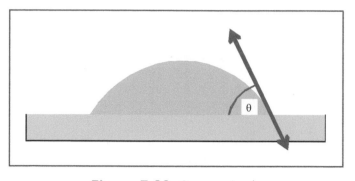

Figure 7.20 Contact Angle

surface (blue double-headed arrow). If the liquid is water and the surface is very hydrophilic, the contact angle drops to nearly zero, and the water spreads evenly across the surface and "wets" the surface. As the surface becomes increasingly nonpolar (e.g., wax paper, Teflon, etc.) the contact angle with water increases and approaches 180°. With glass, the contact angle is typically around 20° to 30°.

We also see the contact angle when water forms a meniscus in a glass tube, such as a pipette or a burette, as illustrated in Figure 7.21. We can envision the air interface as a sphere of radius R (shown in blue). The radius of the tube is r (shown in red). As the radius of the tube decreases, the curvature of the meniscus (R) also decreases.

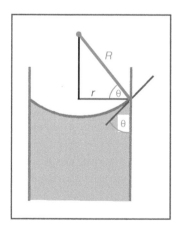

Figure 7.21 Formation of a Meniscus

Capillary rise, or capillary action, is a second effect caused by surface tension. Capillary rise is the tendency of a fluid to rise into a narrow-diameter tube. Figure 7.22 shows this phenomenon. The tube on the far left has the narrowest diameter, so the amount of rise depends on the diameter of the tube.

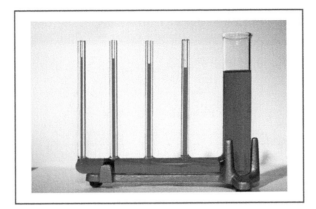

Figure 7.22 Capillary Rise

LaPlace's Law

Let's imagine a very thin film of a liquid, as shown in Figure 7.23. The surface tension in the liquid causes it to curl up. Since we have a two-dimensional surface, we can curl the liquid in only two dimensions. Let's call the radii of curling R_1 and R_2. The black arrows represent the x, y, and z axes of a Cartesian coordinate system.

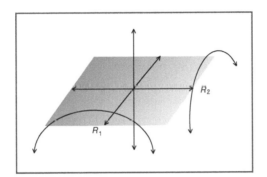

Figure 7.23 Radius of Curvature

The radius of curvature does not necessarily have to describe a circle. However, if R describes a circular curvature, then we have two limiting possibilities, as shown in Figure 7.24. We get a sphere if the radii of curvature are equal, and this is the natural tendency. However, in medicine, a cylinder is useful to consider, because blood vessels are more or less cylindrical.

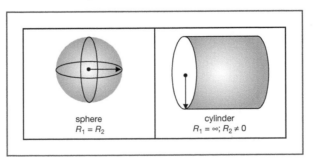

Figure 7.24 Radius of Curvature for a Sphere and a Cylinder

When the liquid is in contact with another surface, this curling creates a pressure difference, which is described by LaPlace's Law. The Greek letter gamma (γ) is the surface tension, P is pressure, and R is the radius of curvature.

$$\Delta P = \gamma\left(\frac{1}{R_1} + \frac{1}{R_2}\right)$$

For a sphere, $R_1 = R_2$, so LaPlace's Law takes the form:

$$\Delta P = \gamma\left(\frac{1}{R_1} + \frac{1}{R_1}\right)$$

which simplifies to:

$$\Delta P = \gamma\left(\frac{2}{R}\right)$$

For a cylinder, $R_1 = \infty$, so LaPlace's Law takes the form:

$$\Delta P = \gamma\left(\frac{1}{R}\right)$$

While we can't divide 1 by infinity (∞), the mathematical limit of dividing 1 by an increasingly large number is zero. Thus, the $1/R_2$ drops out.

We can use LaPlace's Law to calculate the distance a fluid will rise in a capillary. In this system, we have two opposing pressures: the upward pressure due to LaPlace's Law and the downward pressure due to gravity. The fluid will rise until these pressures are equal. We will begin with a model in which the meniscus is a sphere of air in contact with water in a capillary tube, as shown in Figure 7.25.

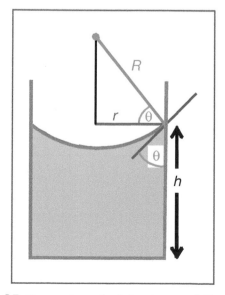

Figure 7.25 Formation of a Mensicus and Capillary Rise

According to LaPlace's Law, the upward pressure is equal to:

$$\Delta P = \gamma \left(\frac{2}{R}\right)$$

We know the surface tension of water is 0.073 N/m, so we need to find the radius of curvature in order to calculate the upward pressure. The radius R is a function of the contact angle and the radius of the tube, r. Since r and R form two sides of a right triangle, we can use trigonometry to relate these two quantities. If you are rusty with trig functions, don't worry. The cosine function is just another mathematical operator on your calculator. Anyway, R is given by:

$$R = \frac{r}{\cos(\theta)}$$

So, if our capillary has a radius of 1.0×10^{-4} m (0.10 mm) and the contact angle is 20°, $R = 1.1 \times 10^{-4}$ m (or 0.11 mm).

$$R = \frac{1.0 \times 10^{-4}\,\text{m}}{\cos(20°)} = \frac{1.0 \times 10^{-4}\,\text{m}}{0.940} = 1.1 \times 10^{-4}\,\text{m}$$

So the upward pressure is:

$$\Delta P = 0.073\,\frac{\text{N}}{\text{m}}\left(\frac{2}{1.1 \times 10^{-4}\,\text{m}}\right) = 1400\,\frac{\text{N}}{\text{m}^2}$$

As we saw in Chapter 5, the downward pressure in a liquid of a given height h is given by:

$$P = \rho g h$$

Since the two pressures are equal, we solve this equation for h by substituting in the previous pressure:

$$h = \frac{1400\,\text{N/m}^2}{(1000\,\text{kg/m}^3)(9.8\,\text{m/s}^2)} = 0.14\,\text{m}$$

In a medical setting, LaPlace's Law explains why the "surface tension" on a blood vessel wall depends on the radius of the vessel. Figure 7.26 shows a (gray) blood vessel of radius R.

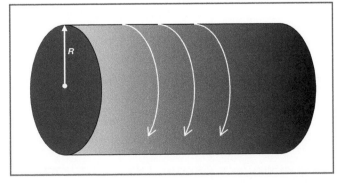

Figure 7.26 Effect of Radius on Blood Vessel Tension

The curved white arrows represent the tension across the interface between the blood and the blood vessel wall.

If we rearrange the LaPlace equation for a cylinder and solve for surface tension, we get:

$$\gamma = \Delta P \cdot R$$

In the case of a blood vessel, we can interpret "surface tension" as the tension on the surface of the vessel. So, we see the tension is proportional to the radius of the vessel. So if the blood pressure remains constant, an artery with a radius of 10^{-3} m must withstand a pressure 100 times greater than a capillary with a diameter of 10^{-5} m, because the radius of the artery is 100 times greater than the capillary.

Aneurysms are approximately spherical, so the tension on the wall of the aneurysm rises as the radius divided by 2:

$$\gamma_{aneurysm} = \Delta P \left(\frac{R}{2} \right)$$

Aneurysms, while not a good diagnosis, represent a physical means for lowering the tension on a blood vessel.

Surfactants

Video 7.2

Surfactants improve a solvent's ability to be a solvent by reducing the surface tension of the solvent. Applications of surfactants include removing greasy stains from clothing, helping premature babies breathe, and killing bacteria. Surfactants are commonly known as soaps or detergents. The word "surfactant" is derived from *surf*ace *act*ive ag*ent* (spelling not withstanding). To see how that happens, we need to consider the chemical structure of a surfactant molecule.

Soaps are derived directly from fats or oils (triglycerides) through a chemical process called *saponification* (Figure 7.27). A triglyceride is a triester derived from glycerin and three fatty acids. A fatty acid is a long-chained carboxylic acid having 12 to 18 carbon atoms (where *n* represents 10 to 16 CH$_2$ units). Heating a fat or oil with excess sodium hydroxide gives three moles of a fatty acid salt, plus one mole of glycerin. A soap is the salt of a fatty acid ($CH_3(CH_2)_nCOO^- Na^+$).

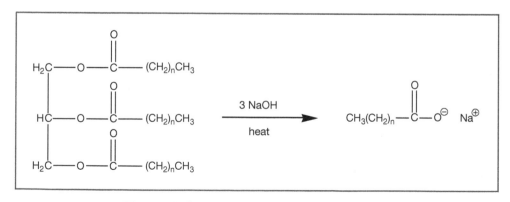

Figure 7.27 Saponification of a Triglyceride

A detergent is chemically synthesized, most commonly from soaps. The molecular structure of detergents can be tweaked to optimize the physical properties of the detergent. Detergents can be positively charged (cationic), negatively charged (anionic), or uncharged (nonionic). Some examples of detergents are shown in Table 7.2.

Table 7.2 SURFACTANTS

ANIONIC SURFACTANT	CATIONIC SURFACTANT	NONIONIC SURFACTANT
$CH_3(CH_2)_9{-}O{-}\overset{\overset{O}{\|\|}}{\underset{\underset{O}{\|\|}}{S}}{-}O^{\ominus}\ Na^{\oplus}$	$CH_3(CH_2)_n{-}\overset{\overset{CH_3}{\|}}{\underset{\underset{CH_3}{\|}}{N}}{\overset{\oplus}{}}{-}CH_3 \quad Cl^{\ominus}$	$C_8H_{17}{-}\bigcirc{-}O{-}(CH_2CH_2{-}O)_n{-}H$
Sodium dodecylsulfate, aka sodium lauryl sulfate (Tide®)	Alkyltrimethyl-ammonium chloride (shampoo)	Octylphenylethoxylate (Triton X-100®, various industrial applications)

The structural feature common to soaps and detergents is that both have a polar (hydrophilic) head and a nonpolar (hydrophobic, greasy) tail. A schematic of a surfactant molecule is shown in Figure 7.28. The red circle represents the polar head, and the wavy violet line represents the greasy (hydrophobic) tail.

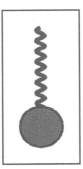

Figure 7.28 Generic Surfactant Molecule

Because *like dissolves like*, when placed into water the surfactant molecule will strive to get the greasy, nonpolar hydrocarbon tail away from the polar water molecules. There are three potential strategies for achieving this outcome: formation of monolayers, formation of bilayers, and formation of micelles.

Monolayers

In forming a monolayer, the surfactant molecules stick their polar heads onto the water, while their greasy tails stick out of the surface of water. In Figure 7.29 the blue spheres represent water molecules.

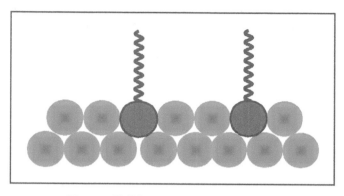

Figure 7.29 Surfactant Monolayer

Adding a surfactant dramatically decreases the surface tension of the water, because the surfactant molecules get in between the water molecules and disrupt the hydrogen bonding between them. The decrease in surface tension finds application in treating a premature baby's lungs. By adding a detergent, the water in the lungs becomes less cohesive, and the surfaces in the lungs are less likely to stick together.

Bilayers

The tails of the surfactant molecules can dissolve in each other to form a double layer, called a *bilayer*, as shown in Figure 7.30.

In this arrangement, the nonpolar tails interact with each other, with the polar heads on the outside interacting with the water molecules. Of course, you recognize that this is the basic structure of cell membranes. This strategy offers the added advantage of creating a "no man's

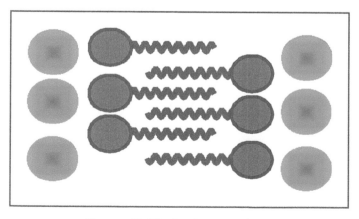

Figure 7.30 Surfactant Bilayer

land" in the center of the membrane. Polar solutes, such as ions, will not enter the nonpolar region unless carried or forced through. Likewise, nonpolar molecules, such as certain proteins and cholesterol, are very happy to find themselves in a nonpolar environment.

Soaps and detergents have bactericidal properties. These nonnative surfactants can work their way into the bacterium's cell membrane and disrupt the integrity of the cell membrane, causing the bacterium to die. Of course, fatty acids are also high-energy food sources, so soaps and detergents lose this bactericidal property at low concentrations.

Micelles

The greasy tails can dissolve in each other, forming a spherical structure called a *micelle*. This is how soaps work in washing your hands or doing dishes. Figure 7.31 shows a two-dimensional slice out of a micelle. Formation of a micelle creates a nonpolar microenvironment in the water. So, when you are scrubbing your hands, or washing dishes or doing laundry, the agitation of the sample causes any greasy, water-insoluble material to migrate into the greasy interior of the micelle, where it can be carried away from your hands, plate, or shirt.

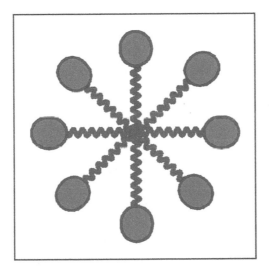

Figure 7.31 A Micelle

Viscosity

Viscosity increases with increasing intermolecular forces. Viscosity is a measure of a fluid's resistance to flow. A more detailed treatment of viscosity is covered in Chapter 5. Let's consider a couple of examples of increasing intermolecular forces resulting in increased viscosity. Water, a hydrogen-bonding species, has a viscosity of 1.0×10^{-3} Pa·s, while nonpolar hexane (C_6H_{14}) has a viscosity of 3.2×10^{-4} Pa·s. Octane (C_8H_{18}), another nonpolar molecule capable only of London forces, has a viscosity of 5.4×10^{-4} Pa·s. Notice that octane is larger than hexane, so octane has more London forces and is more viscous than the smaller molecule (hexane). Molasses, perhaps the poster child for highly viscous liquids, has a viscosity ranging between 5 and 10 Pa·s. Table 7.3 lists the viscosities for some typical liquids.

Table 7.3 VISCOSITIES

SUBSTANCE	PRINCIPAL IM FORCE	VISCOSITY (Pa·s)
Water	Hydrogen bonding	1.0×10^{-3}
Hexane, C_6H_{14}	London forces	3.2×10^{-4}
Octane, C_8H_{18}	London forces	5.4×10^{-4}

Vapor Pressure

Substances having greater intermolecular forces have lower vapor pressures. The most energetic molecules in a liquid have sufficient kinetic energy to overcome the intermolecular forces binding them into the liquid state and escape into the gas phase. Once the molecules are in the gas phase, they are free to zip around, colliding with each other, the surface of the liquid, and the walls of the container. When the gaseous molecules collide with the walls of the container, they give the wall a little push. The forces from all of these collisions, averaged over the area of the walls of the container, result in a pressure. This pressure is called the vapor pressure of the liquid.

So (once again), let's imagine a glass of water at a constant temperature. Let's also put some plastic wrap tightly over the top of the glass, so that the water cannot evaporate out into the surrounding room and escape our system. Some of the water molecules in the liquid sample will vaporize into the gas phase, and some of the gaseous water molecules will condense into the liquid phase. Eventually, the system will reach a state of dynamic equilibrium.

A dynamic equilibrium exists in a system comprised of (at least) two states when the populations of the two states are constant, even though the members of the system are constantly changing from one state to another. Dynamic equilibria occur in many settings, including acid–base chemistry, distribution coefficients for fat-soluble medications, and loading and unloading oxygen from hemoglobin. We will talk more about dynamic equilibria later.

Now, back to our glass of water. At the surface of a liquid, the most energetic molecules can escape into the gas phase. Since the glass is sealed by the plastic wrap, the water molecules are confined to our system and zip around with the other gas molecules, bouncing off the sides of the glass, the plastic wrap, and the surface of the water. At the same time, the slower, less energetic water molecules in the gas phase near the surface of a liquid can be captured by intermolecular forces back into the liquid state. If the rates of vaporization and condensation of the water molecules were not in balance, the level of water in the glass would change. Since everyday experience shows us that the level of water in a sealed container remains constant, we can infer that the rates of evaporation and condensation are equal. In this system, there is a balance between vaporization and condensation, and we have a state of *dynamic equilibrium*.

As temperature increases, the vapor pressure of a liquid increases, as does the volatility. Volatility is a nonquantitative term that describes the tendency of a liquid to evaporate. More volatile liquids evaporate more readily than less volatile liquids. More volatile liquids have a higher vapor pressure than less volatile liquids.

We can use the Boltzmann distribution in Figure 7.32 to explain why vapor pressure increases with increasing temperature. The dark blue line represents the distribution of kinetic energies for a gas at some low temperature. The light blue line represents the distribution

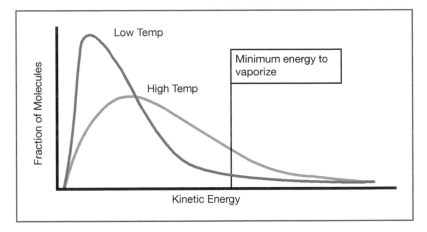

Figure 7.32 Distribution of Kinetic Energies

of kinetic energies for the same amount of the same gas at a higher temperature. There is some minimum kinetic energy necessary to escape into the gas phase. As we can see (in this depiction of an imaginary substance), about one-third of the molecules at the higher temperature have enough energy to vaporize, whereas a much smaller fraction of the molecules at the lower temperature have sufficient energy to vaporize. Therefore, more of the molecules at the higher temperature vaporize, resulting in both a greater vapor pressure and a greater degree of volatility.

Figure 7.33 shows the vapor pressure of water and ethanol over a temperature range of 10°C to 100°C. We see that ethanol is more volatile than water at all temperatures. Each water molecule can form two hydrogen bonds, while ethanol (CH_3CH_2-O-H) can form only one hydrogen bond per molecule. Therefore, we would expect water to have a lower vapor pressure, because water molecules are more tightly bound in the liquid state.

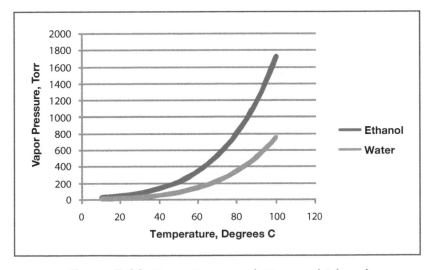

Figure 7.33 Vapor Pressure of Water and Ethanol

The Clausius–Clapeyron equation relates the vapor pressure of a liquid, P, to the Kelvin temperature, T.

$$\ln(P) = -\frac{\Delta H_{vap}}{R}\frac{1}{T} + \text{constant}$$

The heat of vaporization ΔH_{vap} is the amount of energy necessary to liberate one mole of liquid at its boiling point into the gas phase. R is the universal gas constant. The *constant* term is a mathematical artifact from calculus used to derive this equation, and we will ignore it.

Many of the terms in the Clausius–Clapeyron equation are constants, so we can recast the equation into a simpler form.

$$\log P = A + \frac{B}{T}$$

The values of A and B depend on the liquid. For enflurane, $A = 7.967$ torr and $B = -1678$ torr·K. What is the vapor pressure of enflurane at 25°C? The units on these constants tell us we need to express pressure in torr and temperature in kelvins. Let's substitute in the known values.

$$\log P = 7.967 + \frac{-1678}{298\ \text{K}} = 2.34$$

Now, how do we solve for P, since it is tied up in the log term? Remember that a logarithm is a mathematical operator. To free a quantity from an operator, we need to apply the inverse of that operator. So, if we want to know the value of the pressure, we need to apply the antilog operator to each side of the equation. On your calculator, the antilog button is "10^x". So, we find the vapor pressure on enflurane under these conditions is 217 torr.

$$\text{antilog}(3.24) = 10^{2.34} = 217\ \text{torr}$$

The vapor pressure of a volatile liquid determines the maximum mole fraction of that substance in a gas phase in contact with that volatile liquid. Remember that with gases, mole fraction is the same as volume fraction and pressure fraction. This is the principle that guides an anesthetic vaporizer (Figure 7.34).

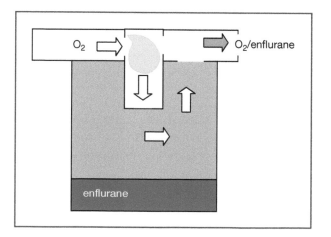

Figure 7.34 Vaporizer

If oxygen gas is sent through a closed container containing liquid enflurane, the oxygen and gaseous enflurane will mix and exit the vaporizer. The composition of the effluent gas will depend on the temperature of the vaporizer (which determines the vapor pressure of the enflurane) and the pressure of the oxygen. The amount of oxygen that is diverted into the vaporization chamber also influences the composition of the effluent, but that is a kinetic (rate determined) effect, not a thermodynamic (what's energetically possible) effect.

Suppose we run oxygen at 750 torr through a vaporizer in such a manner that the oxygen becomes saturated with enflurane. In this case, the partial pressure of enflurane in the effluent will equal the vapor pressure of enflurane. The pressure of the effluent gas remains at 750 torr, but the effluent gas is now saturated with enflurane. The mole fraction of enflurane is given by Dalton's Law:

$$\chi_{enflurane} = \frac{\text{vapor pressure of enflurane}}{\text{total pressure}} = \frac{217 \text{ torr}}{750 \text{ torr}}$$

$$\chi_{enflurane} = 0.29$$

Boiling Point

Compounds with more intermolecular forces have higher boiling points. Do not fall into the trap of defining "boiling point" as the temperature when the liquid boils. First of all, using a term to define itself is a circular definition and is not acceptable. Second, boiling is not a uniquely time-bound event, and, therefore, "when" has no place in the definition. Correctly stated, the boiling point of a liquid substance is the temperature at which the vapor pressure is equal to the ambient pressure. The normal boiling point is the temperature at which the vapor pressure of a liquid is equal to 1 atm of pressure.

Molecules in the liquid state have sufficient intermolecular attractions to keep them from flying away from each other into the gas state, but the intermolecular forces are not strong enough to prevent the molecules from sliding around past each other. In the gas state, there are essentially no intermolecular interactions. Gaseous molecules are "free" from each other. As we add energy to a liquid sample below its boiling point by heating it, the temperature increases, and the average kinetic energy of the molecules in the sample increases. Eventually, there comes a point where the molecules have enough energy so that they are no longer constrained to remain in contact with the other molecules. Vaporization occurs as the molecules escape from the liquid state into the gas state. The boiling point is the temperature that corresponds to the average molecular kinetic energy sufficient to overcome the intermolecular forces that hold the molecules in the liquid state.

Figure 7.35 shows the vapor pressure of water as a function of temperature. We see the vapor pressure of water at 100°C is 760 torr, and we know that water boils at 100°C, at least when the ambient pressure is 760 torr. The boiling point of water when the ambient pressure is 600 torr drops to 93°C.

Because gases respond strongly to pressure, the boiling point of a liquid is highly dependent on pressure. The boiling point of a liquid increases as the ambient pressure increases. The Clausius–Clapeyron equation, which we discussed earlier in the context of vapor pressure, can also be used to quantify the relationship between boiling point and temperature. In

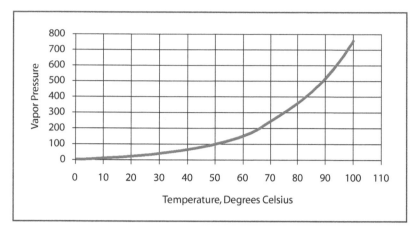

Figure 7.35 Vapor Pressure of Water

this expression of the equation, we have two states (1 and 2). State 1 is some reference state, whereas state 2 is some new state.

$$\ln\frac{P_1}{P_2} = \frac{\Delta H_{vap}}{R}\left(\frac{1}{T_2} - \frac{1}{T_1}\right)$$

For example, what is the boiling point of water at a pressure of 200 kPa? The heat of vaporization of water is 40.7 kJ/mol. We can use the normal boiling point of water of 100°C (373K) at 101 kPa as a reference state. Let's put in the values we know.

$$\ln\frac{1.01 \times 10^5 \text{ Pa}}{2.00 \times 10^5 \text{ Pa}} = \frac{4.07 \times 10^4 \text{ J/mol}}{8.31 \text{ J/mol/K}}\left(\frac{1}{T_2} - \frac{1}{373 \text{ K}}\right)$$

Okay, this is pretty ugly. Let's start by dividing the pressure terms and extracting the natural log of this ratio. Let's also simplify the fractions on the right:

$$-0.6832 = 4898 \text{ K}\left(\frac{1}{T_2} - 0.002681 \text{ K}^{-1}\right)$$

Okay, now let's divide both sides by 4898 and then add 0.002681 to both sides of the equation.

$$\frac{1}{T_2} = -0.0001395 + 0.002681 = 0.002542 \text{ K}^{-1}$$

Now, if we take the reciprocal of both sides, we find that water boils at a temperature of 393K (or 120°C) when the pressure is increased to 200 kPa, or 29 psi.

$$T_2 = \frac{1}{0.002542 \text{ K}^4} = 393 \text{ K}$$

An autoclave takes advantage of this phenomenon. Water in the sealed autoclave is heated to its boiling point and begins to vaporize. Since the autoclave is sealed, the water vapor cannot escape, and the increasing number of moles of gas in the container causes the pressure to increase. The increased pressure causes the water to boil at a higher temperature, because a higher temperature is required to drive the vapor pressure of water up to the increased pressure. Properly controlled by regulation of the heating element and/or pressure relief valves, an autoclave can easily increase the boiling point of water from 100°C to 120°C (or more). Without regulation of the heat source and/or pressure release valves, however, an autoclave would eventually explode.

Melting Point

Compounds with more intermolecular forces have higher melting temperatures. The melting point, or melting temperature, is the temperature at which the solid state reversibly passes into the liquid state. This temperature is also known as the freezing point. At the melting point, thermodynamics does not favor one state over the other, or demand a phase change. The system is in equilibrium. Now, as we said before, the solid state is characterized by a definite shape and volume, which is due to strong intermolecular forces that lock the component molecules (or ions or atoms) into a crystalline array. Even so, the component particles of compounds in the solid state have energy and are vibrating in the crystalline lattice. In the liquid state, the component molecules have sufficient intermolecular attractions to keep them from flying away from each other into the gas state, but the intermolecular forces are not strong enough to prevent the molecules from sliding around past each other.

As we add energy to a solid sample below its melting point by heating it, the temperature increases, and the average kinetic energy of the molecules in the sample increases. Eventually, there comes a point where the molecules have enough energy so that they are no longer constrained to remain in the crystalline lattice. As they begin to break free and slide around, melting has occurred and the sample begins to transition into the liquid state. The melting point is the temperature that corresponds to the average molecular kinetic energy sufficient to overcome the intermolecular forces that hold the molecules in the solid state.

Because ionic bonds are very strong, ionic compounds have very high melting points. With only a handful of exceptions, all ionic compounds are solids at room temperature and pressure.

ENERGY CHANGES AND CHANGES OF STATE

Heat of Fusion and Heat of Vaporization

The molar enthalpy of fusion ($\Delta H°_{fus}$) is the heat necessary to convert one mole of a solid into a liquid at its normal melting point. The molar enthalpy of vaporization ($\Delta H°_{vap}$) is the heat required to convert one mole of a liquid to a gas at its normal boiling point. When melting or vaporization occurs at constant pressure, it is acceptable to use "heat" instead of "enthalpy." This is because heat and change in enthalpy are equal to each other under constant pressure conditions. The interested student should consult any physical chemistry textbook for more details. Both ΔH_{fus} and ΔH_{vap} are inherently endothermic and represent an amount of energy that must be *added* to the sample in order for the phase transition to occur. The heat of fusion represents the amount of energy necessary to overcome the intermolecular forces to the point that the molecules can start to move around each other. The heat of vaporization represents the amount of energy necessary to overcome all intermolecular forces so that the molecules can escape into the gas phase.

For water, the heat of vaporization (41 kJ/mol) is much larger than the heat of fusion (6 kJ/mol). This is true for all substances. Water has relatively large heats of vaporization and fusion because of the strong hydrogen bonds that form among water molecules. Table 7.4 lists the heats of vaporization of several substances and the principal intermolecular forces operating in each.

Table 7.4 HEATS OF VAPORIZATION AND FUSION FOR VARIOUS SUBSTANCES

SUBSTANCE	PRINCIPAL INTERMOLECULAR FORCE	$\Delta H°$VAP, kJ/mol	$\Delta H°$FUS, kJ/mol
H_2O	Hydrogen bonding	40.7	6.02
HCl	Dipole–dipole	16.2	1.99
Cl_2	London forces	20.4	6.41
Br_2	London forces	30.0	10.8

Imagine an 18-g sample of water (i.e., 1.0 mole) as ice at an initial temperature of –40°C. Let's record the temperature of our sample as we add energy by a heat process. Figure 7.36 shows this graphically.

Initially, the temperature rises until the sample reaches the melting temperature. Because the energy added at this point is absorbed in softening up the intermolecular forces so that the transition from solid to liquid occurs, no increase in temperature is observed. Once all of the solid

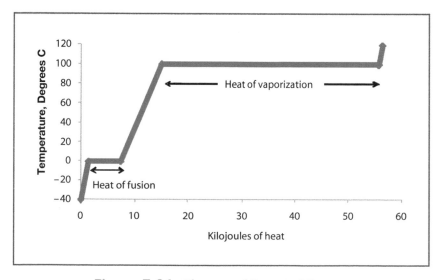

Figure 7.36 Changes of State and Energy

water has melted, added energy once again causes an increase in temperature. When the liquid sample reaches the boiling point, once again the temperature remains constant, because the added energy is used to overcome the remaining intermolecular forces as the liquid water molecules escape into the gas phase. When the sample is completely vaporized, additional energy once again effects an increase in temperature.

The slopes of the lines where the solid, liquid, and gas phases are increasing in temperature are significant. One definition of slope is "rise over run." In this case, the slope has units reflecting a temperature change over energy added (for a fixed amount of sample). You might recognize that this is the reciprocal of heat capacity. So 1/slope is the heat capacity of each phase. Notice that the line in the gas region has the steepest slope. This means the gas has the smallest heat capacity and changes temperature most easily per unit energy added. Thus, the gas phase is the best thermal conductor.

Evaporation has a cooling effect, because vaporization is an endothermic process. Endothermic processes cause a decrease in temperature because they absorb heat from the surroundings. We see this when we sweat (or glisten, depending on your gender). Sweat pores deposit water on our skin. As the water evaporates, it must absorb an amount of energy equal to the heat of vaporization. We can also rationalize this temperature change at a molecular level. When water evaporates, molecules escape from the liquid state into the gaseous state. The molecules most likely to escape are those having the highest kinetic energy. As the sample loses the most energetic molecules, the average kinetic energy of the remaining molecules decreases. The temperature is dependent on the average kinetic energy of the molecules in a sample, so the temperature of the sample decreases.

If vaporization has a cooling effect and absorbs energy, then condensation must have a heating effect, because it releases energy. When gaseous molecules settle into the liquid state, they surrender a great deal of their internal energy to the surroundings. We perceive this release of energy as "heat" coming out of the sample. Unfortunately, many of you are familiar with the devastating burns caused by steam. Even if you're not, you all surely have cooked spaghetti. It doesn't hurt that much to get a drop or two of boiling water on your skin. However, have you ever drained the pasta and gotten a bunch of steam on your wrist? That really hurts. When water at 100°C contacts your skin, the temperature begins to drop immediately as the water cools to your skin temperature. However, when steam at 100°C contacts your skin, it remains at that temperature while releasing the heat of vaporization onto your skin as the gas converts to a liquid. Then, you have water at 100°C on your skin which releases more energy as it cools to your skin temperature.

Let's try a quantitative illustration. Imagine you splash one drop of water (0.05 g) at 100°C on your wrist. The water releases energy into your skin until the temperature of the water reaches skin temperature (37°C). We can calculate this amount of energy based on the specific heat of water.

$$Q = C_p \cdot m \cdot \Delta T$$

The specific heat of water is 4.18 J/g/°C. The change in temperature is the final temperature minus the initial temperature, or 37°C – 100°C = –63°C. So the energy released from the water is 13 J.

$$Q = (4.18 \text{ J/g/°C})(0.050 \text{ g})(-63°C) = -13 \text{ J}$$

The negative sign reminds us that, from the water's point of view, the heat was an exother-mic process. Now, what if an equivalent mass of steam at 100°C contacts your wrist? Now we have two processes. First the steam will undergo a phase transition to liquid water at 100°C, and then the hot water will release energy onto your skin as it cools to skin temperature. The heat of vaporization of water is 40.7 kJ/mol, so first we need to convert the mass of water into moles of water.

$$0.050 \text{ g} \left(\frac{1 \text{ mol}}{18 \text{ g}} \right) = 0.0028 \text{ mol}$$

Now, we can use the heat of vaporization as a conversion factor to calculate the energy trans-ferred from the steam as it condenses. The negative sign is included with the heat of vaporiza-tion because condensation is an exothermic process.

$$0.0028 \text{ mol} \left(\frac{-40.7 \text{ kJ}}{\text{mol}} \right) = -0.113 \text{ kJ}$$

So, the condensation of steam releases 0.113 kJ or 113 J. By the time the steam condenses and cools, it delivers 113 J + 13 J = 126 J of energy. This is nearly 10 times the amount of energy the hot water released. So, steam has the capacity to cause much more devastating burns than hot water.

Desflurane has a heat of vaporization of 45 cal/g. How much heat does desflurane absorb from the surroundings when 2 g vaporize? To solve this problem, just use the heat of vapori-zation as a conversion factor.

$$2 \text{ g} \left(\frac{45 \text{ cal}}{\text{g}} \right) = 90 \text{ cal}$$

PHASE DIAGRAMS

A phase diagram shows the combined effects of temperature and pressure on the state of matter (see Figure 7.37). There are three regions. The lower right region, where the temperatures are highest and the pressures are lowest, defines the gas state. The liquid state exists in the middle,

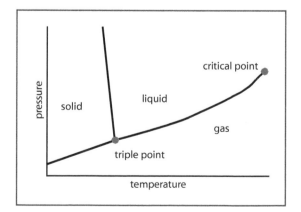

Figure 7.37 Phase Diagram for a Compound Such as Water

where pressures and temperatures are intermediate. The upper left region, where the temperatures are lowest and the pressures are highest, represent conditions that favor the solid state. Any temperature and pressure combination that lands in one of these regions determines the state of the substance.

The lines represent phase equilibrium boundaries. Crossing one of these lines by changing pressure or temperature results in a phase transition (or a change of state). Temperature and pressure combinations that lie on one of these lines allow for two phases to coexist in equilibrium with each other. The triple point of the substance is the single temperature and pressure combination where all three phases can exist in equilibrium with each other. The solid–liquid equilibrium line defines the melting point of the solid. Notice that the melting point doesn't change much with pressure, because the volumes of solids and liquids don't change much with increasing pressure. This line continues upward indefinitely. The liquid–gas equilibrium line defines the boiling point of the liquid at various pressures. The temperatures on this line correspond to the boiling point of the liquid at various pressures.

The liquid–gas equilibrium line terminates at a point known as the critical point. The temperature and pressure that define the critical point are known as the critical temperature and the critical pressure. For example, nitrous oxide has a critical temperature of 36°C and a critical pressure of 72.45 bar (1051 psi). When the temperature and pressure exceed these critical values, the system becomes a supercritical fluid. Supercritical fluids have the flow properties of gases but densities similar to liquids and supercritical fluids have no surface tension. Therefore, supercritical fluids are terrific solvents. For example, supercritical carbon dioxide is an excellent solvent for extracting caffeine from coffee without resorting to more toxic organic solvents like dichloromethane.

Summary

- There are three common states of matter: solids, liquids, and gases. The solid state is characterized by having both a definite volume and a definite shape. The liquid state of matter is characterized by a definite volume, but liquids adopt the shape of their container. Gases have neither a definite volume nor a definite shape. There is a fourth state of matter, plasma, but that is rarely encountered under normal temperature and pressure conditions.

- Intermolecular forces operate among molecules (or atoms or ions) in a sample. If the intermolecular forces are sufficiently strong, the molecules are held in place with respect to each other, and we have the solid state. If the intermolecular forces are sufficiently strong to hold the molecules in contact with each other, but not strong enough to prevent molecules from sliding past each other, we have the liquid state. If the intermolecular forces are not strong enough to keep the molecules in contact with each other, the molecules fly off in random directions, and we have the gaseous state.

- Solids and liquids are called condensed states of matter because the molecules are in contact with each other. For this reason, solids and liquids are not very compressible.

- Intermolecular forces result from a combined effect of the chemical bonding in a molecule, as well as the molecular geometry. Atoms tend to gain, lose, or share electrons so as to achieve an octet of electrons, or a total of eight valence electrons. This electron configuration fills an electron energy shell. This is a particularly stable arrangement that maximizes attractions among the electrons and the nucleus, while minimizing repulsions among the electrons.

- If a neutral atom loses electrons, it acquires a positive charge, and it is called a cation. Representative metals tend to form cations with positive charges equal to the group number.

- If a neutral atom gains electrons, it acquires a negative charge, and it is called an anion. Representative nonmetal elements tend to form anions with charges equal to the group number minus eight.

- Ionic bonds result from the attractions among cations and anions. Because this attraction is not localized between a single cation and a single anion, it is incorrect to refer to an ionic molecule. Ion–ion attractions are the strongest of the "intermolecular" forces, so with very few exceptions, ionic compounds are solids under normal conditions.

- Covalent bonds form between atoms that are nonmetals. Covalent bonds result from sharing electrons, which allows both atoms in the bond to have access to an octet of electrons. The bond results from buildup of electron density between the two nuclei. A covalent bond has an optimal bond length that maximizes attractions among the bonding electrons and the two atomic nuclei. The bond dissociation energy is a measure of how strong the bond is, and this value depends on the type of atoms in the bond.

- Lewis structures are used to represent bonding in a molecule. A line represents a shared pair of electrons, and lone pairs of electrons are represented by dots. To draw a Lewis structure, you must first know the atom connectivity and the number of valence electrons. The atom connectivity is learned by experience and example, so you'll generally be provided atom connectivities at this point. Representative elements have a number of valence electrons equal to the group number. Divide the total number of valence electrons by 2, since electrons tend to form bonds in pairs. Connect each atom in the structure with a line (pair of electrons). Then use the remaining electrons so as to give every atom an octet. If an octet is not possible with the number of valence electrons, use double or triple bonds.

■ The Valence Shell Electron Pair Repulsion (VSEPR) Theory is used to predict the geometry around a given atom in a Lewis structure. VSEPR is based on the repulsion among all of the bonding electrons and the nonbonding electrons (aka lone pairs of electrons). Atoms with two ligands have a linear geometry with bond angles of 180 degrees. Atoms with three ligands have a trigonal planar geometry with bond angles of 120 degrees. Atoms with four ligands have a tetrahedral geometry with bond angles of 109.5 degrees. If the central atom has lone pairs of electrons, the lone pairs influence the geometry by repelling the bonding pairs of electrons. However, lone pairs are not counted as ligands in assigning geometry. An atom with two ligands and two lone pairs has a bent geometry with bond angles close to the tetrahedral angle, 109.5 degrees. An atom with three ligands and one lone pair of electron has a pyramidal geometry, with angles between the ligands close to 109.5 degrees.

■ Electronegativity is a measure of an atom's affinity for electrons. Fluorine is the most electronegative atom, and atoms closer to fluorine on the periodic table are more electronegative.

■ If two atoms of different electronegativity share a pair of electrons in a covalent bond, the "shared" electron pair spend more time around the more electronegative atom. This leads to formation of a partial positive charge on the less electronegative atom and a partial negative charge on the more electronegative atom. This type of bond is called a polar covalent bond.

■ Whether or not a molecule is polar depends on two factors. First, it must have polar covalent bonds. Second, the polar covalent bonds must have a geometry so that the bond dipoles don't cancel each other out. For example, carbon dioxide is a linear molecule that is nonpolar because the two polar carbon–oxygen bond dipoles are pointed in exactly opposite directions, so the bond dipoles cancel each other out. Water is polar because water has a bent geometry, so the two oxygen–hydrogen bond dipoles don't cancel each other out.

■ There are four principal intermolecular forces. Hydrogen bonding is possible only when a hydrogen atom is bonded directly to oxygen, nitrogen, or fluorine. When hydrogen is bonded to these highly electronegative atoms, the bond is so polar that the hydrogen is left as a focused point of partial positive charge. The hydrogen bond results from the attraction between the highly polarized hydrogen and partial negative charge on an oxygen, nitrogen, or fluorine atom in another molecule. Dipole–dipole attractions result between the opposite partial charges among polar molecules. London forces result from the instantaneous and temporary dipoles that result from uneven electron distributions in molecules or atoms. Ion–dipole interactions result from the attraction of an ion with the opposite partial charge on a polar molecule.

■ In general, the strength of the intermolecular forces follows this trend:

Hydrogen bonding > dipole–dipole > London forces

- Solubility is strongly affected by intermolecular forces. "Like dissolves like" means that polar substances are more soluble in polar solvents, while nonpolar substances are more soluble in nonpolar solvents.
- Unbalanced intramolecular forces among molecules at the surface of a liquid create a "skin" on the liquid's surface. This phenomenon is called surface tension. If the liquid is placed on a flat surface, the balance of intermolecular forces of the liquid molecules for each other and the surface determines the amount of "beading" the liquid exhibits. A liquid placed into a thin tube experiences capillary action that pulls the liquid into the tube. Capillary action results from a balance between the intermolecular forces operating among the liquid molecules and the intermolecular forces between the liquid molecules and the walls of the tube. LaPlace's Law quantifies the relationship between the pressure in a liquid and the surface tension of a liquid at the surface of the container. In a blood vessel, the surface tension borne by a blood vessel is inversely related to the diameter of the vessel. (Gamma [γ] is the surface tension, P is blood pressure, and R is the radius of the blood vessel.)

$$\Delta P = \gamma \left(\frac{1}{R}\right)$$

- Surfactants, or surface active agents, reduce the surface tension in water. Soaps are obtained by the hydrolysis of triglycerides (i.e., fats and oils), while detergents are synthetic analogs of soaps. All surfactants have a hydrophilic head (that interacts well with water) and a hydrophobic tail (that does not interact well with water). The hydrophobic tail is normally a long chain of carbon atoms. The hydrophobic head allows the surfactant to dissolve in water, but the hydrophilic tail interrupts the intermolecular forces among the water molecules.
- Liquids exert a vapor pressure as the most energetic molecules at the surface of the liquid break free of their intermolecular forces and fly free as gaseous molecules. Collisions of the gas molecules with the walls of a container result in the pressure. Vapor pressure increases as intermolecular forces decrease. Vapor pressure increases with increasing temperature. The Clausius–Clapeyron equation relates the vapor pressure of a pure liquid (P) to the temperature (Kelvin temperature, T). R is the universal gas constant, and ΔH_{vap} (heat of vaporization) is the energy needed to vaporize one mole of the liquid at its boiling point. C is a constant characteristic of the liquid in question.

$$\ln(P) = -\frac{\Delta H_{vap}}{R}\left(\frac{1}{T}\right) + C$$

- Gathering the constant values affords a simplified form of the equation:

$$\log P = A + \left(\frac{B}{T}\right)$$

where A and B are constants unique to each gas.
- The boiling point of a liquid is the temperature at which the vapor pressure of the liquid equals the ambient pressure. The normal boiling point is the temperature at which the vapor pressure of a liquid equals 760 torr. The boiling point of a liquid increases as

the pressure increases. Liquids with higher boiling points have greater intermolecular forces.

■ The heat of fusion (ΔH_{fus}) and the heat of vaporization (ΔH_{vap}) are the energies necessary to melt one mole of a solid or to vaporize one mole of a liquid, respectively.

■ A phase diagram relates the state of a substance as we change temperature and pressure. The triple point is the temperature and pressure at which all three states of matter can coexist in equilibrium. The critical point defines the temperature and pressure, beyond which a substance becomes a supercritical fluid, which has the density of a liquid but the flow properties of a gas.

Review Questions for States of Matter and Changes of State

1. List and describe the three principal intermolecular forces.

2. List and describe the three states of matter.

3. Do you expect more ideal behavior from helium gas or water vapor, and why?

4. Rationalize the defining characteristics of the states of matter in terms of intermolecular forces and the kinetic molecular theory of matter.

5. What is the difference between an ionic bond and a covalent bond?

6. What is the VSEPR Theory, and why is it so useful?

7. What are the Lewis structures and geometries for CO_2, H_2O, $CHCl_3$, CCl_4, and SO_2?

8. Explain why water is a liquid at room temperature while carbon dioxide is a gas.

9. Explain why NaCl dissolves in water but not in a nonpolar organic solvent.

10. Which has a higher boiling point, $CHCl_3$ or CHI_3? Why?

11. How does vapor pressure depend on temperature? On intermolecular forces?

12. What geometry do you expect for an AX_2E molecule, such as sulfur dioxide?

13. What geometry do you expect for nitrous oxide? The atom connectivity is N–N–O.

14. What is the vapor pressure of enflurane at 37°C? For enflurane, $A = 7.967$ torr and $B = -1678$ torr·K.

15. Explain why orange growers spray their trees with water when there is a threat of freezing temperatures.

16. What is a surfactant? How do they break surface tension? Give examples of when this is useful.

17. An Entonox cylinder has a total pressure of 137 bar. The cylinder is stored at 0.0°C, and a mixture of 80% N_2O and 20% O_2 liquefies in the bottom of the cylinder. The vapor pressure of the nitrous oxide under these conditions is 17,000 torr. This means that, as long as this liquid phase is present in a closed container, the partial pressure of nitrous oxide remains 17,000 torr.

   a. What is the pressure in the tank in units of torr? 1.013 bar = 1 atm and 760 torr = 1 atm.

   b. What is the mole fraction N_2O in the gas phase in the cylinder?

   c. What is the partial pressure of the oxygen in units of torr?

   d. If the gas phase is removed from the cylinder until the total pressure falls to $5\overline{0}$ bar, what is the mole fraction nitrous oxide? Assuming that oxygen behaves ideally, the percent oxygen in the liquid phase will have fallen to 7.3% and the vapor pressure of nitrous oxide will have risen to 21,000 torr.

18. If a mixture of nitrous and oxygen is bubbled through a spirometer having a large volume of water, which gas is more likely to be absorbed (dissolve in) the water? Explain.

19. Do you expect nitrous oxide to be more soluble in the adipose (fatty, nonpolar tissue) or in the blood? Explain.

20. A certain material has the phase diagram shown below. If a sample of this material were in a beaker in front of you right now, what state would it be in?

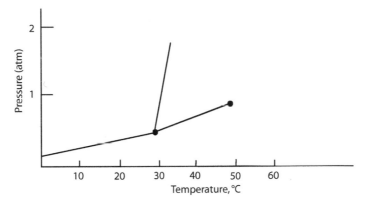

   a. Solid
   b. Liquid
   c. Gas
   d. Cannot determine

21. If the material described in the previous problem were maintained at a constant temperature of 20°C while the pressure was reduced from 1.0 atm to 0.001 atm, what phase transition would have occurred?
   a. Melting
   b. Sublimation
   c. Vaporization
   d. Condensation

22. Which state of matter (solid, liquid, or gas) do you expect to be the least dense? Explain.

23. Give the Lewis structure for each of these compounds. The atom connectivities are given as shown in the partial structures.
   a. H-O-Cl (hypochlorous acid)
   b. H-O-Cl-O (chlorous acid)

       H—O—Cl—O
              |
   c.   O   (chloric acid)

              O
              ‖
       H—O—Cl—O
              |
   d.   O   (perchloric acid)

$$
\begin{array}{c}
O \\
\| \\
S{-}O \\
\| \\
O
\end{array}
$$

e. 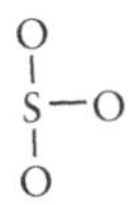 O   (sulfur trioxide)
f. F-O-F (oxygen difluoride)
g. H-O-O-H (hydrogen peroxide)
h. O-C-C-C-O (carbon suboxide)
i. I-Br (iodine monobromide)
j. H-C-N (hydrogen cyanide)
k. H-N-C-O (isocyanic acid)
l. H-O-N-C (cyanic acid)
m. H-C-N-O (fulminic acid)

24. Predict the geometry around each nonterminal atom in each compound in the previous question.

25. Give the strongest intermolecular force you expect to find in each of these compounds.
   a. IBr
   b. I_2
   c. CO_2
   d. H_2O
   e. N_2O
   f. F_2
   g. HF
   h. HCl
   i. NH_3
   j. CH_3CH_2OH
   k. CH_3OCH_3
   l. $CH_3CH_2CH_3$

26. Which compound in the following pairs of compounds has the higher boiling point? Cite which intermolecular force is responsible for the higher boiling point.
   a. CH_3CH_2OH or CH_3OCH_3
   b. CH_3OCH_3 or $CH_3CH_2CH_3$
   c. HCl or HF
   d. Br_2 or IBr
   e. CO_2 or SO_2
   f. NaCl or HCl
   g. Na_2O or H_2O
   h. He or Ar

27. Use LaPlace's Law to estimate the tension on the surface of an aorta having a radius of 0.01 m if the blood pressure is 16 kPa.

28. Use LaPlace's Law to estimate the tension on the surface of a femoral artery having a radius of 0.005 m if the blood pressure is 16 kPa.

29. Use LaPlace's Law to estimate the tension in an aneurism having a radius of 0.01 m (if the blood pressure is 16 kPa). Assume the aneurysm is spherical, so be sure to use the correct form of LaPlace's Law.

30. The constants in the modified Clausius–Clapeyron equation for sevoflurane are A = 8.083 torr and B = –1726 torr·K. Remember that temperature is expressed in kelvins.

$$\log(P) = A + B/T$$

   a. Calculate the vapor pressure at a temperature of 20°C.
   b. Calculate the vapor pressure at a temperature of 25°C.
   c. Calculate the vapor pressure at a temperature of 35°C.
   d. Calculate the boiling point of sevoflurane when the ambient pressure is 760 torr.
   e. Calculate the boiling point of sevoflurane when the ambient pressure is 700 torr.
   f. Calculate the boiling point of sevoflurane when the ambient pressure is 800 torr.

31. The boiling points of sevoflurane and desflurane are 59°C and 23°C, respectively.
   a. Which compound has the greater amount of intermolecular forces?
   b. Which compound do you believe is more polar?
   c. Based on your answer above, which compound is more soluble in a polar solvent, like water? (Neither is very soluble, but which one is *more* soluble?)

32. The vapor pressure of isoflurane at various temperatures is given in the table below. Graph the log of vapor pressure (*y*-axis) versus the reciprocal of the Kelvin temperature (*x*-axis) and determine the slope and intercept of the line. Given the modified Clausius–Clapeyron equation, log(*P*) = *A* + *B*/*T*, which constant, A or B, does the slope of the line represent? How about the intercept?

VAPOR PRESSURE, TORR	TEMP °C	LOG *P*	1/T, K⁻¹
238	20	2.377	0.003411
295	25	2.470	0.003354
367	30	2.565	0.003299
450	35	2.653	0.003245
760	48.5	2.881	0.003109

33. Use your data from the previous question to determine the vapor pressure of isoflurane at 22°C.

34. The triple point of nitrous oxide occurs at a temperature of 182K and 12.9 psi. The critical temperature for nitrous oxide is 310K, and the critical pressure is 1050 psi. On the phase diagram of nitrous oxide, assume the solid–liquid phase boundary line is vertical

up from the triple point and the liquid–gas phase boundary line is approximately a straight line from the triple point to the critical point. Match each of the temperature and pressure combinations with a state of matter for nitrous oxide.

1. Solid	a. T = 182K and P = 12.9 psi
2. Liquid	b. T = 250K and P = 10 psi
3. Gas	c. T = 275K and P = 1000 psi
4. Solid, liquid, gas	d. T = 350K and P = 1500 psi
5. Supercritical fluid	e. T = 150K and P = 25 psi

Solutions and Their Behavior

Chemists almost always work with solutions of analytical reagents, rather than the pure materials. Likewise, the majority of injectable and intravenous medications are delivered as solutions. The reasons for both scenarios are very similar. Using relatively dilute solutions allows for the use of larger and more precisely measured volumes, and dilute solutions are frequently better behaved than the pure reagent. For example, no one would want to do a titration with concentrated sulfuric acid because the sulfuric acid reacts almost explosively with basic analytes. A 1% to 2% solution of lidocaine stings enough: imagine the pain of introducing solid lidocaine hydrochloride under the skin of a patient! Finally, some medications are solids, and it is necessary to dissolve them in an appropriate solvent (usually water) in order to deliver them to a patient in a controlled manner. Therefore, let us consider what exactly a solution is, how concentrations of solutions are expressed, and how the concentration of a solute affects the chemical properties of a solution.

SOME BASIC TERMINOLOGY

A solution is a homogeneous mixture that consists of one or more solutes uniformly dispersed at the molecular or ionic level throughout a medium known as the solvent. A homogeneous mixture means that it is not possible to discern phase boundaries between the components of the mixture. A phase boundary separates regions of a mixture where the chemical or physical properties of the mixture change. Even with the most powerful light microscope, you cannot see where the ions in a normal saline solution are, or where there is pure water. Normal saline is uniform throughout the sample. The solute is the material that got dissolved, and the solvent is the material that does the dissolving, although these are probably unacceptably casual definitions. More practically, the solute is the component of the solution present in the smaller quantity. For example, normal saline is a solution containing sodium chloride and water. Since the sodium chloride comprises 0.89% of the mixture and water comprises 99.11% of the mixture, NaCl is the solute and water is the solvent.

Solutions aren't necessarily liquids. Air is a solution of nitrogen, oxygen, and a few other minor gases. Mixtures of oxygen and nitrous oxide are also solutions. Dental fillings used to be made of silver amalgams, which are solid solutions of silver and mercury.

SOLUTION CONCENTRATIONS

In order to deliver the necessary dose of a medication in solution form, the concentration of the solutes must be considered. Typically, concentrations in medical applications are expressed as percentages. However, there are several concentration units encountered in chemistry, and different units are more applicable than others in different situations. Therefore, we will consider several of the most common concentration units.

Molarity

Molarity, or molar concentration, is defined as moles of solute per liter of solution.

$$M = \frac{\text{mol of solute}}{\text{L of solution}}$$

Because virtually all stoichiometric calculations involve moles (abbreviated mol) of material, molarity is probably the most common concentration unit in chemistry. If we dissolved 1.0 mol of glucose *in enough water to give a total volume of 1.0 L*, we would obtain a 1.0 molar solution of glucose. Molarity is abbreviated with a capital M. Notice that, because molarity has units of moles per liter, molar concentrations are conversion factors between moles of material and liters of solution.

Notice our word choice in the preceding example. We said enough water to give a total volume of 1.0 L of "solution," not 1.0 L of "water." If we dissolved 1.0 mol of glucose in 1.0 L of water, is the molarity of the resulting solution greater than, equal to, or less than 1.0 molar? What will happen to the volume of 1 L of water if we add 180 g (1.0 mol) of glucose? It will increase, of course. Therefore, the molarity will be less than 1.0 M.

You have probably administered IV solutions of D5W, which is a 5% solution of dextrose[1] in water. Let's calculate the molarity of a solution of D5W, which can be prepared by dissolving 1.0 g of glucose ($C_6H_{12}O_6$) in enough water to give a total volume of 20 mL. To solve a problem like this, you need to find the moles of solute (glucose in this case) and the liters of solution. The molecular mass of glucose is 180 g/mol, so we find the moles of glucose as:

$$1.0 \text{ g}\left(\frac{1 \text{ mol glucose}}{180 \text{ g}}\right) = 5.56 \times 10^{-3} \text{ mol}$$

The volume of the *solution* is given as 20 mL, which is equal to 0.020 L. Now we can calculate the molarity of the solution by plugging these values into the definition of molarity.

$$\frac{5.56 \times 10^{-3} \text{ mol glucose}}{0.020 \text{ L of solution}} = 0.28 \text{ M}$$

---

[1] Dextrose is the old name for glucose. Dextrose is so named because a solution of glucose rotates plane polarized light to the right. Dextro is derived from the Latin *dexter*, on the right side. Of course your left-handed authors are not offended that Latin for "on the left" gives us the English word *sinister*, or that the French word for "left" gives us *gauche*.

We can also use molarity as a conversion factor between moles and liters. For example, what volume of the D5W solution is needed to give a patient 2.8 mol of glucose? This amount of glucose contains about 2000 kcals, a day's worth of food energy. Once again, let us stress that memorization is not needed here. Our solution is 0.28 M, or 0.28 mol/L, or 1 liter of D5W contains 0.28 mol of glucose. We are given a number of moles and asked to find a volume in liters. Let the units guide you through this calculation!

$$2.8 \text{ mol glucose}\left(\frac{1 \text{ L of D5W solution}}{0.28 \text{ mol glucose}}\right) = 10 \text{ L D5W solution}$$

Molality

Molality, or molal concentration, expresses concentration in terms of moles of solute per kilogram of solvent. Molality can be used as a conversion factor between moles of solute and kilograms of solvent. Molality is abbreviated with a lower case m. Notice we said kilograms of *solvent*, not kilograms of *solution*. If we dissolved 1.0 mol of NaCl in 1.0 kilogram of water, we would obtain a 1.0 molal solution of NaCl.

$$m = \frac{\text{mol of solute}}{\text{of sol ent}}$$

Let's calculate the molality of a solution prepared by dissolving 1.0 g of glucose in 20 g of water. Again, we start by finding the moles of glucose.

$$1.0 \text{ g}\left(\frac{1 \text{ mol}}{180 \text{g}}\right) = 5.56 \times 10^{-3} \text{ mol}$$

The mass of the solvent is 20 g or 0.020 kg, so we calculate the molality of the solution by plugging these values into the definition of molality.

$$\frac{5.56 \times 10^{-3} \text{ mol glucose}}{0.020 \text{ kg of solvent}} = 0.28 \text{ m}$$

Molality and molarity are each very useful concentration units, but it is very unfortunate that they sound so similar, are abbreviated so similarly, and have such a subtle but crucial difference in their definitions. Because solutions in the laboratory are usually measured by volume, molarity is very convenient to employ for stoichiometric calculations. However, since molarity is defined as moles of solute per liter *of solution*, molarity depends on the temperature of the solution. Most things expand when heated, so molar concentration will decrease as the temperature increases. Molality, on the other hand, finds application in physical chemistry, where it is often necessary to consider the quantities of solute and solvent separately, rather than as a mixture. Also, mass does not depend on temperature, so molality is not temperature dependent. However, molality is much less convenient in analysis, because quantities of a solution measured out by volume or mass in the laboratory include both the solute and the solvent. If you need a certain amount of solute, you measure the amount of solution directly, not the amount of solvent. So, when doing stoichiometry, molality requires an additional calculation to take this into account.

Molality is never equal to molarity, but the difference becomes smaller as solutions become more dilute. So, which solution is more concentrated: 0.28 m glucose or 0.28 M glucose? From the way the preparation of these two solutions was described, you might recognize that less water was added in the molarity example: "enough water to give a total volume of 20 mL," as opposed to 20 g of water. Remember the density of water at 25°C is 1.0 g/mL. So, as long as we are near room temperature, you can consider the mass of water in grams to be equal to the volume of water in milliliters. So, 0.28 M glucose is more concentrated than 0.28 m glucose.

But we can also answer this question by converting molarity to molality. So, what is the molal concentration of a 0.28 molar solution of glucose? To convert between molality and molarity, we need to know the density of the solution. The density of a D5W solution is 1.0157 g/mL. We also need to be very careful about the definitions of molarity and molality, and keep in mind whether we are dealing with liters of solutions or kilograms of solvent.

Step 1: Think about our goal. To find the molality of the solution, we need to find the number of moles of glucose and the kilograms of water in a sample of the solution. That is, we need values for x and y.

$$\text{molality of D5W solution} = \frac{x \text{ mol glucose}}{y \text{ kg of solvent}}$$

Step 2: Choose a sample size. Since concentration is an intensive physical property, we can choose any sample size we want. A 1-L sample is convenient, because that would contain 0.28 mol of glucose, which is one of the numbers we need to know. We're halfway there!

$$\text{m of solution} = \frac{0.28 \text{ mol glucose}}{y \text{ kg of solvent}}$$

Step 3: Now we need to find the kilograms of water in our 1-L sample of D5W solution. Here is where we need the density of the solution. Using density, we can convert the volume of the solution into the mass of the solution. The mass of a solution is equal to the mass of the solvent plus the mass of the solute(s). Therefore, if we subtract the mass of the glucose from the mass of the solution, we get the mass of the water.

$$\text{mass of solution} = 1000 \text{ mL}\left(\frac{1.0157 \text{ g}}{\text{mL}}\right) = 1015.7 \text{ g}$$

$$\text{mass of solute} = 0.28 \text{ mol}\left(\frac{180 \text{ g}}{\text{mol}}\right) = 50 \text{ g glucose}$$

$$\text{mass of water} = \text{mass of D5W} - \text{mass glucose}$$

$$\text{mass of water} = 1015.7 \text{ g} - 50 \text{ g} = 966 \text{ g}$$

Step 4: Now we just need to plug in the values for the moles of glucose and kilograms of water.

$$\text{molality of D5W solution} = \frac{0.28 \text{ mol glucose}}{0.966 \text{ kg of solvent}} = 0.29 \text{ m}$$

Percent

You are certainly familiar with percent concentrations, but you might not know that the term *percent* is somewhat ambiguous. Of course, percent means *parts per 100*. However, there are three common variations on percent, depending on whether you use mass or volume to measure the amount of solute and solution.

Percent by Weight to Volume (% w/v)

This is percent concentration you encounter in a clinical setting, because you want to measure out a volume of medicine in a syringe and be able to calculate the mass of the drug you will give the patient. Formally, percent weight to volume is defined as grams of solute per 100 mL of solution. Oops, we have a language problem at this point. Grams are a unit of mass, not weight, so we are really talking about percent by mass to volume, not percent by weight to volume. However, the terminology of percent by weight to volume is so solidly ingrained in the language of chemistry, with all due apologies to the physics folks, we will say percent by weight.

As stated previously, percent by weight to volume is defined as grams of solute per 100 mL of solution. There are two mathematically equivalent statements of this definition:

$$\%(w/v) = \frac{\text{grams of solute}}{100 \text{ mL of solution}} \quad \text{or} \quad \%(w/v) = \frac{\text{grams of solute}}{\text{mL of solution}} \cdot 100\%$$

The first formulation is useful as a conversion factor between grams of solute and milliliters of solution. The second is more useful to calculate the concentration of a solution. For example what is the percent (w/v) of a solution prepared by dissolving 25 g of glucose in enough water to give a total volume of 500 mL?

$$\%(w/v) = \frac{25 \text{ g of glucose}}{500 \text{ mL of solution}} \cdot 100\% = 5.0\% \ (w/v)$$

We can also use percent concentrations as conversion factors. For example, how many liters of D5W are required to deliver 100 g of glucose? Well, D5W is 5%(w/v) glucose, so we can interpret 5% as 5 g of glucose/100 mL D5W solution. That's a conversion factor, and now the units can take us to the answer.

$$100 \text{ g glucose}\left(\frac{100 \text{ mL solution}}{5 \text{ g glucose}}\right) = 2000 \text{ mL D5W}$$

Percent by Weight (% w/w)

Percent by weight is commonly reported by chemical analysts. The definition of percent by weight is exactly analogous to the definition of percent weight to volume, except the denominator expresses the quantity of solution in terms of grams, not milliliters.

$$\%(w/v) = \frac{\text{grams of solute}}{100 \text{ g of solution}} \quad \text{or} \quad \%(w/w) = \frac{\text{grams of solute}}{\text{g of solution}} \cdot 100\%$$

For example, what is the percent by weight concentration of glucose in a solution prepared by dissolving 25 g of glucose in 475 g of water? As with all solutions problems, we need to be very careful about the definitions. The mass of solute is obviously 25 g, but what about the grams of solution? The solution consists of both glucose and water, so the total solution mass is 500 g, not 475 g. So the percent concentration is:

$$\%(w/w) = \frac{25 \text{ g of glucose}}{500 \text{ g of solution}} \cdot 100\% = 5.0\%(w/w)$$

To relate percent by weight to percent weight to volume, we again need to employ the density of the solution. For example, what is the percent by weight of a D5W solution? The density of D5W is 1.0157 g/mL. Let's see: we need to convert 5.0 g per 100 mL into grams per 100 g. That is, we need to convert the denominator from 100 mL into 100 g. Density is the quantitative relationship between mass and volume. So, let's apply the density as a conversion factor to convert the amount of solution from milliliters into grams. You will find this solution is 4.9% (w/w) glucose.

$$\frac{5 \text{ g glucose}}{100 \text{ mL solution}} \left(\frac{1 \text{ mL}}{1.0157 \text{ g}} \right) = \frac{4.9 \text{ g glucose}}{100 \text{ g solution}}$$

Now, just in case you're thinking: "That's not all that different. Why are we making such a big deal about it?" Well, our reply is this. Surely at some point in your critical care experience, you have encountered a situation where "pretty close" just isn't good enough. Besides, we believe that you all have the experience and capability so that you can begin to probe the subtleties of chemistry and physics, and don't have to rely on crude approximations.

Percent by Volume (% v/v)

Percent by volume is never used in an analytical laboratory, because volumes are not additive. I know, that sounds impossible, but it's true. If you mix 50 mL of water with 50 mL of ethanol, you do not get 100 mL of solution! Because of strong intermolecular interactions between ethanol and water, the volume of the solution is less than 100 mL.

Normality and Equivalents

Normality and equivalents were once standard topics in chemistry courses, but they have largely been abandoned, although they still appear in the medical community. An equivalent (abbreviated Eq) is analogous to a mole, and normality is analogous to molarity. Specifically, one equivalent of a substance contains one mole of chemical reactivity. For example, one equivalent of an acid could deliver one mole of H^+ ions. One equivalent of an oxidizing agent could accept one mole of electrons. Normality is equal to the equivalents of solute per liter of solution. It was kind of a convenient system, because the context of the chemistry, as well as a significant part of the stoichiometric calculations, was contained in the concentration unit. The problem is that, unless the context of the chemistry is specified, normality is ambiguous. For example, nitric acid (HNO_3) has one unit of chemical activity per mole as an acid, but two or three units of chemical activity as an oxidizing agent. So a 0.10-M solution of nitric acid could be correctly labeled as 0.10 N or 0.20 N or 0.30 N, depending on the application to which it is used. So normality has largely disappeared from the chemical literature.

However, when you see a patient's blood work, you might see a calcium ion content of say, 40 mEq/L. What does that mean? When expressing the equivalency of electrolyte cations, the number of equivalents in a mole is equal to the charge on the cation. So, an Na^+ cation has 1 Eq/mol, and a Ca^{2+} cation has 2 Eq/mol. Did you notice this relationship is set up as a conversion factor? So let's play with that calcium ion concentration. First, let's change the concentration into equivalents per liter.

$$\frac{40 \text{ mEq}}{L}\left(\frac{10^{-3} \text{ Eq}}{\text{mEq}}\right) = \frac{0.040 \text{ Eq}}{L}$$

Now, how many moles per liter is this?

$$\frac{0.040 \text{ Eq}}{L}\left(\frac{1 \text{ mol}}{2 \text{ Eq}}\right) = \frac{0.020 \text{ mol}}{L}$$

What about milligrams per liter?

$$\frac{0.020 \text{ mol}}{L}\left(\frac{40 \text{ g}}{\text{mol}}\right)\left(\frac{\text{mg}}{10^{-3} \text{ g}}\right) = \frac{800 \text{ mg}}{L}$$

Parts Per Million

The concentration of extremely dilute solutions is sometimes expressed as parts per million. For example, one supplier of ultrahigh purity nitrous oxide lists the concentration of carbon monoxide as less than 0.1 ppm. A ppm concentration is analogous to a percent concentration, except you are comparing the amount of solute to a million parts of solution, rather than 100 parts. Formally, this definition has the following mathematical form.

$$\text{ppm} = \frac{\text{grams solute}}{1 \times 10^6 \text{ grams solution}} = \frac{\text{grams solute}}{\text{grams solution}} 1 \times 10^6 \text{ ppm}$$

Let's try an example. You know that chromium is a micronutrient that is an essential part of our diets. You might also remember that chromium picolinate got a lot of attention several years ago as a magic weight-loss supplement. The Centers for Disease Control and Prevention lists normal chromium levels in blood as around 2.5 µg per 100 mL of whole blood. What is the concentration of chromium in parts per million? You will need to know that the density of whole blood is 1.06 g/mL.

Step 1: Let's look at the goal to see where we are and where we need to go. From the definition of parts per million, we need to determine the mass of a sample and the mass of chromium in that sample. That is, we need to find numerical values for x and y, and plug them into the definition.

$$\text{ppm of Cr in blood} = \frac{x \text{ g of Cr}}{y \text{ g blood}} 1 \times 10^6 \text{ ppm}$$

Step 2: If we take a 100-mL sample size, we would have 2.5 µg (2.5×10^{-6} g) of chromium.

$$\text{ppm of Cr in blood} = \frac{2.5 \times 10^{-6} \text{ g of Cr}}{y \text{ g blood}} 1 \times 10^{6} \text{ ppm}$$

Step 3: Now we need to express our 100-mL sample size as a mass in grams. For this, we need to use the density.

$$100 \text{ mL}\left(\frac{1.06 \text{ g}}{1 \text{ mL}}\right) = 106 \text{ g}$$

Step 4: We're at the end. If we plug in 106 g as the mass of the solution and do some arithmetic, we find that the chromium concentration in whole blood is 0.024 ppm.

$$\frac{2.5 \times 10^{-6} \text{ g of Cr}}{106 \text{ g of blood}} 1 \times 10^{6} \text{ ppm} = 0.024 \text{ ppm}$$

SOLUBILITY

While few things are absolutely insoluble, some solutes are much more soluble in a given solvent than others. The solubility of a solute is the amount of the solute that will dissolve in a given amount of solvent at a given temperature. For example, sodium chloride is quite soluble in water and has a solubility of 39.5 g per 100 mL of water at 25°C. We'll talk later about the effect of temperature on solubility. Oxygen, on the other hand, is not very soluble in water, with a solubility of 42 mg per 100 mL of water at 25°C and a pressure of 1 atm. We'll talk later about the effect of temperature and pressure on the solubility of gases.

A saturated solution contains the maximum amount of a solute, as defined by its solubility. No more solute will dissolve in a solution saturated with that solute. If the solution is not saturated, more solute will dissolve in that solution. Sometimes, a solution will become supersaturated with a solute. A supersaturated solution contains more solute than allowed by the solubility of the solute. This is not a stable system, because there is more solute dissolved in the sample than the solvent can accommodate. In this case, the excess solute will come out of solution crystallizing as a solid, separating as a liquid, or bubbling out as a gas. For example, when blood or urine in the kidneys becomes supersaturated with calcium oxalate or calcium phosphate, a kidney stone can form. If the solute is a gas in liquid solvent, you would see bubbles forming in the solution. Perhaps you've seen this phenomenon when you open a bottle of beer or soda pop.

Two liquids are miscible if they are soluble in each other in all proportions. Alcohol and water are miscible with each other. Oil and water, on the other hand, are immiscible with each other.

One of the great "rules of thumb" in chemistry is "like dissolves like." That means that polar solutes are more soluble in polar solvents, while nonpolar solutes are more soluble in nonpolar solvents. We have discussed molecular polarity in a previous chapter, but you can consider most organic compounds to be nonpolar. The most polar species are, obviously, ionic compounds, followed by species that can form hydrogen bonds, such as water and ethanol (CH_3CH_2OH). Therefore, you would expect ionic compounds to be soluble in water, but not very soluble in an organic solvent such as ether or hexane. On the other hand, you would expect an organic compound, like the vast majority of the pure form of injectable medications, to be relatively insoluble in a water-based medium such as blood.

What about the solutions of lidocaine and ketamine you have doubtlessly given to patients? Didn't they look like solutions in the syringe? The answer to this puzzle is that you have never given a patient lidocaine, but rather a solution of lidocaine *hydrochloride*. Like most organic compounds, lidocaine is not very soluble in water. Lidocaine belongs to the class of organic compounds called amines, and amine functional groups contain a nitrogen atom (see Figure 8.1). Amine nitrogen atoms are basic, and that means they will accept a hydrogen ion from a strong acid, such as HCl. When the nitrogen undergoes this chemical process, it not only acquires a new chemical bond to a hydrogen atom, but it also picks up the positive charge that was carried by the H^+ ion. The negatively charged chloride ion is necessary in order to maintain electrical neutrality. Lidocaine (a nonpolar, water-insoluble organic compound) reacts with hydrochloric acid to give an ionic salt, which is called lidocaine hydrochloride. Because lidocaine hydrochloride has an ionic bond, it is readily soluble in water.

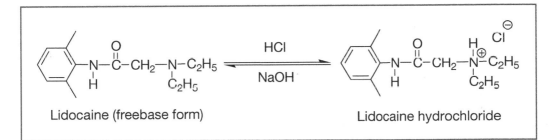

Figure 8.1 Conversion of Lidocaine Into Lidocaine Hydrochloride

When lidocaine is treated with a base, such as sodium hydroxide, the hydrochloric acid is absorbed from the lidocaine salt, leaving lidocaine as its water-insoluble "freebase" form. This freebasing process can be a problem, because the pH of blood is rigidly buffered. Depending on the medication, some hydrochloride salts are converted back into the water-insoluble freebase form.

You might have heard about freebasing in the context of drugs like cocaine, and the chemical process is identical. Cocaine hydrochloride is a highly crystalline, water-soluble species suitable for direct introduction into the user. However, cocaine hydrochloride, like all ionic compounds, has a very low vapor pressure and is not suitable for smoking. Therefore, a base such as ammonia or sodium bicarbonate (baking soda) is added to convert the salt to the more volatile freebase form. If the process doesn't include removing the inorganic by-products, the result is crack cocaine.

ENERGY CHANGES AND THE SOLUTION PROCESS

When a solute dissolves in a solvent, there is an associated energy change, and there is often times a noticeable change in the temperature of the solution. You exploit this phenomenon when you use chemical hot packs or cold packs. This energy change is called the heat of solution or the enthalpy of solution. Formally, the enthalpy of solution, ΔH_{soln}, is defined as the energy change that accompanies dissolving exactly one mole of solute in a given solvent. Remember that enthalpy H is equal to the heat Q as long at the pressure remains constant. This is an extremely subtle difference, and chemists typically say "heat of solution" when they really mean "enthalpy of solution," although this is not strictly correct. This energy change may be endothermic or exothermic. Energy flows into the system during an endothermic process, whereas energy flows out of the system during an exothermic process.

If the solution process is exothermic, energy flows out of the system (solvent and solute) into the surroundings (where you are). The system will have lost energy. As an observer who is part of the surroundings, you will note that the temperature of the solution increases. This makes sense, of course, because energy is exiting the system. If the solution process is endothermic, energy flows from the surroundings into the system, and the energy of the system increases. In this case, you will observe a temperature decrease in the solution.

Now, wait just a minute. If energy flows *into* the system, how can the temperature *decrease*? This example strongly illustrates why we shouldn't talk about heat as energy. Heat is a process, not a product. Heat is the transfer of energy into or out of a system caused by a difference in temperature. If you add energy to a system, like, say heating a pot of water on the stove, the influx of energy goes into increasing the average kinetic energy of the molecules, which is manifested as an increase in temperature. However, the added energy can also be absorbed by the system in the form of potential energy. One way for a chemical system to store potential energy is in how the molecules in the system interact with one another. So, energy can be added to a chemical system without increasing the average kinetic energy of the molecules, and you will not observe an increase in temperature.

To help understand why the heat of solution is sometimes endothermic and sometimes exothermic, let's imagine splitting the heat of solution into two processes. Remember that enthalpy is a state function, so it is perfectly legitimate to split it into two processes, as long as our imaginary route starts at the same point as the observed process and takes us to the same endpoint. We are starting with an ionic substance in the solid state, so let's first imagine pulling all of the ions away from each other. To accomplish this, we have to overcome the lattice energy, because all of the cations are attracted to all of the anions by Coulomb's Law (opposite charges attract). Of course, pulling oppositely charged ions away from each other requires energy, so this is an endothermic process.

We end up with separated ions. So now, let's add the water molecules to our isolated ions. Now, an ion is the ultimate polar species. Also recall that water is a polar molecule. The oxygen end of the water molecule has a partial negative charge, and the hydrogen end of the water molecule has a partial positive charge. So, how is water going to interact with ions?

Of course, cations will be surrounded by clusters of the water molecules, with the negatively charged end of water's molecular dipole pointing toward the cation. Anions will be surrounded by clusters of water molecules, with the positively charged end of water's molecular dipole pointed toward the anion. This process is called *solvation*. Since solvation involves opposite charges coming toward each other, solvating an ion is an inherently exothermic process. Figure 8.2 illustrates a chloride ion solvated by a sphere of water molecules.

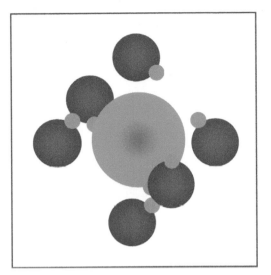

Figure 8.2 Solvated Chloride Ion

The figure is an illustration to help you visualize the solvation process. However, solvation is not a crystalline and orderly process. There are not a fixed number of solvating molecules, nor is it a highly organized process, nor is there a single sphere of solvation molecules. In short, solvation is a very messy process, and whoever comes up with a workable model will certainly win the Nobel Prize in chemistry.

Whether or not the heat of solution is an endothermic or an exothermic process depends on the relative magnitudes of the lattice energy and the heat of solvation. If tearing the ions apart requires more energy than is released by solvation, then ΔH_{soln} is going to be positive (endothermic). If the energy released by solvation is greater than the energy required to tear the ions apart, ΔH_{soln} is going to be exothermic.

An instant cold pack has a vial of ammonium chloride inside a bag that contains water. To activate the cold pack, the inner vial of ammonium chloride is broken, and the ammonium chloride dissolves in the water. Since the heat of solution for ammonium chloride is endothermic, this causes the system to get colder.

Let's consider the energy diagram (Figure 8.3) to explain how the solution process for ammonium chloride in water can be endothermic. We start with solid ammonium chloride and first go to separated ions in the gas phase. To do this, we need to add enough energy to the system to overcome the lattice energy, and the energy of the system rises. When we add the water molecules, the ions are solvated, and this exothermic process releases energy from the system. Since the magnitude of the lattice energy is greater than the magnitude of the solvation energy, in going from solid ammonium chloride to a solution of ammonium chloride our system has actually risen in energy. That is, it has undergone an endothermic change.

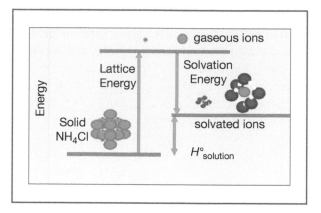

Figure 8.3 Lattice and Solvation Energy

FACTORS THAT AFFECT SOLUBILITY

Effect of Pressure on Solubility

Pressure has a dramatic effect on the solubility of gaseous solutes in liquid solvents. As pressure increases, the solubility of a gaseous solute in a liquid solvent increases. You have no doubt observed this phenomenon when opening a carbonated beverage. Have you noticed that when you open a bottle of champagne (or soda pop, or beer, or club soda), small bubbles of carbon dioxide gas start to form? That "pop" you hear when removing the cork is caused by the escape of excess carbon dioxide gas from the bottle. When the gas escapes, the pressure in the container decreases. With less pressure, the solubility of the carbon dioxide dissolved in the wine decreases. As the carbon dioxide comes out of the solution, it forms those tiny (wonderful) little bubbles. Since solids and liquids are not very compressible, at least not compared to gases, pressure has very little effect on the solubility of solid and liquid solutes.

The quantitative relationship between pressure and solubility is given by Henry's Law:

$$S = k_H P_{gas}$$

where S = solubility, k_H = Henry's constant, P_{gas} = partial pressure of the gas.

If you think about it for a moment, Henry's Law makes sense. If we increase the partial pressure of a gas, that means there are more of those gaseous molecules (in a given volume, at a given temperature) zipping around near the surface of the liquid. Doesn't it make sense that more of the gas molecules will be trapped by the liquid solution? Also, if the dissolved gas molecules were to escape into the gas phase, the pressure would increase. So, assuming that the pressure remains constant, the would-be escapees are crushed back into solution.

The only tricky part of this equation is the pressure term. P_{gas} is the partial pressure of the gas in question, not the total pressure. Imagine a glass containing 1 L of water in a hyperbaric chamber that is filled with pure oxygen at a pressure of 1 atm. At room temperature, about 42 mg of oxygen will dissolve in 1 L of water (S = 42 mg/L). Now, let's double the pressure in the chamber by pumping in pure oxygen gas. Of course, the amount of oxygen that dissolves in water will double to 84 mg/L. But what if we had pressurized the chamber by pumping in a mixture of 50% oxygen and 50% nitrogen? Now the pressure due to just the oxygen gas molecules is only increased by 50%. There are only 50% more oxygen molecules zipping along near the surface of the water, not twice as many. The other half of the molecules are nitrogen. So, the concentration of oxygen only increases by 50%, to 63 mg/L. There is a somewhat imperfect analogy that you have no doubt seen innumerable times in clinical settings. When you administer oxygen to a patient who is having trouble breathing, the pressure of the gas is near atmospheric pressure, but the concentration of oxygen in pure oxygen gas is nearly five times the concentration of oxygen in air. Therefore, you expect the blood to become more quickly saturated with oxygen.

Let us consider an example. The Henry's Law constant for oxygen in water is 0.042 g/L/atm at 25°C. What is the solubility (in mg/L) of O_2 in pure water when the atmospheric pressure is 740 torr? In air, the mole fraction of oxygen is 0.21.

Since the value of the constant has pressure units of atmospheres, we need to convert torr into atm.

$$P_{total} = 740 \text{ torr}\left(\frac{1 \text{ atm}}{760 \text{ torr}}\right) = 0.974 \text{ atm}$$

Since oxygen comprises 21% of air, oxygen constitutes 21% of the total pressure:

$$P_{O_2} = \chi_{O_2}P_{total} = (0.21)(0.974 \text{ atm}) = 0.204 \text{ atm}$$

Therefore, the solubility of oxygen under these conditions is:

$$S_{O_2} = \left(0.042\frac{\text{g}}{\text{L} \cdot \text{atm}}\right)(0.204 \text{ atm}) = 0.0086 \frac{\text{g}}{\text{L}}$$

So, the solubility of oxygen gas at a pressure of 740 torr and 25°C is 8.6 $\frac{\text{mg}}{\text{L}}$.

Effect of Temperature on Solubility

The solubility of solid and liquid solutes in liquid solvents generally increases with increasing temperature. There are a few exceptions to this rule, but not many. This is why there are a few medications that are stored at room temperature. For example, a concentrated mannitol solution will crystallize if stored at low temperatures. Warming the solution will cause the crystallized solid to redissolve. Temperature has the opposite effect on the solubility of gaseous solutes in liquid solvents. As the temperature increases, the vapor pressure of the gaseous solutes increases to the point that they can escape the solvent into the gas phase. There is a simple experiment you can do to verify this: Try opening a hot can of soda pop.[2]

---

[2] This is also true of opening a bottle of champagne. Holding the bottle by the neck increases the temperature of the wine in the neck, and you will lose more of the wine to frothing. If you hold the bottle at the bottom, or use a towel to insulate the bottle from your warm hand, frothing will be minimized.

COLLIGATIVE PROPERTIES OF SOLUTIONS

As solutes are added to a solution, the physical properties of the solution change. Some properties, such as electrical conductivity, depend, at least to some extent, on the identity of the solute. However, some properties do not depend on the identity of the solute, but only on the number of solute particles in the solution. The properties of a solution that depend only on the number of solute particles, and not the identity of the solute particles, are called the *colligative properties* of solutions.

There are four commonly cited colligative properties:

1. The vapor pressure of a solution decreases with increasing solute concentration.
2. The boiling point of a solution increases with increasing solute concentration.
3. The freezing point of a solution decreases with increasing solute concentration.
4. The osmotic pressure of a solution increases with increasing solute concentration.

Vapor Pressure

As we have previously discussed, the vapor pressure of a liquid results from the most energetic molecules near the surface of the liquid escaping into the gas phase. The most likely escape sites for the liquid molecules are at or near the surface of the liquid. As we begin to introduce solute molecules, some of these escape sites are occupied by the solute molecules. Therefore, fewer solvent molecules can escape into the gas phase. Therefore, the vapor pressure of the solution is less than the vapor pressure of the pure solvent.

Raoult's Law gives the quantitative relationship between vapor pressure and solute concentration. Raoult's Law states the vapor pressure of a volatile component of a solution (P) is equal to the vapor pressure of the pure substance ($P°$) times the mole fraction (χ) of that substance.

$$P = \chi P°$$

For example, we can calculate the vapor pressure of a 5.0% (by mass) sucrose solution at 45°C. At this temperature, the vapor pressure of pure water is 71.9 torr. Since sugar is essentially nonvolatile, we have to concern ourselves only with the vapor pressure of the water.

First, we need to calculate the mole fraction of the water. Since the solution is 5.0% (*w/w*), a 100-g sample of the solution would contain 95 g of water and 5 g of sucrose. To find the mole fraction, we need find the moles of each compound. Recall we calculated mole fractions in Chapter 6.

$$95 \text{ g H}_2\text{O}\left(\frac{1 \text{ mol}}{18 \text{ g}}\right) = 5.28 \text{ mol water}$$

$$5 \text{ g C}_{12}\text{H}_{22}\text{O}_{11}\left(\frac{1 \text{ mol}}{342 \text{ g}}\right) = 0.0146 \text{ mol sucrose}$$

The mole fraction of water is:

$$\chi \text{H}_2\text{O} = \frac{5.28 \text{ mol water}}{5.28 \text{ mol water} + 0.0146 \text{ mol sucrose}} = 0.997$$

Since water molecules comprise a mole fraction of 0.997 of the molecules in the solution, the vapor pressure of the solution is only 99.7% that of pure water:

$$P = (0.997)(71.9 \text{ torr}) = 71.7 \text{ torr}$$

We can also use Raoult's Law to calculate the vapor pressure of a mixture of volatile substances. For example, what is the total vapor pressure of a solution comprised of 0.75 mol of benzene and 0.45 mol hexane at 60°C? ($P°_{\text{benzene}} = 390$ torr; $P°_{\text{hexane}} = 580$ torr). We can calculate the vapor pressures of each of the components using Raoult's Law.

$$P_{\text{benzene}} = \left(\frac{0.75 \text{ mol benzene}}{0.75 \text{ mol benzene} + 0.45 \text{ mol hexane}}\right) 390 \text{ torr}$$

$$P_{\text{benzene}} = 244 \text{ torr}$$

$$P_{\text{hexane}} = \left(\frac{0.45 \text{ mol benzene}}{0.75 \text{ mol benzene} + 0.45 \text{ mol hexane}}\right) 580 \text{ torr}$$

$$P_{\text{hexane}} = 218 \text{ torr}$$

The total pressure of the solution will then be the sum of the component pressures (Dalton's Law).

$$P_{\text{total}} = P_{\text{hexane}} + P_{\text{benzene}} = 218 \text{ torr} + 244 \text{ torr} = 462 \text{ torr}$$

Boiling Point

Boiling point is defined as the temperature at which the vapor pressure of the material is equal to the ambient pressure. Since the vapor pressure of a solution is decreased by the addition of nonvolatile solutes, a higher temperature is needed to drive the vapor pressure up to the point where it equals the ambient pressure. In other words, the boiling point of a solution increases as the concentration of solute(s) increases. The change in boiling point is directly proportional to the molal concentration of the solute particles.

$$\Delta T_{\text{bp}} = k_{\text{bp}} \cdot m_{\text{total}}$$

where ΔT_{bp} is the number of degrees by which the boiling point increases, k_{bp} is a constant (called the *ebullioscopic constant*) that is characteristic of the solvent, and m_{total} is the molal concentration of all solute particles.

This effect is fairly modest. For example, we all know that you cook pasta in salted water. Do we add salt to increase the boiling point of the water or to season the pasta? Well, a typical recipe might call for 4 L of water (which has a mass of 4 kg) and 2 tablespoons of salt, or about 40 g. The ebullioscopic constant of water is 0.512 °C/molal.

First, we need to calculate the molality of the sodium chloride:

$$m_{\text{NaCl}} = \frac{\text{mol NaCl}}{\text{kg water}} = \frac{40 \text{ g NaCl}\left(\frac{1 \text{ mol NaCl}}{58.5 \text{ g}}\right)}{4 \text{ kg water}} = 0.17 \text{ m}$$

Here's where you have to be careful. The colligative properties depend on the concentration of solute particles, whether they are molecules or ions. When NaCl dissolves in water, it separates into a sodium ion and a chloride ion. Thus, the concentration of particles is 0.34 molal.

$$m_{particles} = \frac{0.17 \text{ mol NaCl}}{\text{kg water}} \left(\frac{2 \text{ moles of ions}}{1 \text{ mol NaCl}} \right) = 0.34 \text{ m}$$

Therefore, the temperature of the salted water will increase by 0.18°C.

$$\Delta T_{bp} = (0.512°C/molal) \cdot (0.34 \ m_{total}) = 0.18°C$$

So, we add the salt primarily to season the pasta.

This is a good time to bring in a concept that you probably have not encountered before: *activity*. Activity is the *effective concentration* of a solute, and activity is always less than molality. You need to become increasingly concerned about the difference between molality and activity when the concentration rises above 0.1 molal, especially if you are dealing with ionic solutes. In the previous example, we assumed that sodium chloride dissolves in water to give completely separated ions that have no interactions with each other. Is this reasonable? The answer is, "Not really." To be sure, there is considerable separation of the ions, or else a salt water solution would not conduct electricity, sodium chloride would not be an electrolyte, and your heart wouldn't beat. However, as the concentration of ions increases, the solvating water molecules will have increasing difficulty in keeping the positively charged cations separated from the negatively charged anions. This "clumping together" of ions results in a smaller concentration of particles in the solution. So, will the boiling point elevation for 1 m NaCl be twice as much as a 1-m solution of sugar? No. Almost, but not quite, because at a concentration of 1 molal NaCl, there will be significant clustering of the ions, and the total concentration of particles in solution will be less than 2 molal. The Debye–Hückel Theory attempts to quantify this phenomenon, but is beyond the scope of our present conversation.

Freezing Point

The freezing point (or melting point) of a sample is the temperature at which the liquid phase of the material is in equilibrium with the solid phase. In order to enter into the solid state, the molecules (or ions or atoms) of the sample need to settle into an orderly, crystalline lattice structure. The presence of solute particles interferes with this process by getting in the way. So it is necessary to cool the sample to lower temperatures, thereby lowering the kinetic energy of the molecules even further, before they will settle into the solid phase.

The relationship that quantifies the degrees of freezing point depression has an identical form to boiling point elevation.

$$\Delta T_{fp} = k_{fp} \cdot m_{total}$$

where ΔT_{fp} is the number of degrees the freezing point decreases, k_{fp} is a constant (called the *cryoscopic constant*) that is characteristic of the solvent, and m_{total} is the molal concentration of all of the solute particles.

Once again, this effect is fairly modest, although it is generally larger than boiling point elevation. For example, adding "antifreeze" to a car's radiator lowers the freezing point of the coolant liquid. Many commercial antifreezes are about 95% ethylene glycol. The manufacturers recommend automobile cooling systems be filled with 50% antifreeze and 50% water. Let's calculate the freezing point of a 50:50 mixture of ethylene glycol and water. To simplify the calculations, let's use percent by mass. The cryoscopic constant for water is 1.86 °C/molal.

Let's start by calculating the molality of the solute. Ethylene glycol (HO-CH_2CH_2-OH, or EG for short) is an organic alcohol with a molecular mass of 62 g/mol. Since it is a molecular substance, it is a nonelectrolyte.

$$m_{EG} = \frac{mol\ EG}{kg\ water} = \frac{50\ g\ EG\left(\frac{1\ mol\ EG}{62\ g}\right)}{0.050\ kg\ water} = 16\ m$$

Therefore, the freezing point of this solution will decrease by 30°C, from 0 to –30°C.

$$\Delta T_p = (1.86°C/molal) \cdot (16\ m_{total}) = 30°C$$

The observed freezing point of this mixture is –34°C, which represents about a 12% error, but we can easily rationalize this discrepancy. We made some tremendous assumptions in the previous example. First, we are ignoring the other components of antifreeze, which will also lower the temperature of the solution. Second, we are assuming that this solution will exhibit ideal behavior. This second assumption is the more problematic. In the first place, these laws don't apply well to highly concentrated solutions. Moreover, both ethylene glycol and water can form hydrogen bonds, which leads to significant solvent–solute interactions. All of the colligative properties postulate there are no solvent–solute interactions, and that is why colligative properties do not depend on the identity of the solute.

Osmotic Pressure

As you doubtlessly learned in physiology, osmosis is diffusion of water through a semipermeable membrane. The semipermeable membrane allows water to move through it, but most solute particles are either too big or too polar to make it across the membrane. The relative concentration of solutes in osmotic systems is called the *tonicity*. Two solutions are isotonic if they contain equal concentrations of particles. If the concentrations are not equal, the one with the greater concentration is the hypertonic solution, and the one with the lower concentration is the hypotonic solution. It is critically important to notice that tonicity is a comparative concept, and it makes no sense to call a solution hypertonic without indicating to which solution you are comparing it. For example, is a 5% NaCl solution hypotonic or hypertonic? You are probably tempted to say "hypertonic," because you are mentally comparing this solution to normal saline, which is 0.89% (w/w) NaCl. So, 5% NaCl is hypertonic to normal saline. However, 5% NaCl is hypotonic to 10% NaCl and isotonic with another solution of 5% NaCl.

If we separate a 5% NaCl solution from a 1% NaCl solution by a semipermeable membrane, as illustrated in Figure 8.4, in which direction will osmosis occur? Osmosis is the diffusion of water, and diffusion always spontaneously occurs in the direction from an area of high concentration to an area of low concentration. Since the concentration of water in the hypotonic solution is greater than the concentration of water in the hypertonic solution, osmosis always spontaneously occurs from the hypotonic solution to the hypertonic solution.

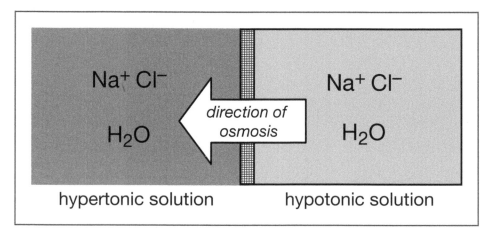

Figure 8.4 Osmosis

The Second Law of Thermodynamics provides a theoretical framework to understand the spontaneous direction of osmosis. The second law states that the entropy of the universe is continually increasing. Therefore, spontaneous processes almost always have increasing entropy. Recall that entropy is loosely defined as disorder or randomness in a system. When two solutions have different concentrations, they appear to be slightly organized, at least through the lens of the second law. More solute particles are sorted out into one solution. Therefore, entropy demands that osmosis occur between two solutions of unequal tonicity until the concentrations of the two solutions are equal.

Osmotic pressure (symbolized as capital pi, Π) results from the potential drive for the concentration of water to equalize. Osmotic pressure is a colligative property, and the osmotic pressure of a solution increases with increasing solute concentration. The relationship between osmotic pressure and concentration is given by:

$$\Pi V = nRT$$

Yes, this does strongly resemble the ideal gas law. If the volume is in liters and we divide both sides of the relationship by volume, we get:

$$\Pi = MRT$$

where Π is the osmotic pressure, M is the molarity of the solute particles, R is the ideal gas constant, and T is the absolute temperature.

Of all of the colligative properties, osmotic pressure shows the most dramatic effect. For example, a 1.0-M solution of glucose exerts an osmotic pressure of 24.5 atm at 25°C. A pressure of 24.5 atm is equivalent to 360 psi!

$$\Pi = (1 \text{ mol/L})(0.08205 \text{ L} \cdot \text{atm/mol/K})(298 \text{ K})$$

$$\Pi = 24.5 \text{ atm}$$

For example, osmosis is at least partially responsible for moving water up to the top of trees. What glucose concentration is needed to supply an osmotic potential (compared to pure water) necessary to push water to the top of a 25-ft (7.6-m) tree when the temperature is 25°C? First, recall that one atmosphere of pressure can support a column of mercury 0.76 m in height (or 760 mm) or a column of water 10.3 m in height.

$$7.6 \text{ m of water}\left(\frac{1 \text{ atm}}{10.3 \text{ m}}\right) = 0.74 \text{ atm}$$

If we solve the osmotic pressure equation for molarity and insert our known values:

$$M = \frac{0.74 \text{ atm}}{(0.08205 \text{ l} \cdot \text{atm} \cdot \text{mol}^{-1} \cdot \text{K}^{-1})(298 \text{ K})}$$

$$M = 0.030 \frac{\text{mol glucose}}{L}$$

The concentration of sugar in maple tree sap, for example, is about 0.06 M.

You know that a red blood cell placed into pure water will burst. What is the total force exerted on a single red blood cell due to osmotic pressure? Cells are isotonic with normal saline, which is 0.89% (*w/v*) NaCl. A red blood cell is approximately a cylinder with a diameter of 6 μm and a total surface area of 140 μm². While this is not a difficult problem, it is easy to make a mistake if we don't decide on units early in the process. The safest route is to stick with SI units: volume in cubic meters, pressure in pascals, and force in newtons.

First, let's calculate the total particle concentration.

$$\frac{0.89 \text{ g NaCl}\left(\frac{1 \text{ mol}}{58.5 \text{ g}}\right)}{0.10 \text{ L solution}} = 0.15 \frac{\text{mol NaCl}}{L}$$

But 1 mol of NaCl dissociates into 2 mol of ions, so the total particle concentration is 0.30 M. The pressure exerted by an osmosis gradient of 0.30 M NaCl is 7.4×10^5 Pa.

$$\Pi = (0.30 \text{ mol/L})(0.08205 \text{ L} \cdot \text{atm/mol/K})(298 \text{ K}) = 7.3 \text{ atm}$$

$$7.3 \text{ atm}\left(\frac{101{,}325 \text{ Pa}}{\text{atm}}\right) = 7.4 \times 10^5 \text{ Pa}$$

The surface area of a red blood cell, expressed in square meters, is:

$$140 \ \mu m^2\left(\frac{10^{-6} \text{ m}}{\mu m}\right)^2 = 1.4 \times 10^{-10} \text{ m}^2$$

Recalling that 7.4×10^5 Pa means 7.4×10^5 N/m^2, the force exerted on the red blood cell is:

$$1.4 \times 10^{-10} \text{ m}^2\left(\frac{7.4 \times 10^5 \text{ N}}{\text{m}^2}\right) = 1.0 \times 10^{-4} \text{ N}$$

While this might not initially seem like much force, we're talking about a single cell. For example, if the cell were the size of a basketball, which has a surface area of 0.179 m^2, the total force would be 30,000 pounds!

COLLOIDS

Colloids are similar to solutions in that they consist of one phase uniformly dispersed in a second phase. There are many common examples of colloids, including milk, blood, paint, and jelly. However, colloids are not true solutions because the particles in the dispersed phase are not the size of molecules or ions. The particles in a colloid range in size from 10 to 200 nm. The dispersed particles might be supersized molecules (e.g., proteins) or aggregates of ions. While these particles are typically too small to be discerned, even by a microscope, they are much larger than, say, a sodium ion, which has a diameter of about 0.1 nm.

Colloidal particles cannot be filtered and do not settle out of solution. Colloids can be stable for years if they are stored under controlled conditions. Colloids exhibit the Tyndall effect, whereas solutions do not. You have seen the Tyndall effect when you turn on the headlights on your car and see the "beams" of light shooting out ahead of you. The particles in a colloid are large enough to scatter light passing through, so you can see the beam of light shooting through a colloid. The particles in a true solution are too small to scatter visible light, so solutions do not exhibit the Tyndall effect.

Summary

- A solution is a homogeneous mixture comprised of one or more solutes distributed at a molecular or ionic level throughout a medium called the solvent. Homogeneous means there are no visible phase boundaries, even under a microscope.
- Solutes and solvents can be solids, liquids, or gases.
- The molarity (*M*) of a solution is defined as the moles of solute per liter of *solution*. The molarity is temperature dependent. As the temperature increases, molarity (generally) decreases because the volume of solvent expands with higher temperatures.

$$M = \frac{\text{moles of solute}}{\text{liter of solution}}$$

- The molality (m) of a solution is defined as the moles of solute per kilogram of *solvent*. Since molality is based on mass, not volume, molality does not depend on temperature.

$$m = \frac{\text{moles of solute}}{\text{kg of solvent}}$$

- Percent weight-to-volume (% *w/v*) is defined as grams of solute per 100 mL of solution.

$$\%(m/v) = \frac{\text{grams of solute}}{\text{mL of solution}} 100\% = \frac{\text{grams of solute}}{100 \text{ mL of solution}}$$

- Percent by weight (% *w/w*) is defined as grams of solute per 100 g of solution.

$$\%(w/w) = \frac{\text{grams of solute}}{\text{g of solution}} 100\% = \frac{\text{grams of solute}}{100 \text{ g of solution}}$$

- Percent volume-to-volume (% *v/v*) is not generally used for precise analytical work because volumes are not necessarily additive.
- Normality (N) is defined as the equivalents (Eq) of solute per liter of solution. An equivalent is defined as one unit of chemical reactivity, so you have to know the chemical context for the term *equivalent* to have an exact meaning. In physiology, equivalents of a metal ion are equal to the moles of the metal ion times the ionic charge.

$$N = \frac{\text{equivalents of solute}}{\text{liters of solution}}$$

- Equivalents of a metal ion = (moles of metal ion) × (ionic charge)
- Parts per million (ppm) is defined as grams of solute per one million grams of solvent. For dilute aqueous solutions, the parts per million of a solute gives the mg of solute per liter of solution.

$$ppm = \frac{\text{grams of solute}}{\text{grams of solution}} 1 \times 10^6 \, ppm \approx \frac{\text{mg of solute}}{\text{liters of solution}}$$

- Solubility is defined as the mass of solute that will dissolve in a given amount of solvent. A saturated solution contains a solute at a concentration equal to its solubility.

A supersaturated solution contains a concentration of solute greater than its solubility limit. Supersaturated solutions are not stable, and the excess solute will precipitate out of solution.

■ The solution process may be endothermic or exothermic, depending on how the solute particles interact with each other and how the solute particles interact with the solvent. These interactions are based on intermolecular forces (hydrogen bonding, dipole–dipole, London, or ion–dipole). The attraction of solute particles for each other is called the lattice energy. When the solute particles are in solution, they are surrounded by an ill-defined solvation layer of solvent molecules. The attraction between the solute particles and the surrounding solvent molecules is called the solvation energy. If the absolute value of the lattice energy is greater than the absolute value of the solvation energy, then the solution process will be endothermic, and dissolution of the solute causes the solution to become colder. If the absolute value of the lattice energy is less than the absolute value of the solvation energy, the solution process will be exothermic, and dissolution of the solute causes the solution to become warmer.

■ The solubility of solid and liquid solutes is not greatly affected by changing the pressure. However, the solubility of solid and liquid solutes tends to increase with increasing temperature.

■ The solubility (S) of a gaseous solute in a liquid solvent increases with the partial pressure (P_{gas}) of the gas. The relationship is given by Henry's Law. The proportionality constant (k_H) is called the Henry's Law constant for the gas and is unique to each gas/solvent combination.

$$S = k_H P_{gas}$$

■ The solubility of a gas in a liquid solvent decreases with increasing temperature.

■ The colligative properties of solutions are a set of physical properties that are affected by the concentration of solute particles, but do not depend on the identity of the solute particles. The colligative properties include:

   ■ Vapor pressure decreases with increasing solute concentration
   ■ Boiling point increases with increasing solute concentration
   ■ Freezing point decreases with increasing solute concentration
   ■ Osmotic pressure increases with increasing solute concentration

■ Raoult's Law states the vapor pressure of a solution (P) is equal to the vapor pressure of the pure solvent ($P°$) times the mole fraction (χ) of the solvent.

$$P = \chi P°$$

■ The mole fraction of a solute is equal to the moles of solute in a sample divided by the total number of moles of substances in a sample.

$$\chi_i = \frac{\text{mol}_i}{\text{mol total}}$$

■ The boiling point equation states the increase in boiling point (ΔT_{bp}) is equal to the boiling point constant (k_{bp}) times the molal concentration of all the solute particles (m_{total}).

$$\Delta T_{bp} = k_{bp} \cdot m_{total}$$

■ The freezing point equation states the decrease in freezing point (ΔT_{fp}) is equal to the freezing point constant (k_{fp}) times the molal concentration of all the solute particles (m_{total}).

$$\Delta T_{fp} = k_{fp} \cdot m_{total}$$

■ The osmotic pressure (Π) is equal to the molarity of all solvent particles (M) times the ideal gas constant (R) times the Kelvin temperature (T).

$$\Pi = MRT$$

■ Osmosis is the diffusion of water across a semipermeable membrane. The direction of spontaneous osmosis is from the hypotonic solution to the hypertonic solution. As osmosis progresses, removal of water from the hypotonic solution causes its concentration to increase. Likewise, osmosis of water into the hypertonic solution causes its concentration to decrease. Osmosis continues until the two solutions are isotonic.

■ The tonicity of two solutions describes the concentration of each solution *relative to the concentration of the other*. The hypertonic solution has the *greater* concentration, while the hypotonic solution has the *lesser* concentration. Two solutions that have equal concentrations of solute particles are isotonic with each other. In physiological settings, the tonicity of a solution is compared to normal saline, which is 0.89% (*m/v*) NaCl in water.

Review Questions for Solutions

1. What is a solution? How does it differ from a colloid?

2. Give examples of solutions that have:
   a. A liquid solute in a liquid solvent
   b. A solid solute in a liquid solvent
   c. A gaseous solute in a liquid solvent
   d. A solid solute in a solid solvent

3. List and describe the colligative properties of solutions.

4. Which colligative property has the most profound impact on physical properties?

5. What is the molarity of pure water?

6. What does miscible mean?

7. Why do some compounds have endothermic heats of solution, while others have exothermic heats of solution?

8. You dissolve $9\overline{0}$ g of sucrose (molar mass = 342 g/mol) in $1\overline{0}00$ g of water. What is the molality of the solution? If the cryoscopic constant for water is 1.9 degrees/molal, what is the freezing point of the solution?

9. How does temperature normally affect solubility of solid solutes? Gaseous solutes?

10. If the solubility of oxygen in water is 0.0404 g/L at 1 atm of pressure and 25.0°, what is the solubility at 3.50 atm of pressure and the same temperature?

11. Why is the vapor pressure of a solution less than the vapor pressure of a pure liquid? How does this impact boiling point? What is the quantitative relationship between solute concentration and vapor pressure?

12. Why do you soak potatoes in salt water before making French fries?

13. Why are there no great white sharks in fresh-water lakes?

14. Describe the terms *hypertonic, isotonic,* and *hypotonic.*

15. If an anesthetic (like all other compounds) has a smaller molecular weight, it tends to have smaller intermolecular attractions. Therefore:
    a. A smaller anesthetic molecule will have a higher vapor pressure.
    b. A larger anesthetic molecule will have a higher vapor pressure.
    c. A smaller anesthetic molecule will have a higher boiling point.
    d. A larger anesthetic molecule has a greater tendency to evaporate.

16. The graph below shows the vapor pressure of ethanol as a function of temperature. At what temperature does ethanol boil high in the mountains, where the atmospheric pressure is 650 torr?
    a. 65°C
    b. 75°C
    c. 85°C
    d. 95°C

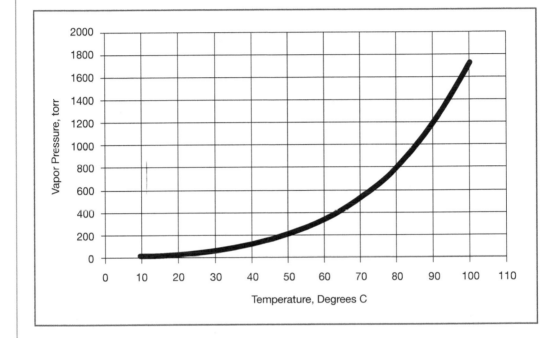

17. The normal boiling point of a liquid is:
    a. The temperature at which a liquid normally boils.
    b. The temperature at which the vapor pressure equals the ambient pressure.
    c. The temperature at which the liquid and vapor phases are in free equilibrium.
    d. None of these.

18. Which of these best explains the cooling effect of evaporation?
    a. The most energetic molecules in a liquid are more likely to escape, thus raising the average kinetic energy of those left behind.
    b. The liquid molecules absorb the heat energy of the escaping gas molecules.
    c. The heat of vaporization is inherently exothermic.
    d. None of these is correct.

19. Which of these are practical applications of the colligative properties of solutions?
    a. Ethylene glycol (EG) is one of the main components of antifreeze. EG has a freezing point of −13°C, whereas a 50:50 mixture of EG and water freezes at −70°F.
    b. Adding sodium dodecylsulfate (SDS, also known as Tide) to a solution of a protein extract causes precipitation of the protein.
    c. A pressure cooker achieves a cooking temperature around 250°F.
    d. Both a and c.

20. Blood cells placed into pure water will:
    a. Swell, because they are hypertonic to pure water.
    b. Shrink, because they are hypertonic to pure water.
    c. Swell, because they are hypotonic to pure water.
    d. Shrink, because they are hypotonic to pure water.

21. Exactly 1 mole of NaCl is placed into 1 L of water. This mixture is then thoroughly mixed.
    a. The molarity of the solution will be equal to 1.
    b. The molarity of the solution will be less than 1.
    c. The molarity of the solution will be greater than 1.
    d. The molarity will equal the molality.

22. Which of these concentration units shows temperature dependence?
    a. Molarity
    b. Molality
    c. Percent (w/v)
    d. Both a and c

23. Surfactant therapy is used to prevent an infant's lungs from collapsing and "sticking together." Which of these is the most reasonable rationale for this effect?
    a. The surfactant establishes a favorable osmotic potential between the lung tissues.
    b. The surfactant cleans the surface of the lungs.
    c. Surfactants reduce both surface tension and adhesive forces in water.
    d. The surfactant lowers the freezing point of the aqueous portion of the lung tissue.

24. If the solubility of oxygen in water is 0.040 g/L under normal pressures, what is the solubility of oxygen in water in an airplane at an altitude of $2\overline{0}00$ feet, where the air pressure is only 350 torr?
    a. 0.018 g/L
    b. 0.22 g/L
    c. 0.087 g/L

25. Which of these is NOT a colligative property of solutions?
    a. Lowering vapor pressure
    b. Increasing boiling point
    c. Lowering surface tension
    d. Increasing osmotic pressure

26. Calculate the missing quantities for a solution of sucrose (MW = 342.34 g/mol).

SOLUTION NUMBER	DENSITY [g/cm³]	% (w/w)	% (w/v)	MOLARITY	MOLALITY
1	1.0021	1.000%			
2	1.0060		2.0120		
3	1.0279			0.2252	
4	1.0381				0.3246
5		20.00%	21.6200		

27. Calculate the missing quantities for these solutions of sodium chloride.

SOLUTION NUMBER	PERCENT BY WEIGHT	DENSITY [g/cm³]	MASS OF NaCl IN 100.0 g SOLUTION	MASS OF NaCl IN 100.0 mL SOLUTION	MASS OF SOLUTION THAT CONTAINS 100.0 g OF NaCl	mL OF SOLUTION THAT CONTAINS 100.0 g OF NaCl
1	1.000%	1.0020				
2	2.500%	1.0078				
3	5.000%	1.0175				
4	10.00%	1.0375				

28. A solution is prepared by dissolving 1.00 g of NaCl in enough water to give a volume of exactly 100 mL of solution. The density of the resulting solution is 1.005 g/mL. What is the concentration of this solution in units of:
    a. %(w/w)
    b. %(w/v)

29. A solution is prepared by dissolving 1.00 g of NaCl in exactly 99 g of water, giving 100 g of solution. The density of the resulting solution is 1.005 g/mL. What is the concentration of this solution in units of:
    a. %(w/w)
    b. %(w/v)

30. A child presents symptoms of lead poisoning. Analysis of the blood shows a lead level of 15 µg/dL. What is the concentration in units of parts per million, or mg of lead/liter of blood. One thousand micrograms equals 1 milligram, and there are 10 dL in 1 liter. If the child has a total blood volume of 3.5 L, what is the total mass of lead in the child, expressed in milligrams?

31. Parts per million with gaseous solutions is defined as $\text{ppm} = \dfrac{\text{mL of gaseous solute}}{\text{cubic meter of gas}}$.
    A tank of nitrous oxide claims a purity of 99.5% but also contains a carbon monoxide impurity at a level of 5.0 ppm. If the cylinder can deliver a total volume of 900 L of gas at room temperature and a pressure of 1.00 atm, how many mL of carbon monoxide will be delivered from the tank?

32. How many equivalents of each ion are contained in one mole of that ion?
    a. Na^+
    b. Ca^{2+}
    c. Al^{3+}

33. The solubility of sodium chloride in water is 359 g/L. How many grams of NaCl will dissolve in 1.50 L of water?

34. The solubility of phenobarbital in water is 1 g/L. What volume of solution will contain 250 mg of phenobarbital?

35. Ammonium chloride is used in instant cold packs. When this compound dissolves in water, the solution becomes colder. What are the relative magnitudes of the lattice energy and the solvation energy for ammonium chloride?

36. Sodium acetate is used in "hand-warmer" packs. These plastic bags contain supersaturated sodium acetate and a metal disc. When you click the metal disc, sodium acetate trihydrate precipitates from solution, and the solution gets warm.
    a. Is this an endothermic or an exothermic process?
    b. How does the absolute value of the solvation energy compare to the lattice energy?
    c. Do you think dissolving solid sodium acetate trihydrate in water will be endothermic or exothermic?

37. Pure diazepam (Valium™) is a solid compound that is not very soluble in water. Diazepam is not very polar and water is very polar. How do you think the solubility will change if:
    a. We increase the temperature?
    b. We increase the pressure?
    c. We change solvents to a less polar solvent, such as ethanol?

38. Nitrous oxide is not very soluble in water, because nitrous oxide is not very polar, and water is very polar. How do you think the solubility will change if:
    a. We increase the temperature?
    b. We increase the pressure?
    c. We change solvents to a less polar solvent, such as ethanol?

39. The Henry's Law constant for oxygen in water is 5.5×10^5 g/L/torr. What is the solubility of oxygen in water when:
    a. $P_{O2} = 150$ torr?
    b. $P_{O2} = 750$ torr?

40. The Henry's Law constant for carbon monoxide is 0.057 g/L/atm. What is the solubility of carbon monoxide in water if:
    a. The partial pressure of carbon monoxide is 25 torr?
    b. The total pressure of the gas is 760 torr and the mole fraction of carbon monoxide is 0.15?

41. The cryoscopic constant for water is 1.85 degrees/molal. Calculate the freezing point of water that contains:
    a. 2.0 m glucose
    b. 2.0 m NaCl (remember that NaCl dissolves to give Na^+ ions and Cl^- ions, so the total solute concentration is 4.0 m)
    c. 1.0 m $CaCl_2$

42. What is the freezing point of a solution that is prepared by mixing 18 g of glucose (MW = 180 g/mol) in 1 kg of water?

43. What molal concentration is needed to lower the freezing point of water to –10°C?

44. The ebullioscopic constant for water is 0.512 degrees/molal. Assuming ideal behavior, what is the boiling point of an aqueous solution of:
    a. 4.0 m glucose
    b. 4.0 m NaCl
    c. 4.0 m $CaCl_2$

45. What osmotic pressure would an aqueous solution of each of these exert against pure water at 298K?
    a. 1.0 M glucose
    b. 1.0 M NaCl
    c. 1.0 M $CaCl_2$

46. The vapor pressure of pure water is 24.0 torr. What will the vapor pressure be if 90 g (0.50 mol) of sugar are dissolved in 180 mL (10 mol) of water?

47. The vapor pressure of ethanol at its normal boiling point is 760 torr. At the same temperature, the vapor pressure of water is 355 torr. If 18 g of water (1.0 mol) is mixed with 46 g of ethanol (1.0 mol):
    a. What is the mol fraction of ethanol?
    b. What is the partial pressure of ethanol in this solution (assume ideal behavior)?
    c. What is the partial pressure of water in this solution (assume ideal behavior)?
    d. What will the total vapor pressure of this mixture be at the normal boiling point of ethanol?
    e. Will this mixture boil as a temperature above or below the boiling point of pure ethanol?

Acids, Bases, and Buffers

You probably studied acid–base chemistry as part of your undergraduate studies. However, acids and bases play such a key role in medicine and physiology that we believe the subject merits another look. The central theme in acid–base chemistry is relatively simple. Acids donate hydrogen ions to bases. The uses of this reaction are myriad in variation and in application. By controlling the acid–base conditions, you can ensure that a medication stays in solution. If the acid content of blood changes by a tiny amount, the patient dies. The acidic and/or basic properties of amino acids located in the active site of an enzyme catalyze a staggering number of chemical transformations which are essential for life.

So, let's proceed with a review of the basics. We'll start with another look at equilibrium, and then the structure and nomenclature of some common acids and bases. We'll next take a look at some basic pH calculations. Then, however, we'll delve into an area that you might not have seen before: ionization equilibria of weak acids and bases. Finally, we'll explore the preparation and properties of buffers.

CHEMICAL EQUILIBRIA

You can't get very far into acid–base chemistry before you run into acid–base equilibria. So, let's get started with acids and bases by first revisiting the concept of dynamic equilibria. As we said in Chapter 7, a dynamic equilibrium exists in a system comprised of (at least) two states when the populations of the two states are constant, even though the members of the system are constantly changing from one state to another. We illustrated this principle with vapor pressure. Now let's consider some chemical examples. Most chemical reactions are reversible. Starting materials combine to give products, and products break back down into starting materials (also called reactants).

$$\text{starting materials} \rightleftharpoons \text{products}$$

These two competing processes occur simultaneously. The double arrows ($\rightleftharpoons$) indicate a reaction is a reversible, equilibrating process.

When the system reaches a state of equilibrium, the forward and reverse reactions continue to take place, but the concentrations of all the products and starting materials remain constant.

Le Châtelier's Principle

Le Châtelier's Principle states that equilibrium is a good thing, and nature strives to attain and/or maintain a state of equilibrium. In physiology, you called this *homeostasis*. If a cell gets away from a balance point, it has to get back to the balance point, or it dies. Let's imagine some chemical reaction that is in a state of equilibrium. The ratio of products to starting materials is some constant value. Now, let's imagine messing with the system by adding and/or removing one or more of the reactants and/or products. This system will always react to undo the change we caused. The system will always strive to reestablish the equilibrium condition.

Changing Concentration

If you *add* products, the equilibrium will shift toward reactants. If you *remove* products, the equilibrium will shift toward products. The system will adjust to counteract whatever change you make.

You all know that hemoglobin in red blood cells carries oxygen from the lungs to the rest of the body. Hemoglobin is a tetrameric protein with a molecular mass of around 68,000 g/mol. Hemoglobin is a quaternary aggregate of four polypeptide (i.e., protein) chains. Each polypeptide carries an organic heme group, which is a cyclic structure that carries a Fe^{2+} ion. Each iron (II) ion can carry one oxygen molecule, so each hemoglobin molecule can carry four oxygen molecules. Since the hemoglobin is such a huge molecule, let's use Hb to stand for hemoglobin. Hb reacts with oxygen in the lungs according to this equation:

$$Hb_{(aq)} + 4\ O_{2\ (g)} \rightleftharpoons Hb(O_2)_{4\ (aq)}$$

Note the "aq" in the above equation means that a particular species is in aqueous solution while "g" signifies a gas. It is a very good thing (for us, anyhow) that this reaction is an equilibrating process and that equilibrating processes respond to Le Châtelier's Principle. In the lungs the concentration of oxygen is high. The system perceives this increased oxygen concentration as adding a starting material. Therefore, the equilibrium shifts toward products (oxyhemoglobin, $Hb(O_2)_4$), trying to undo the increase in oxygen. In cellular tissue, the concentration of oxygen is low, and the system perceives this as removing a reactant. Therefore, the equilibrium shifts toward starting materials, trying to replace the missing reactant, oxygen. So it is in accordance with Le Châtelier's Principle that hemoglobin loads up on oxygen in the lungs and dumps oxygen into the cells.

Changing Temperature

Exothermic reactions evolve energy from the system, and endothermic reactions absorb energy into the system from the surroundings. Therefore, you can think of heat as a product in exothermic reactions and as a reactant in endothermic reactions. Therefore, increasing temperature favors endothermic processes.

Changing Volume and Pressure

Changing volume and/or pressure significantly impacts equilibrium reactions only when at least one of the reactants or products is a gas, because solids and liquids aren't compressible. Decreasing volume increases pressure. Therefore, decreasing the volume or increasing the pressure favors the smaller number of gas particles. Think of squeezing many gas particles into fewer gas particles, which offsets the increase in pressure.

Back to our hemoglobin example. When the partial pressure of oxygen is increased, the equilibrium shifts to the right. This is why giving a patient pure oxygen results in greater oxygen saturation in blood.

$$Hb_{(aq)} + 4\ O_{2\ (g)} \rightleftharpoons Hb(O_2)_{4\ (aq)}$$

The Equilibrium Constant

A system is in a *state of equilibrium* when there is a balance between reactants and products. This balance is defined by thermodynamic parameters, namely, bond strengths and the intermolecular forces between all the molecules in the system. The equilibrium constant (K) is the numerical description of that balance. K is equal to the product of all of the molar concentrations of the products, each raised to the power of their stoichiometric coefficients, divided by the product of the molar concentrations of the reactants, each raised to the power of their stoichiometric coefficients. This sounds a lot worse than it is. For example, with this model reaction:

$$a\,A + b\,B \rightleftharpoons c\,C + d\,D$$

The equilibrium constant is given by:

$$K = \frac{[D]^d [C]^c}{[A]^a [B]^b}$$

The units on K depend on the values of the coefficients, so the units on K are meaningless. Therefore we will ignore units on equilibrium constants.[1]

Subscripts

The equilibrium constant (K) is often appended with a subscript. The subscript denotes the type of equilibrium reaction.

K_{eq} is a generic equilibrium constant.
K_a is an equilibrium constant governing the ionization of weak acids.

---

[1] Actually, K is a ratio of activities, not molarities, and activities are dimensionless. Therefore, K is technically dimensionless as well. So what we said above is not strictly true.

K_b is an equilibrium constant governing the ionization of weak bases.

K_{sp} is an equilibrium constant governing the solubility of sparingly soluble compounds.

Meaning of K

As K increases, the reaction tends to increasingly favor products. That is, the *forward* reaction becomes more favorable. As K decreases, the reaction tends to increasingly favor starting materials, and the *reverse* reaction becomes more favorable.

For example, K for the reaction below is 56.

$$H_2 + I_2 \rightleftharpoons 2\ HI$$

If equal amounts of hydrogen and iodine are placed into a reaction vessel and allowed to come to equilibrium, will there be more HI or H_2?

Well, K for this reaction is given by:

$$K = \frac{[HI]^2}{[H_2][I_2]} = 56$$

Since $K > 1$, we know the reaction is product-favored. So at equilibrium, there will be more HI.

Carbon monoxide kills by tying up the hemoglobin so that not enough oxygen can be delivered to the body. The equilibrium constant for hemoglobin and carbon monoxide is about 200 times larger than the equilibrium constant for hemoglobin and oxygen. Therefore, when carbon monoxide is present, a greater percentage of the hemoglobin is in the carboxy form, because the equilibrium position is further to the product side. This means there is less free hemoglobin available to bind oxygen.

$$Hb_{(aq)} + 4\ CO_{(g)} \rightleftharpoons Hb(CO)_{4\ (aq)}$$

Solids and Liquids

Pure solids or liquids comprise a different phase from where reactions in aqueous media occur. Furthermore, the concentration of pure solids and liquids is a constant value (or very nearly so), and these constant values for concentration are included with the equilibrium constant value. Water sometimes appears as a reactant or product in reactions occurring in aqueous media. Because the concentration of water is so much greater than the concentrations of any of the reacting species, the concentration of water is also essentially constant. Therefore, the concentrations of solids, liquids, and water (as a solvent) do not appear explicitly in the equilibrium constant expression.

For example, let's write the equilibrium constant expression for the basic ionization of ammonia in water. The equation for the chemical equilibrium reaction is:

$$NH_3 + H_2O \rightleftharpoons NH_4^+ + OH^-$$

The equilibrium constant expression is:

$$K = \frac{[NH_4^+][OH^-]}{[NH_3][H_2O]}$$

This expression can be rearranged to get:

$$K[H_2O] = \frac{[NH_4^+][OH^-]}{[NH_3]}$$

Finally, we can define the new equilibrium constant K_b to be:

$$K_b = K[H_2O] = \frac{[NH_4^+][OH^-]}{[NH_3]}$$

Most texts will write this expression as:

$$K_b = \frac{[NH_4^+][OH^-]}{[NH_3]}$$

Notice the water does not appear in the final equilibrium constant expression, because the concentration of water remains virtually constant in this reaction and is included in the equilibrium constant K_b.

Reversing a Reaction

Since equilibrating reactions are, by definition, reversible, what happens to K when the chemical equation is reversed? Let's consider a model reaction:

$$a\,A + b\,B \rightleftharpoons c\,C + d\,D$$

The equilibrium constant is given by:

$$K_{forward} = \frac{[D]^d[C]^c}{[A]^a[B]^b}$$

Now, let's write the equation for the reaction in reverse.

$$c\,C + d\,D \rightleftharpoons a\,A + b\,B$$

The equilibrium constant for the reverse reaction is:

$$K_{reverse} = \frac{[A]^a[B]^b}{[D]^d[C]^c}$$

So, we see that if we reverse the equation for a chemical reaction, $K_{forward}$ is the reciprocal of $K_{reverse}$.

$$K_{forward} = \frac{1}{K_{reverse}}$$

ACIDS AND BASES

Definition of Acids and Bases

While there are many definitions of acids and bases, probably the Arrhenius definition is the most operational one.

- An acid is a species that increases the hydronium ion (H_3O^+) concentration in an aqueous solution.
- A base is a species that increases the hydroxide ion (OH^-) concentration in an aqueous solution.

However, the Brønsted[2] definition is the most generally useful definition of acids and bases.

- An acid is a species that donates a hydrogen ion (H^+) to a base.
- A base is a species that accepts a hydrogen ion (H^+) from an acid.

Conjugate Acid–Base Pairs

One of the most useful features of the Brønsted system is that is takes into account what happens *after* an acid–base reaction occurs. You all know HCl is an acid, so let's write an equation for it behaving as an acid. That is, let's allow HCl to give away its hydrogen ion.

$$HCl \rightarrow H^+ + Cl^-$$

A few comments are in order here. First, what is an H^+ ion? Well, it's just a proton, and for that reason, we use those terms interchangeably. Second, the chloride ion has a special relationship with the HCl. Imagine the reaction running in reverse. If the chloride ion just picked up that hydrogen ion, it would be back as HCl. But wait. If a chloride ion took a hydrogen ion, that would mean it was acting *as a base*. So, Cl^- is the conjugate base of HCl. Conversely, HCl is the conjugate acid of Cl^-. They form a conjugate acid–base pair.

It is easy to recognize conjugate acid–base pairs. They always differ by exactly one hydrogen ion. The charge on the conjugate acid is always one greater than the charge on its conjugate base. Table 9.1 gives some conjugate acid–base pairs.

When an acid gives away a proton, it always gives it to a base, because a proton is a much too concentrated charge to float around unnoticed and alone in a solution. So, when an acid gives away its hydrogen ion to a base, the acid is converted into its *conjugate base*. Likewise, when a base accepts a proton from an acid, the base is converted into its *conjugate acid*. In generic form, we can express this process as:

$$HA + B \rightarrow A^- + BH^+$$

---

[2] Many older texts refer to this as the Lowry-Brønsted definition. Apparently, though, Lowry has gone the way of Montgomery and Roebuck.

Table 9.1 CONJUGATE ACIDS AND BASES

ACID	CONJUGATE BASE		BASE	CONJUGATE ACID
HNO_3	NO_3^-		NH_3	NH_4^+
H_2SO_4	HSO_4^-		H_2O	H_3O^+
HSO_4^-	SO_4^{2-}		OH^-	H_2O
$HC_2H_3O_2$	$C_2H_3O_2^-$		CO_3^{2-}	HCO_3^-
H_2O	OH^-		HCO_3^-	H_2CO_3

If you, like most people, prefer a concrete example, let's take a look at what really happens when HCl dissolves in water (see Figure 9.1). HCl is an acid and wants to give away its H^+ ion. There is nothing else in the solution, so the only available base is H_2O. Therefore, water in this case behaves as a base and accepts the proton from HCl. In this process, water is converted to its conjugate acid, H_3O^+, and HCl is converted into its conjugate base, Cl^-.

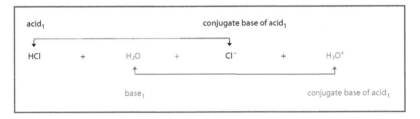

Figure 9.1 Conjugate Acid–Base Pairs (HCl and H_2O)

So, in water, we don't have H^+ floating around; we really have the hydronium ion (H_3O^+). If you like, you can think of a hydronium ion as a proton that has acquired a water of hydration. However, the hydronium ion is not a unique species and can pick up additional waters of hydration to form, for example, $H_5O_2^+$. So, while H_3O^+ is more correct than writing H^+, it is not perfectly correct. Furthermore, we feel that sometimes the chemistry is clarified by simply considering H^+ to be the species in solution. You can take your pick, because most authors tend to use H^+ and H_3O^+ interchangeably. We will most often use H^+, however.

What about a base? Let's consider dissolving sodium bicarbonate in water (see Figure 9.2). Well, the bicarbonate ion is a weak base (more on weak bases in a moment). Bases accept

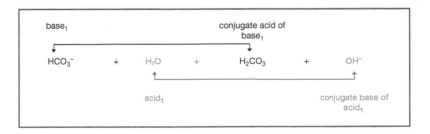

Figure 9.2 Conjugate Acid–Base Pairs (HCO_3^- and H_2O)

hydrogen ions from acids. Again, the only *reacting* partner in this system is water. So, here, water acts as an acid. When the bicarbonate ion takes a hydrogen ion from water, the water is converted into its conjugate base (OH^-), and the bicarbonate ion is converted into its conjugate acid (H_2CO_3).

Amphiprotic Species

An amphiprotic species can behave as either an acid or a base. In the preceding examples, we saw that water is an amphiprotic species. It reacts as a base in the presence of HCl, but it reacts as an acid in the presence of bicarbonate.

Monohydrogenphosphate is an important amphiprotic species that helps maintain the pH balance in blood. That is, monohydrogenphosphate can act as either an acid or a base.

Here, monohydrogenphosphate acts as a base by accepting a hydrogen ion and is converted into its conjugate acid, dihydrogenphosphate.

$$HPO_4^{2-} + H^+ \rightarrow H_2PO_4^-$$

Here, monohydrogenphosphate acts as an acid by donating a hydrogen ion and is converted into its conjugate base, phosphate.

$$HPO_4^{2-} \rightarrow H^+ + PO_4^{3-}$$

Strong Acids

Now, let's explore some of the common acids and bases. Acids donate protons; that's their job. Stronger acids are really determined to foist their proton off onto some base. When a strong acid dissolves in water, it is essentially 100% ionized. That means *essentially* all of the molecular species dissociate into ions. The reaction is *not* an equilibrating process, so all of the starting materials are converted into products. For example, when HCl dissolves in water, the following reaction occurs:

$$HCl + H_2O \rightarrow H_3O^+ + Cl^-$$

So, if 1 mole of HCl dissolves in enough water to give 10.0 L of solution, the molar concentration of [HCl] is essentially zero, while the molar concentrations of [H_3O^+] and [Cl^-] are both 0.10 M. By the way, the concentrations of species in acid–base chemistry are always expressed as molarity. So when you see square brackets—[ ]—around a species, it means "the molar concentration of that species." A bottle of this reagent would likely be labeled as 0.10 M HCl, but this is not strictly true. It is more correct to say the *formal concentration* of HCl is 0.10 M.

Strong acids are relatively rare. Most acids are weak acids and only partially ionize in water. We'll delve into weak acids in greater detail in just a moment. Some common strong acids include:

H_2SO_4	HI	HBr	HCl	HNO_3
Sulfuric acid	Hydroiodic acid	Hydrobromic acid	Hydrochloric acid	Nitric acid

Strong Bases

Bases accept (or take) hydrogen ions. In water, the strongest possible base is the hydroxide ion, OH^-. A strong base ionizes essentially 100% to produce the OH^- ion, so a strong base is a soluble ionic hydroxide. Like all soluble ionic compounds, ionic hydroxides are strong electrolytes and dissociate completely (more or less) into ions. For example, NaOH dissolves in water to give a solution of sodium ions and hydroxide ions.

$$NaOH \rightarrow Na^+ + OH^-$$

So, if 1 mole of NaOH is dissolved in enough water to give 10.0 L of solution, the formal concentration of NaOH is 0.1 M, [NaOH] is 0 M, but [Na^+] and [OH^-] are both 0.10 M.

Not many ionic hydroxides are soluble except for the hydroxides of the Group 1A and Group 2A metals (except for magnesium—Mg(OH)$_2$ is insoluble). So, the commonly encountered strong bases include the following:

LiOH	NaOH	KOH	$Ba(OH)_2$
Lithium hydroxide	Sodium hydroxide	Potassium hydroxide	Barium hydroxide

Weak Acids

Weak acids are able to donate hydrogen ions to bases, but they are less determined to do so than strong acids. So when a weak acid dissolves in water, it establishes a dynamic equilibrium between the molecular form of the acid and the ionized form. For example, acetic acid is a very common weak acid. In water, acetic acid establishes the following equilibrium:

$$HC_2H_3O_2 \rightleftharpoons H^+ + C_2H_3O_2^-$$

If 1 mole of acetic acid is dissolved in enough water to give 10.0 L of solution, 98.7% of the acetic acid remains in the molecular form. That is, at equilibrium, [$HC_2H_3O_2$] = 0.0987 M, while [H^+] and [$C_2H_3O_2^-$] are both 0.0013 M.

Some common weak acids are listed below. Notice that ions as well as molecular compounds can behave as acids.

$HC_2H_3O_2$	H_2CO_3	H_3PO_4	NH_4^+
Acetic acid	Carbonic acid	Phosphoric acid	Ammonium ion

Weak Bases

Weak bases are able to accept hydrogen ions from acids, but they are less determined to do so than strong bases. Weak bases do not completely ionize in water to produce an equivalent concentration of the hydroxide ion, because when a weak base dissolves in water, it establishes a dynamic equilibrium between the molecular form and the ionized form. For example, ammonia is a very common weak base. In water, ammonia establishes the following equilibrium:

$$NH_3 + H_2O \rightleftarrows NH_4^+ + OH^-$$

If 1 mole of ammonia is dissolved in enough water to give 10.0 L of solution, 98.7% of the ammonia remains in the molecular form, NH_3. That is, at equilibrium, $[NH_3] = 0.0987$ M, while $[NH_4^+]$ and $[OH^-]$ are both 0.0013 M.

You might have noticed the numbers for ammonia and acetic acid are the same, but don't be misled to believe these numbers are universal for all weak acids and bases. It is entirely coincidental that the numbers for acetic acid and ammonia are the same. The degree of ionization is dependent on the inherent acid or base strength of the species. This is a topic we will consider shortly.

Some common weak bases are given below. Notice that ions as well as molecular compounds can behave as bases.

NH_3	HCO_3^-	CO_3^{2-}	HPO_4^{2-}
Ammonia	Bicarbonate ion	Carbonate ion	Monohydrogenphosphate ion

Polyprotic Acids

A diprotic acid has two hydrogen ions to donate. So a diprotic acid such as carbonic acid can behave as an acid twice:

$$H_2CO_3 \rightleftarrows H^+ + HCO_3^-$$

$$HCO_3^- \rightleftarrows H^+ + CO_3^{2-}$$

A triprotic acid has three hydrogen ions to donate. For example, phosphoric acid (H_3PO_4) is a triprotic acid.

The number of acidic protons is not necessarily the number of hydrogens in the molecular formula. *Generally* (but not always), acidic hydrogens are listed first in the molecular formula. For example, acetic acid ($HC_2H_3O_2$) is a monoprotic acid (Figure 9.3).

The reason for this dichotomy of hydrogens lies in the molecular structure of acetic acid. The acidic hydrogen is bonded to a highly electronegative oxygen atom. The O–H bond is polarized toward the oxygen to a point that a base can snatch away the hydrogen as an H^+ ion from the acetic acid molecule. The other three hydrogens are bonded to a carbon atom, and carbon and hydrogen have almost identical electronegativities. Therefore, those bonds are nearly nonpolar, and the hydrogens have no tendencies to be removed as H^+ ions.

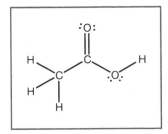

Figure 9.3 Lewis Structure of Acetic Acid

Acid and Base Strength: K_a and K_b

Some acids are stronger than others. A stronger acid is more determined to give its proton to some base, and a stronger base is more determined to take a proton from some acid. There are some guidelines for predicting acid and base strength, and interested readers are encouraged to read more on this topic in any general or inorganic chemistry textbook.

First, let's divide acids into two categories: strong and weak. The strong acids are a much smaller group and include H_2SO_4, HI, HBr, HCl, and HNO_3. When a strong acid dissolves in water, it dissociates completely. For example, nitric acid dissolves in water to give a hydrogen ion and a nitrate ion. This reaction goes to completion and is not an equilibrating process.

$$HNO_3 \rightarrow H^+ + NO_3^-$$

All other acids (that we'll mention) are weak acids. The vast majority of acids fall into this category. However, weak acids show a range of acid strength. The relative strength of a weak acid is quantified by the equilibrium constant K_a governing the ionization of the weak acid.

Let's write an equation for the ionization of acetic acid and the corresponding equilibrium constant. To write an equation for the ionization of the acid, just remember that acids give away hydrogen ions.

$$HC_2H_3O_2 \rightleftarrows H^+ + C_2H_3O_2^-$$

The equilibrium constant for this reaction is:

$$K_a = \frac{[H^+][C_2H_3O_2^-]}{[HC_2H_3O_2]} = 1.8 \times 10^{-5}$$

If you're panicking at this point over memorizing numerical values for equilibrium constants, stop it. No one remembers all of these numbers. They are available in tables from many reference sources. Do not waste your time memorizing these numbers.

A larger value of K_a means a stronger acid. On the following list, phosphoric acid is the strongest acid, while the ammonium ion is the weakest.

ACID	$HC_2H_3O_2$	H_3PO_4	$H_2PO_4^-$	NH_4^+
K_a	1.8×10^{-5}	7.1×10^{-3}	6.3×10^{-8}	5.6×10^{-10}

Phosphoric acid merits a closer look. The K_a of H_3PO_4 is 7.1×10^{-3}, while the K_a of dihydrogenphosphate is 6.3×10^{-8}. You probably noticed that dihydrogenphosphate is the conjugate base of the triprotic phosphoric acid. Additionally, ionization of dihydrogenphosphate represents the loss of the second proton from phosphoric acid. For polyprotic acids, K_a values indicate acid strength for donating just one hydrogen ion, not all of them.

$$H_3PO_4 \rightleftharpoons H^+ + H_2PO_4^- \quad K_a = 7.1 \times 10^{-3}$$

$$H_2PO_4^- \rightleftharpoons H^+ + HPO_4^{2-} \quad K_a = 6.3 \times 10^{-8}$$

The first equation shows phosphoric acid donating its first proton, while the next equation shows phosphoric acid donating its *second proton*. Notice that phosphoric acid is a much stronger acid than dihydrogenphosphate, by a factor of $7.1 \times 10^{-3}/5.6 \times 10^{-8} = 100,000$. This is true in general. For polyprotic acids, the first proton is donated much more readily than each successive proton. This makes sense, of course, because opposite charges attract each other. It is easier to drag an H^+ ion away from $H_2PO_4^-$, which has a smaller negative charge than the doubly negative HPO_4^{2-}.

When dealing with polyprotic acids, the exact solution for pH calculations quickly becomes very daunting. However, Le Châtelier's Principle can come to our rescue. The first ionization is much more complete than the second because the first K_a is larger than the second. The hydrogen ion produced by the first ionization is also a product of the second ionization. Thus, the increase in hydrogen ion concentration pushes the second equilibrium back to the left, suppressing the acid ionization of $H_2PO_4^-$. Therefore, with polyprotic acids, we can usually just consider the first acid ionization and ignore the subsequent acid ionizations.

$H_3PO_4 \rightleftharpoons H^+ + H_2PO_4^- \quad K_a = 7.1 \times 10^{-3}$: more complete; dominates pH

$H_2PO_4^- \rightleftharpoons H^+ + HPO_4^{2-} \quad K_a = 6.3 \times 10^{-8}$: suppressed by first step; less important in pH

Weak bases also show a range in base strength. The strength of a base is quantified by K_b.

BASE	NH_3	HCO_3^-	CO_3^{2-}	HPO_4^{2-}
K_b	1.8×10^{-5}	2.3×10^{-8}	1.8×10^{-4}	1.6×10^{-7}

Let's write an equation for the carbonate ion behaving as a base. To write this equation, remember that bases take hydrogen ions from water, giving hydroxide ions and the conjugate acid of the base.

$$CO_3^{2-} + H_2O \rightleftharpoons HCO_3^- + OH^-$$

The equilibrium constant expression is:

$$K_b = \frac{[OH^-][HCO_3^-]}{[CO_3^{2-}]} = 1.80 \times 10^{-4}$$

Acid/Base Strength of Conjugate Acid–Base Pairs

Now we are ready to examine another great rule of thumb in chemistry: *The stronger the acid, the weaker its conjugate base.* For example, let's take a close look at carbonic acid.

ACID	K_a	CONJUGATE BASE	K_b
H_2CO_3	4.3×10^{-7}	HCO_3^-	2.3×10^{-8}
HCO_3^-	5.6×10^{-11}	CO_3^{2-}	1.8×10^{-4}

Carbonic acid is a stronger acid than bicarbonate. In fact, bicarbonate is such a weak acid that it actually is a base in water. Therefore, carbonate is a stronger base than bicarbonate, because it is derived from the weaker conjugate acid.

$$H_2CO_3 \rightleftharpoons H^+ + HCO_3^-$$

$$HCO_3^- \rightleftharpoons H^+ + CO_3^{2-}$$

The inverse of this rule is also true. The stronger the base, the weaker the conjugate acid. So, for example, ammonia is a stronger base than the acetate ion.

$$NH_3 + H_2O \rightleftharpoons NH_4^+ + OH^- \quad K_b = 1.8 \times 10^{-5}$$

$$C_2H_3O_2^- + H_2O \rightleftharpoons HC_2H_3O_2 + OH^- \quad K_b = 5.8 \times 10^{-10}$$

Since K_b for ammonia is larger than K_b for the acetate ion, ammonia is a stronger base. Therefore, acetic acid ($HC_2H_3O_2$) is a stronger acid than the ammonium ion (NH_4^+).
Here are some general guidelines.

1. The conjugate base of a really strong acid has no base strength. HCl is a really strong acid. That means that its conjugate base, the chloride ion, has no base strength. A solution containing NaCl is neutral in terms of acid/base level.
2. The conjugate base of a weak acid has base strength. Acetic acid is a weak acid. Therefore, the acetate ion has base strength. You probably noticed the stumbling language here. We didn't say, "Since acetic acid is a weak acid, the acetate ion is a strong base," because that isn't true. However, a solution of sodium acetate will be basic.
3. The conjugate acid of a weak base has acid strength. Ammonia is a weak base, so the ammonium ion behaves as an acid in water. Therefore, a solution of ammonium chloride will be acidic.

ACID–BASE REACTIONS

Acid–base reactions involve transfer of a hydrogen ion from the acid to the base. In order to predict the products of an acid–base reaction, you first have to identify which is the acid and which is the base. Then, you just move an H^+ ion from the acid to the base. This converts the acid into its conjugate base and the base into its conjugate acid. So, any acid–base reaction has two acids and two bases. There is one acid and one base on the reactant side, and the conjugate acid and the conjugate base on the product side.

If you don't recognize which is the acid and which is the base, the base almost always has a lower (more negative) charge than the acid. Also, hydrogen is almost always the first atom listed in the formula of an acid.

The reaction equilibrium always favors the formation of the weaker acid. So, you need to look up the K_a's for the reacting acid and the conjugate acid of the reacting base.

For example, let's consider the reaction of the bicarbonate ion with the dihydrogenphosphate ion. This is about the hardest example, because it is not immediately obvious which is the acid and which is the base. Both ions have a −1 charge, and both start with H. So, let's consider both possibilities.

First, let's let bicarbonate be the acid, so we need to move an H^+ from bicarbonate to dihydrogenphosphate, as shown in Figure 9.4. We see the product acid, phosphoric acid, is a stronger acid than bicarbonate. Therefore, this equilibrium reaction will favor the starting materials.

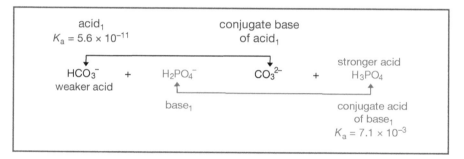

Figure 9.4 Predicting Acid–Base Reactions

Well, let's try letting bicarbonate be a base and take a hydrogen ion from dihydrogenphosphate (see Figure 9.5).

Carbonic acid is a stronger acid than dihydrogenphosphate. So, this reaction favors reactants as well.

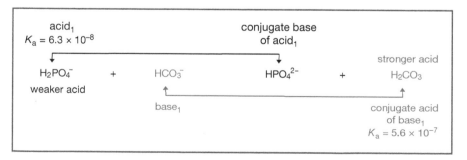

Figure 9.5 Predicting Acid–Base Reactions

We see that dihydrogenphosphate will not react with bicarbonate in an acid–base reaction. This also means they are compatible with each other in solution (e.g., blood). While these ions serve other purposes (e.g., dihydrogen phosphate is the *inorganic phosphate* in metabolism), both of these ions have acid–base properties and they assist us in maintaining a constant pH.

MEASURING ACIDITY: THE pH FUNCTION

Video 9.1

You are doubtless familiar with the pH scale. On this scale, solutions with a pH less than 7 are acidic, and solutions greater than 7 are basic or alkaline. Solutions with a pH of exactly 7 are pH neutral, neither acidic nor basic. You know certain bodily fluids have characteristic pHs. Blood has a pH of 7.4. Stomach juice has a pH of around 1. What does the little "p" mean? Why is 7 the midpoint? How do we calculate pH?

The p-Function

The *p* in pH is a mathematical operator. We have been dealing with several operators, including addition, subtraction, and square roots. The p-function operator isn't that different, but it looks a little strange. The p-function operator means *the negative logarithm of*. The *H* in pH means hydrogen ion concentration. So we have a definition of pH:

$$pH = -\log [H^+]$$

Chemists use the p-function operator to express the concentrations of many ions. pOH for hydroxide ion, pCa for Ca^{2+}, and so forth. The meaning of the p is the same in every case—take the logarithm of the concentration and then change the algebraic sign.

Now, there are very practical reasons for defining the p-function. First, the hydrogen ion concentration varies from 10^0 M to 10^{-14} M in solutions commonly encountered in the laboratory. That is 14 orders of magnitude—a factor of a hundred million million. Imagine what a pH meter would look like without the logarithm part of the p-function definition. The scale would have a hundred million million divisions. A logarithm function is a way to map a vast range of values onto a much smaller set of values. In this case, the logarithm function maps a range from 0 to 10^{-14} onto a range from 1 to 14. The negative sign in the definition is one of convenience. Nobody likes working with negative numbers. The negative sign simply ensures that the pH values will be positive numbers.

To apply the p-function, you simply take the logarithm of the $[H^+]$ concentration and then change the sign. For example, what is the pH of a solution when the $[H^+]$ is 1.0×10^{-3} M?

$$pH = -\log (1.0 \times 10^{-3} \text{ M}) = 3.00$$

A couple of comments are in order here. First, did you notice the value of the pH is the same as the absolute value of the exponent? This will always be true when the first part of the scientific notation is exactly 1. The second comment relates to significant figures. There are two significant figures in the molarity measurement of 1.0×10^{-3} M. There are also two significant figures in the pH value of 3.00. Finally, pH values have no intrinsic units. Logarithms represent "pure numbers," and as such, have no units.

Wait a minute. As a normal measurement, 3.00 has three significant figures. However, as a logarithmic measurement, it has only two. This is because the digits to the left of the decimal place (those greater or equal to 1) are called the *characteristic* and tell you where the decimal place in the molarity measurement was. The digits to the right of the decimal place (those less than 1) are called the *mantissa* and represent the significant figures in the measurement. So, with pH measurements, only the decimal places represent significant figures.

Now, let's try one more. What is the pH of a solution whose $[H^+]$ concentration equals 5.0 $\times 10^{-3}$ M?

$$pH = -\log (5.0 \times 10^{-3} \text{ M}) = 2.30$$

Another comment is in order. This second example had a solution in which the hydrogen ion concentration was five times greater than the first solution, but the pH only changed by 0.70 pH units. Logarithms map a huge range of values onto a much smaller range. Each change of 1 pH unit means the hydrogen ion concentration is changing by a factor of 10. This means that small changes in pH correspond to much larger changes in acidity level. That is why small changes in blood pH have such devastating effects on a patient's well-being.

Removing the p-function is accomplished by changing the sign of pH and then taking the antilog. On your calculator the antilog button is the "10^x" key. So that you aren't forced to just memorize another equation, let's do this stepwise. First we change the signs on both sides of the equation that defines pH.

$$pH = -\log [H^+]$$

$$-pH = \log [H^+]$$

Then we take the antilog of both sides of the equation.

$$\text{antilog } (-pH) = \text{antilog } (\log [H^+])$$

The log and antilog operators are inverses of each other, just like the x^2 operator and the square root operator. Therefore, the operators on the right side of the equation cancel out.

$$\text{antilog } (-pH) = [H^+]$$

or

$$[H^+] = 10^{-pH}$$

So, what is the hydrogen ion concentration in a solution whose pH is 2.30?

$$[H^+] = 10^{-2.30} = 5.0 \times 10^{-3} \text{ M}$$

Depending on the syntax of your calculator, you either enter –2.30 first and then push the anti-log key, or the other way around. Take some time to play with your calculator to make sure you can do these calculations. By the way, we are sure you noticed we used the same numerical values in these examples. There is a reason beyond laziness for this. It illustrates using a model system to test an unknown mathematical operation. If you apply the p-function to a number, and then remove the p-function, you should get the same number back, right? So, if you're not positive that your calculator is giving you the correct answers, try applying the p-function to some number and then removing it. If you get the original number back, you are doing the calculations correctly.

Self-Ionization of Water

The acid–base chemistry we will encounter occurs in aqueous media. Water itself has some very weak acid–base properties, and therefore sets some limits on the parameters of the pH scale. A tiny fraction of water molecules dissociates or ionizes into a hydrogen ion and a hydroxide ion:

$$H_2O \rightleftharpoons H^+ + OH^-$$

The equilibrium constant that governs this equilibrium is called K_w. At 25°C, $K_w = 1.0 \times 10^{-14}$. Of course, pK_w is 14.00. Since water is a pure liquid, the concentration of water does not appear in the equilibrium constant.

$$K_w = [H^+][OH^-] = 1.0 \times 10^{-14} \text{ at } 25°C$$

So, let's find $[H^+]$ and $[OH^-]$ in pure water. If we take a look at the equation for the ionization of water, we see that for each water molecule that falls apart, exactly one H^+ and one OH^- are produced. Therefore, *in pure water*, the concentrations of these two ions are equal.

$$[H^+] = [OH^-] \text{ in pure water}$$

Let's substitute the hydrogen ion concentration $[H^+]$ for the hydroxide ion concentration $[OH^-]$ into the K_w expression.

$$K_w = [H^+][H^+]$$

or

$$[H^+]^2 = 1.0 \times 10^{-14}$$

If we take the square root of both sides, we find the hydrogen ion concentration is 1.0×10^{-7} M.

$$[H^+] = \sqrt{1.0 \times 10^{-14}} = 1.0 \times 10^{-7} \text{ M}$$

So, if the $[H^+]$ is 1.0×10^{-7} M, the pH is 7.00.

So, the pH of pure water is 7.00 because that is the negative logarithm of the hydrogen ion concentration that is generated by the self-ionization of water. It is a value based on the inherent acid–base properties of water.

You should now be able to prove to yourself that, in pure water, the $[OH^-]$ concentration is 1.0×10^{-7} M, and the pOH is 7.00.

Relationship Between pH and pOH

Because pH and pOH are derived from the ionization of water, there is a fixed relationship between them. Let's see if we can figure it out. First, let's start with the K_w expression for the ionization of water.

$$K_w = [H^+][OH^-]$$

Now, let's apply the p-function to both sides of the equation.

$$-\log (K_w) = -\log ([H^+][OH^-])$$

Well, the left side is easy enough. The negative log of K_w is pK_w and is equal to 14.00. The right side is more interesting. Most people are rusty with the laws of logarithms. One key law is *the log of a product is equal to the sum of the logs*. This means we can separate the terms on the right side, to give.

$$14.00 = (-\log ([H^+]) + (-\log [OH^-])$$

Now, we see the right side can easily be recast as pH and pOH.

$$14.00 = pH + pOH$$

Therefore, the pH plus the pOH of any aqueous solution (at 25°C) always adds up to 14. So, if you know the pH, just subtract the pH from 14, and you have the pOH.

pK_a and pK_b

The equilibrium constants for acids and bases are commonly presented as p-functions as well. For example, the K_a and pK_a for carbonic and acetic acid are given below.

Uh-oh. The negative sign in the p-operator plays a dirty trick here. Acetic acid is a stronger acid than carbonic acid, and acetic acid has a larger K_a than carbonic acid. However, acetic acid has a smaller pK_a than carbonic acid. So a stronger acid has a larger K_a, but a smaller pK_a. We see the pK_a of acetic acid is 4.74. The conjugate base of acetic acid is the acetate ion. So, what is the pK_b of the acetate ion? The pK_a and pK_b of a conjugate acid–base pair sum to give 14:

$$pK_a + pK_b = 14$$

So, the pK_b of acetate is 9.26.

ACID	K_a	pK_a
Acetic acid	1.8×10^{-5}	4.74
Carbonic acid	4.3×10^{-7}	6.37

CALCULATING THE pH OF SOLUTIONS

The Key Relationships

To solve pH calculation problems, you need to be conversant with six key relationships:

$$pH = -\log [H^+]$$
$$pOH = -\log[OH^-]$$
$$[H^+][OH^-] = K_w = 1.00 \times 10^{-14}$$
$$pH + pOH = pK_w = 14.00$$
$$pK_a + pK_b = 14.00$$
$$K_a K_b = K_w$$

Now, let's see how these are applied in the following examples.

Calculating the pH of a Strong Acid Solution

Strong acids completely ionize in water. This means every mole of strong acid falls apart into an equal number of moles of hydrogen ions. Therefore, $[H^+]$ is equal to the formal concentration of the acid.

So, what is the pH of a 0.15-M solution of HCl? Well, HCl is a strong acid; therefore, if the formal concentration of HCl is 0.15 M, then $[H^+] = 0.15$ M, and the pH is 0.82.

Calculating the pH of a Strong Base Solution

Strong bases completely ionize in water. This means every mole of strong base falls apart into an equal number of moles of hydroxide ions, except for $Ba(OH)_2$ which gives two moles of hydroxide ions per mole. Therefore, $[OH^-]$ is equal to the formal concentration of the base.

So, what is the pH of a 0.15-M solution of NaOH? Well, NaOH is a strong base; therefore, if the formal concentration of NaOH is 0.15 M, then $[OH^-] = 0.15$ M, so we can easily find the pOH.

$$pOH = -\log [OH^-] = -\log [0.15 \text{ M}] = 0.82$$

The sum of pH and pOH is 14, so we find the pH by subtracting pOH from 14.

$$pH = 14 - 0.82 = 13.18$$

You have to be very careful with calculating the pH of a basic solution, because the first number that comes out of your calculator is going to be pOH, not pH. Just remember to ask yourself if the answer is reasonable. If you are dealing with a basic solution, you know the pH has to be greater than 7.

Calculating the pH of a Weak Acid Solution

Calculating the pH of weak acids and bases is more challenging, because they do not ionize completely. So, in a 0.10-M solution of acetic acid, the $[H^+]$ concentration is not 0.10 M. We must use the equilibrium constant expression to find the pH of weak acid solutions. So, what is the pH of a 0.10-M solution of acetic acid? The K_a of acetic acid is 1.8×10^{-5}.

Let's start with the ionization equilibrium.

$$HC_2H_3O_2 \rightleftharpoons H^+ + C_2H_3O_2^-$$

The equilibrium constant for this reaction is:

$$K_a = \frac{[H^+][C_2H_3O_2^-]}{[HC_2H_3O_2]} = 1.8 \times 10^{-5}$$

Now the K_a expression has three concentration terms, or three variables: the concentration of the hydrogen ion, the concentration of the acetate ion, and the concentration of the acetic acid. We can solve only algebraic expressions that have one variable. So, we have to express each concentration in terms of a single variable. It makes sense to us to use $[H^+]$ as a variable, rather than x, because it keeps us focused on our goal: finding the pH. So, let's see if we can express the concentrations of acetic acid and the acetate ion in terms of the hydrogen ion concentration.

First, we see that for every acetic acid that dissociates, we get exactly one hydrogen ion and one acetate ion. Therefore, in a solution that contains only acetic acid, we know that the concentrations of these two ions have to be equal.

$$[H^+] = [C_2H_3O_2^-]$$

Now, we need to deal with the acetic acid concentration. You might be tempted to say the molar concentration of acetic acid is 0.10 M, because that is the value cited in the problem. However, 0.10 M is the *formal concentration*. The formal concentration is the molarity that is calculated on the basis of the amount of material that we used to prepare the solution. So, if we dissolve 0.10 moles of acetic acid in enough water to give 1.0 L of solution, the formal concentration of acetic acid is 0.10 M:

$$c_{\text{acetic acid}} = 0.10 \text{ M}$$

The K_a expression has the *equilibrium molarity* of acetic acid in the denominator, not the formal concentration. The equilibrium molarity is the molar concentration of acetic acid that is in the solution after equilibrium is established. As soon as the acetic acid dissolves in water, some of it dissociates. So the equilibrium molarity is less than the formal concentration.

If we look at the ionization equilibrium, we see that one hydrogen ion appears as a product for each acetic acid that dissociates.

$$HC_2H_3O_2 \rightleftharpoons H^+ + C_2H_3O_2^-$$

Therefore the equilibrium concentration of acetic acid, $[HC_2H_3O_2]$, is equal to the formal concentration of the acetic acid minus the hydrogen ion concentration.

$$[HC_2H_3O_2] = c_{\text{acetic acid}} - [H^+]$$

Next, we see the value of K_a is very small, 0.000018. This means the equilibrium position greatly favors the acetic acid over the ions. This also means that the acetic acid concentration doesn't change significantly. So, the equilibrium concentration of acetic acid remains about 0.10 M, even though some of it falls apart into ions. This assumption greatly simplifies the upcoming calculations and will allow us to avoid needing to use the quadratic formula.

$$[HC_2H_3O_2] = C_{HC_2H_3O_2} - [H^+] \approx 0.010 \text{ M}$$

We can summarize these concentrations in a table that shows the initial and equilibrium concentrations. The initial concentrations are the concentrations before the ionization equilibrium was established. So, since there are no other significant sources of hydrogen ions or acetate ions, their initial concentrations are zero. After equilibrium was established, we expressed the concentration of these ions in terms of $[H^+]$.

	$HC_2H_3O_2$	$\rightleftharpoons$	H+	+	$C_2H_3O_2^-$
Initial concentration	$c = 0.10$ M		0 M		0 M
Equilibrium concentration	0.10 M – $[H^+]\approx$ 0.10 M		$[H^+]$		$[H^+]$

Now, let's make these substitutions into the equilibrium constant expression.

$$K_a = \frac{[H^+][C_2H_3O_2^-]}{[HC_2H_3O_2]} = 1.8 \times 10^{-5}$$

$$\frac{[H^+][H^+]}{[0.10]} = 1.8 \times 10^{-5}$$

$$\frac{[H^+]^2}{[0.10]} = 1.8 \times 10^{-5}$$

Now, if we solve this expression for $[H^+]$, we get:

$$[H^+] = \sqrt{0.10(1.8 \times 10^{-5})} = \sqrt{1.8 \times 10^{-6}} = 0.0013 \text{ M}$$

So, at equilibrium, the concentration of all species is:

	$HC_2H_3O_2$	$\rightleftharpoons$	H+	+	$C_2H_3O_2^-$
Initial concentration	$c = 0.10$ M		0 M		0 M
Equilibrium concentration	0.10 M – 0.0013 M $\approx$ 0.10 M		0.0013 M		0.0013 M

Notice that our assumption about the acetic acid concentration was valid. If we subtract the hydrogen ion concentration from the formal concentration, the difference, expressed to the correct number of significant figures, remains at 0.10 M.

If we take the negative log of the hydrogen ion concentration, we find the pH of a 0.10-M acetic acid solution is 2.87.

Calculating the pH of a Weak Base Solution

Calculating the pH of a weak base involves exactly the same process as calculating the pH of a weak acid. So, what is the pH of a 0.10 M solution of ammonia? K_b for ammonia is 1.8×10^{-5}.

	NH_3	+	H_2O	$\rightleftharpoons$	NH_4^+	+	OH^-
Initial concentration	0.10 M				0 M		0 M
Equilibrium concentration	$c_{NH_3} - [OH^-]$ ≈ 0.10 M				$[OH^-]$		$[OH^-]$

Let's start with the ionization equilibrium.

$$NH_3 + H_2O \rightleftharpoons NH_4^+ + OH^-$$

We then write the equilibrium constant expression.

$$K_b = \frac{[OH^-][NH_4^+]}{[NH_3]} = 1.8 \times 10^{-5}$$

Just as with the weak acid example, we can summarize the equilibrium concentration in a table. We need to use a single variable, so let's express the equilibrium concentrations of each species in terms of the hydroxide ion.

If the only source of hydroxide ions and ammonium ions is from the basic ionization of ammonia, the molar concentrations of the two products must be equal. Therefore, in a solution that contains only ammonia, we know that concentrations of these two ions have to be equal.

$$[OH^-] = [NH_4^+]$$

Next, we see the value of K_b is very small, 0.000018. This means the equilibrium position greatly favors the ammonia over the ions. Therefore, the molar concentration of ammonia remains about 0.10 M, even though some of it falls apart into ions.

$$[NH_3] = c_{NH_3} - [OH^-] \approx 0.10 \text{ M}$$

Now, let's make these substitutions into the equilibrium constant expression.

$$\frac{[OH^-][OH^-]}{[0.10]} = 1.8 \times 10^{-5}$$

$$\frac{[OH^-]^2}{[0.10]} = 1.8 \times 10^{-5}$$

If we solve this expression for $[OH^-]$, we get:

$$\frac{[OH^-][OH^-]}{[0.10]} = 1.8 \times 10^{-5}$$

$$\frac{[OH^-]^2}{[0.10]} = 1.8 \times 10^{-5}$$

Therefore, the pOH of a 0.10-M ammonia solution is 2.87, and the pH is 11.13.

pH of Salt Solutions

Acids and bases react to give salts and (usually) water. That is one definition of a salt. So, when we say salt solution, we are really talking about solutions that contain the conjugate acid or base of some other acid or base. The pH of a salt solution depends on the acid/base strength of the acid or base from which it was derived. There are three permutations on the problem: salts of strong acids/bases, salts of weak acids, and salts of weak bases. Let's consider them each in turn.

Salts of Strong Acids/Bases

Strong acids include HCl and HNO_3. Because these are very strong acids, their conjugate bases (Cl^- and NO_3^-) have no base strength. Therefore these ions do not change the pH of a solution.

Strong bases like NaOH create salts containing the sodium ion. So, we could say that Na^+ is the salt of a strong base. Like the chloride ion, Na^+ does not participate in acid–base chemistry. It is purely a spectator ion.

So, the pH of a NaCl solution is 7 or, at least very close to 7. Anytime you are dealing with ions in an aqueous solution, activity is going to rear its ugly head. We think that you should be aware of activity, and that it causes solutions to behave nonideally. However, a detailed treatment of activity is a topic beyond the scope of our goals. So, for our purposes here, the pH of a NaCl solution is 7.

What about sulfuric acid? Well, H_2SO_4 is certainly a strong acid. In fact, it is a much stronger acid than HCl. Like all diprotic acids, sulfuric acid dissociates in a stepwise manner. The first proton is very, very acidic. K_a for the first ionization is very large, something on the order of 10^{10}, so the first ionization of sulfuric acid in water is 100% complete.

$$H_2SO_4 \rightarrow H^+ + HSO_4^- \quad K_a = \text{very large}$$

The second acid ionization has a much smaller K_a. Therefore, bisulfate is a weak acid, albeit a very strong one. Bisulfate is one of the strongest of the weak acids. So, the salt of sulfuric acid, bisulfate, is still an acid.

$$HSO_4^- \rightleftharpoons H^+ + SO_4^{2-} \quad K_a = 0.010$$

However, bisulfate is not a strong acid. The salt, or conjugate base, of bisulfate is a base, albeit a very weak one. Therefore, a solution of sodium sulfate would be very slightly alkaline.

Salts of Weak Acids

Salts of weak acids are bases. Most acids are weak acids. Therefore the conjugate bases of these acids have some appreciable base strength. That means that the vast majority of anions increase the pH of a solution. So, what is the pH of a 0.10-M solution of sodium acetate?

Well, in case you hadn't noticed, sodium ions don't seem to do much in chemistry. They are almost always spectator ions because they don't participate in any of the chemical reactions. Their job is to provide a charge balance to the anions in solution. So, in calculating the pH of sodium acetate, we ignore sodium. The acetate ion, however, is the conjugate base of the weak acid, acetic acid. Therefore, the acetate ion is a base, and we can write this ionization equilibrium equation.

$$C_2H_3O_2^- + H_2O \rightleftharpoons HC_2H_3O_2 + OH^-$$

The equilibrium constant expression for this equilibrium is:

$$K_b = \frac{[OH^-][HC_2H_3O_2]}{[C_2H_3O_2^-]}$$

A table of initial and equilibrium concentrations can be constructed that expresses all concentrations in terms of the hydroxide ion.

	$C_2H_3O_2^-$	+	H_2O	$\rightleftharpoons$	$HC_2H_3O_2$	+	OH^-
Initial concentration	0.10 M				0 M		0 M
Equilibrium concentration	$c_{acetate} - [OH^-] \approx 0.10$ M				$[OH^-]$		$[OH^-]$

This looks familiar. If the only source of hydroxide ions and acetic acid is from the basic ionization of the acetate ion, the molar concentrations of the two products must be equal.

$$[OH^-] = [HC_2H_3O_2]$$

Now, what's the value of K_b? The base constants for conjugate bases of weak acids are usually not included in reference tables because they are easily calculated from the K_a's of the weak acid. Let's start with the K_a and K_b expressions for acetic acid and the acetate ion.

$$K_a = \frac{[H^+][C_2H_3O_2^-]}{[HC_2H_3O_2]}$$

$$K_b = \frac{[OH^-][HC_2H_3O_2]}{[C_2H_3O_2^-]}$$

Let's try multiplying these two expressions.

$$K_a K_b = \left(\frac{[H^+][C_2H_3O_2^-]}{[HC_2H_3O_2]} \right) \left(\frac{[OH^-][HC_2H_3O_2]}{[C_2H_3O_2^-]} \right)$$

Notice the concentrations of acetic acid and acetate cancel out, leaving us with:

$$K_a K_b = [H^+][OH^-] = K_w$$

So, the product of multiplying K_a times K_b is K_w. So, we can find K_b for the acetate ion.

$$K_b = \frac{K_w}{K_a} = \frac{1.00 \times 10^{-14}}{1.8 \times 10^{-5}} = 5.6 \times 10^{-10}$$

So, we see that K_b for the acetate ion is very, very small, and only a tiny amount reacts with water. Therefore, the molar concentration of the acetate ion remains about 0.10 M.

$$[C_2H_3O_2^-] = c_{acetate} - [OH^-] \approx 0.10 \text{ M}$$

Now, let's make these substitutions into the equilibrium constant expression.

$$K_b = \frac{[OH^-][HC_2H_3O_2]}{[C_2H_3O_2^-]} = 5.6 \times 10^{-10}$$

$$\frac{[OH^-][OH^-]}{[0.10]} = 5.6 \times 10^{10}$$

$$\frac{[OH^-]^2}{[0.10]} = 5.6 \times 10^{-10}$$

If we solve this expression for [OH⁻], we get:

$$[OH^-] = \sqrt{0.10 (5.6 \times 10^{-10})} = \sqrt{5.6 \times 10^{-11}} = 7.45 \times 15^{-6} \text{ M}$$

Therefore, the pOH of a 0.10 M sodium acetate solution is 5.13, and the pH is 8.87.

Salts of Weak Bases

Salts of weak bases are acids. This is probably the more significant salt solution pH calculation, because many medications are organic amines, or alkaloids. Alkaloids are (naturally occurring) weakly basic organic compounds. Since these organic compounds are generally not very soluble in water, they are converted to their conjugate acid form by reacting with a strong acid (such as HCl). The ionic lidocaine hydrochloride is much more soluble in water than lidocaine, as shown in Figure 9.6.

Since lidocaine is a weak base, the salt of that weak base will be an acid. The pK_a of lidocaine hydrochloride is 7.9 ($K_a = 1.3 \times 10^{-8}$). The pH of a solution of lidocaine hydrochloride will be slightly acidic.

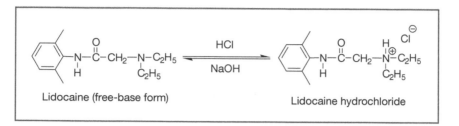

Figure 9.6 Acid–Base Reaction of Lidocaine

Just like sodium ions, chloride ions are spectator ions in acid–base chemistry. Their job is to provide a charge balance to the cations in solution. So, in calculating the pH of lidocaine hydrochloride we ignore the chloride ion. Now we could draw out the structure or write the molecular formula of lidocaine and its conjugate acid, but it is tedious to do so. Let's do what most chemists do, and postulate a temporary abbreviation for these species. How about using L for lidocaine and HL^+ for its conjugate acid? Now, we can write an equation for the acid ionization equilibrium reaction.

$$HL^+ \rightleftharpoons H^+ + L$$

The equilibrium constant expression for this equilibrium is:

$$K_a = \frac{[H^+][L]}{[HL^+]}$$

A table of initial and equilibrium concentrations can be constructed that expresses all concentrations in terms of the hydrogen ion.

	HL^+	$\rightleftharpoons$	H^+	+	L
Initial concentration	0.10 M		0 M		0 M
Equilibrium concentration	$c - [H^+] \approx 0.10$ M		$[H^+]$		$[H^+]$

This looks familiar, and that's very good. Solving pH problems for weak acids and bases is really applying the same strategy to different situations. That is why the presentations have been so repetitive. We want you to see that there is one problem-solving strategy that works for many of these applications.

If the only source of hydrogen ions and lidocaine is from the acidic ionization of the HL^+ ion, the molar concentrations of the two products must be equal.

$$[H^+] = [L]$$

Since K_a for HL^+ is small, the molar concentration of the HL^+ ion remains about 0.10 M.

$$[HL^+] = c - [H^+] \approx 0.10 \text{ M}$$

Now, let's make these substitutions into the equilibrium constant expression.

$$\frac{[H^+][H^+]}{[0.10]} = 1.3 \times 10^{-8}$$

$$\frac{[H^+]^2}{[0.10]} = 1.3 \times 10^{-8}$$

If we solve this expression for $[H^+]$, we get:

$$[H^+] = \sqrt{0.10(1.3 \times 10^{-8})} = \sqrt{1.3 \times 10^{-9}} = 3.6 \times 10^{-5} \text{ M}$$

Therefore, the pH of a 0.10 M lidocaine hydrochloride solution is 4.4.

Extreme Concentrations

Thus far, the examples have used concentrations that are reasonably high, but not so high that activity must be considered. Values for equilibrium constants have been small. There are no hard and fast rules about when the magnitudes of concentrations or K_a values will cause you problems, but you should be wary when concentrations get below, say 10^{-3} molar, or when the concentrations aren't that different from the K_a values. What happens when we stray out of the safe ranges we have been exploring?

For example, what is the pH of 1.0×10^{-8} M HCl? Well, if you whip out your calculator, and take the negative log of 1.0×10^{-8}, you get an answer pH = 8.0. Does that make sense? If you have a liter of pure water at a pH of 7.0, and you add 1.0×10^{-8} moles of HCl, will the pH go up? Of course not. This example underscores that you cannot be totally complacent about these kinds of calculations, especially when the concentrations become very dilute.

Well, since we brought it up, what is the pH of 1.0×10^{-8} M HCl? The answer to this conundrum lies in the fact that there are two sources of hydrogen ion. The first is from the HCl, and the second is from the ionization of water. The position of this equilibrium reaction is governed by K_w.

$$H_2O \rightleftharpoons H^+ + OH^-$$

$$K_w = [H^+][OH^-]$$

Let's see: In pure water the hydrogen ion concentration is 1.0×10^{-7} M. Okay, you say. So let's just add the 1.0×10^{-8} M from the HCl to the 1.0×10^{-7} M from the water. But that doesn't work, because the introduction of H^+ from HCl impacts the self-ionization of water. According to Le Châtelier's Principle, the position of the equilibrium will be shifted to the left, because we are adding a product. Many biological systems are coupled equilibria, so if you change one, you change them all. If you want to solve one system, you have to solve all of them simultaneously, because they are all interconnected.

Well, back to our problem. We know that the product of $[H^+]$ times $[OH^-]$ equals K_w. Before we added the HCl, the concentrations of these two ions were equal. Since we're going to change the concentration of [H⁺] by adding HCl, let's use the hydroxide ion concentration as our variable.

	H_2O	$\rightleftharpoons$	H^+	$+$	OH^-
Initial concentration			[OH⁻]		[OH⁻]
Equilibrium concentration			[OH⁻] + 1.0 × 10⁻⁸		[OH⁻]

We have both concentrations expressed in a single variable, so let's plug these into the equilibrium constant expression.

$$K_w = ([OH^-] + 1.0 \times 10^{-8})[OH^-]$$

If we multiply the right side through by the concentration of the hydroxide ion, and rearrange the resulting quadratic equation, we obtain:

$$[OH^-]^2 + 1.0 \times 10^{-8}\,[OH^-] - 1.0 \times 10^{-14} = 0$$

To solve this equation, we need to use the quadratic formula:

$$x = \frac{-b \pm \sqrt{b^2 - 4ac}}{2a}$$

where our x is [OH⁻], the linear coefficient b is 1.0×10^{-8}, and the constant term c is -1.0×10^{-14}. The quadratic coefficient a is 1. Making these substitutions, we obtain:

$$[OH^-] = \frac{\left(-(1.0 \times 10^{-8}) \pm \sqrt{(1.0 \times 10^{-8})^2 - 4(1)(-1.0 \times 10^{-14})}\right)}{2(1)}$$

Okay, it's painful, but not impossible, to work through the algebra to find:

$$[OH^-] = 9.5 \times 10^{-8}\ M$$

We discard the other root ([OH⁻] = −1.1 × 10⁻⁷ M) because, while this is a mathematically valid solution, a negative concentration has no meaning. Therefore, the pOH of the solution is 7.02 and the pH is 6.98. So adding a miniscule amount of HCl caused a tiny drop in the pH, just as you would have expected.

OTHER ACIDIC SPECIES

Nonmetal oxides dissolve in water to give acid solutions. The most physiologically important example is carbon dioxide. Of course, you already know this. Buildup of carbon dioxide in the blood results in acidosis. This is not necessarily a bad thing. In cellular tissue, where the carbon dioxide concentration is relatively high, the increased acidity slightly alters the structure of hemoglobin and facilitates the release of oxygen.

Carbon dioxide is a nonmetal oxide because it is a compound comprised of a nonmetal and oxygen. Nonmetal oxides are sometimes called acid anhydrides because they are produced by stripping water from an acid.

When carbon dioxide dissolves in water, it combines with a water molecule to give carbonic acid. Notice that the reverse direction corresponds to stripping a water molecule away from carbonic acid, leaving carbon dioxide.

$$CO_2 + H_2O \rightleftharpoons H_2CO_3$$

When the carbonic acid forms, it dissociates according to its acid strength.

$$H_2CO_3 \rightleftharpoons H^+ + HCO_3^-$$

So, when CO_2 dissolves in water, the pH drops.

BUFFERS

A pH buffer is a solution that resists changes in pH. A pH buffer is a solution that contains a weak acid (HA) and its conjugate base (A$^-$), or a weak base and its conjugate acid. Now, buffer solutions *resist* change in pH. So, when a small amount of a strong acid or base is added to the solution, the pH will change, but not nearly as much as with an unbuffered solution. The chemistry for minimizing this pH change is pretty straightforward. If a strong base is added to a buffered solution, the weak acid in the buffer HA reacts with the hydroxide ion to give water and the weak base A$^-$.

$$HA + OH^- \rightleftharpoons H_2O + A^-$$

This results in converting a strong base OH$^-$ into a weak base A$^-$. So, the pH will increase, but not by much.

If a strong acid is added to a buffered solution, the weak base in the buffer, A$^-$, reacts with the H$^+$ ion to give HA.

$$A^- + H^+ \rightleftharpoons HA$$

This results in converting a strong acid H$^+$ into a weak acid HA. So, the pH will decrease, but not by much.

Calculating the pH of a Buffer

It is very tempting to guess that buffers maintain a neutral pH of 7, but this is not true. The pH of a buffer is determined by the acid strength of HA. As the acid strength of HA increases, the pH range maintained by a buffer system based on HA decreases. This means that we can fine tune our buffer systems to maintain any desired pH by careful selection of the weak acid and its conjugate base.

But what is the relationship between the pH of a buffer and the strength of HA? Well, as with all equilibrating processes, let's start with an equation for the acid ionization of HA.

$$HA \rightleftarrows H^+ + A^-$$

The equilibrium constant expression for this equilibrium is:

$$K_a = \frac{[H^+][A^-]}{[HA]}$$

But can we use this expression for a buffer? Yes. Up to this point, all of the equilibrium calculations we have considered began with a pure substance, just HA or just A^-. But the equilibrium state can be achieved from an infinite number of starting points. So, if we start with pure HA, it will ionize until the concentrations of $[H^+]$, $[A^-]$, and $[HA]$ satisfy the K_a expression. But if we start from a mixture of HA and its conjugate base A^-, each species will undergo its acid–base reactions simultaneously until the concentrations of all species reach the equilibrium state. Remember: Equilibrium is a good thing, and systems will strive to reach the equilibrium state, regardless of the starting point.

So, the K_a expression holds for buffers. Since we want to find the pH of a buffer, let's solve the K_a expression for $[H^+]$.

$$[H^+] = K_a \frac{[HA]}{[A^-]}$$

Now, if we want pH, let's apply the p-function to both sides.

$$-\log[H^+] = -\log K_a \frac{[HA]}{[A^-]}$$

Well, the negative log of the hydrogen ion concentration is pH, but what about the right-hand side of the equation? Remember that the log of a product is equal to the sum of the logs. So, we can separate the K_a constant from the concentration terms.

$$pH = -\log K_a + \left(-\log K_a \frac{[HA]}{[A^-]}\right)$$

Okay, the negative log of K_a is pK_a. Furthermore, we can get rid of the pesky negative sign in the remaining log term by recalling that a constant times a logarithm is equal to the logarithm raised to that power.

$$a\log b = \log b^a$$

So the negative log of $[HA]/[A^-]$ is equal to the log of $[HA]/[A^-]$ raised to the minus one power.

$$-\log \frac{[HA]}{[A^-]} = \log \left(\frac{[HA]}{[A^-]}\right)^{-1}$$

Of course, any number raised to the minus one power (except zero) is the reciprocal of that number. So, we can eliminate the minus sign by reciprocating the [HA]/[A⁻] ratio.

$$pH = pK_a + \log\frac{[A^-]}{[HA]}$$

This is the Henderson–Hasselbalch equation, or simply the buffer equation. We can use the buffer equation to calculate the pH of a buffer. We can also use it to determine the ratio of weak acid to conjugate base at a given pH.

So, what is the pH of a buffer that contains acetic acid at a concentration of 0.10 M and sodium acetate at a concentration of 0.15 M? Well, acetic acid is the weak acid, and its concentration is given. The conjugate base of acetic acid is the acetate ion, and we are told the concentration of sodium acetate is 0.15 M. Once again, you need to recognize that sodium ions are spectators, and you can just ignore them. To calculate the pH of a buffer, we need to know the pK_a of the acid. The K_a of acetic acid is 1.8×10^{-5}, so the pK_a is 4.74. Now, we have values for all the terms in the buffer equation, and we can solve for the pH:

$$pH = 4.74 + \log\frac{0.15\ M}{0.10\ M}$$

$$pH = 4.74 + 0.18 = 4.92$$

The value of the pH relative to the pK_a is significant. The concentration of the conjugate base is greater than the concentration of the weak acid. Therefore, the pH is on the basic side of pK_a.

Let's try another one. What is the pH of a buffer containing 0.25 M ammonia and 0.75 M ammonium chloride? Well, the weak acid in this case is the ammonium ion. The chloride ion is a spectator to be ignored. Ammonia is a weak base and the conjugate base of the ammonium ion. So, since this solution contains a weak conjugate acid–base pair, it is a buffer, and we can calculate the pH using the Henderson–Hasselbach equation. The Henderson–Hasselbach equation calls for the pK_a of the acid, so in this case we need the pK_a for the ammonium ion. The pK_b for ammonia is 4.74, so the pK_a for the ammonium ion is 9.26. If we substitute this value and the values for the concentrations into the buffer equation, we find the pH of this solution is 8.78.

$$pH = 9.26 + \log\frac{0.25\ M}{0.75\ M}$$

$$pH = 9.26 + (-0.48) = 8.78$$

Two comments: First, notice the pH is on the acidic side of the pK_a, because the concentration of the weak acid (NH_4^+) was greater than the concentration of its conjugate base (NH_3). Second, notice how different the pH of an ammonia/ammonium chloride buffer system is from the pH of an acetic acid/sodium acetate buffer system. The pH of a buffer depends on the acid strength of the weak acid from which it is composed.

One last example, and this one is special. What is the pH of a buffer that contains acetic acid and sodium acetate both at a concentration of 0.30 M? When we put the numbers into the buffer equation, we see the ratio of acetate ion to acetic acid is 1. The logarithm of 1 is 0, so the pH is equal to the pK_a. This is true of any buffer system. When the concentrations of the weak acid and its conjugate base are equal, the pH equals the pK_a.

$$pH = 4.74 + \log\frac{0.30 \text{ M}}{0.30 \text{ M}}$$

$$pH = 4.74$$

Dilution of Buffer Solutions

One of the really interesting things about buffers is that the pH depends on the ratio of weak base to its conjugate acid, not the numerical value of each. This means a pH buffer can be diluted with water without changing the pH of the solution. For example, if we have 1.0 L of 0.10 M HCl and we add 9.0 L of water, what happens to the pH? Well, the initial pH is 1.0. When we add the water, the concentration of HCl drops to 0.010 M, so the pH rises to 2.0.

But what if we had a buffer solution containing acetic acid and sodium acetate, each at 0.10 M? The previous example shows us the pH of this buffer is equal to the pK_a. If we dilute 1.0 L of this buffer to a new volume of 10.0 L, the concentration of each component falls to 0.010 M. However, the ratio of the two components remains the same, so the pH remains at 4.74. Since blood is a buffered solution, you can administer several liters of normal saline without changing the pH of blood.

Buffer Capacity

To be an effective buffer, the pH must be within one pH unit of the pK_a of the weak acid.

$$pH_{effective} = pK_a \pm 1$$

A quick look at the buffer equations explains this. The pH changes from the pK_a by the log of the ratio of the weak base to the weak acid. If the pH is one unit above the pK_a, the ratio of the weak base to weak acid is 10. This is still an effective buffer because there is still an appreciable concentration of both the weak acid and the weak base.

$$pH - pK_a = 1 = \log\frac{[A^-]}{[HA]}$$

$$10^1 = \frac{[A^-]}{[HA]} = 10$$

If the pH is three pH units above the pK_a, the ratio of the weak base to weak acid is 1000.

$$pH - pK_a = 3 = \log\frac{[A^-]}{[HA]}$$

$$10^3 = \frac{[A^-]}{[HA]} = 1000$$

This will not be an effective buffer because the concentration of the weak acid is so much less than the concentration of the weak base. Since the concentration of HA is so small, there is not enough HA present to react with added OH⁻ and prevent the pH from rising.

To prepare an effective buffer, therefore, we need to identify an acid with a pK_a as close as possible to the desired pH. For example, phosphate buffers are commonly used in the biochemistry laboratory (and in blood). If we want a buffer to maintain a pH of 7.4, which phosphate species do we need and in which concentration?

Well, when we say "phosphate," we have four choices: phosphoric acid, dihydrogenphosphate, monohydrogenphosphate, and phosphate.

$$H_3PO_4 \rightleftharpoons H^+ + H_2PO_4^-\ pK_a = 2.15$$

$$H_2PO_4^- \rightleftharpoons H^+ + HPO_4^{2-}\ pK_a = 7.20$$

$$HPO_4^{2-} \rightleftharpoons H^+ + PO_4^{3-}\ pK_a = 12.35$$

The pK_a of dihydrogenphosphate is the closest to our desired pH. So, the weak conjugate acid–base pair we need is dihydrogenphosphate and monohydrogenphosphate. If we substitute in the pH and pK_a values into the buffer equation, we have:

$$7.4 = 7.20 + \log\frac{[HPO_4^{2-}]}{[H_2PO_4^-]}$$

If we subtract 7.20 from each side of the equation and then take the antilog, we get:

$$\frac{[HPO_4^{2-}]}{[H_2PO_4^-]} = 10^{0.20} = 1.6$$

So, we need the ratio of monohydrogenphosphate to dihydrogenphosphate to be 1.6.

Alpha Plots

An alpha plot for a buffer (or an amino acid) shows the percentage of each component in a buffer system as a function of pH. Alpha for a given species is the percent of all the material that is present in that form. So, when alpha for the acid form is 75%, the ratio of acid to conjugate base is 75:25. Naturally, the sum of the alpha values for the acid and the conjugate base adds up to 100%.

The shape of the curves in an alpha plot is determined by the buffer equation. At low pH, most of the material is present as its acid. As the pH increases, the acid form converts to the conjugate base form. At high pH, most of the material is present in the conjugate base form.

Let's take a look at an alpha plot for lidocaine (Figure 9.7). Lidocaine is a weak base, so a solution of lidocaine and its conjugate acid, lidocaine hydrochloride, will comprise a buffer. We see at low pH's, up to about pH = 6, the alpha value for lidocaine hydrochloride is close to 100%. This means that essentially 100% of the lidocaine is in the hydrochloride, or conjugate acid, form. As the pH rises to the pK_a of lidocaine (pH = 7.9), the alpha values for lidocaine

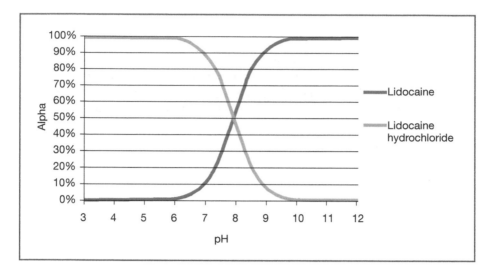

Figure 9.7 Alpha Plot for Lidocaine

hydrochloride and lidocaine are both 50%. Of course, you know that when the concentration of a weak acid and its conjugate base are equal, the pH of a buffer is equal to the pK_a of the acid. As we continue increasing the pH, the amount of the lidocaine hydrochloride falls to zero, as the fraction of the material in the conjugate base form rises to 100%. So, we see that at the blood pH of 7.4, about 76% of the anesthetic is in the ionized form and 24% is in the free-base form.

Summary

■ Few reactions go to completion. Rather, most reactions are in a state of dynamic equilibrium, with a constant interchange between reactants and products.

■ A system is at equilibrium when the concentrations of the products and reactants reach a constant value. The ratio of the concentrations of the products divided by the concentrations of the reactants is equal to the equilibrium constant. The numerical value of K varies widely, depending on the specific reaction. The value of K also depends on the temperature.

■ Systems are driven to reach equilibrium. LeChâtlier's Principle states that if a system at equilibrium is disturbed so that the ratio of products to reactants no longer equals the equilibrium constant, and it will react so as to reestablish the equilibrium state. That is, the system will strive to undo whatever changes are enacted upon it.

■ A large value of K means the equilibrium favors the products. A value of K that is less than one means the equilibrium favors the reactants.

■ The Brønsted theory defines an acid as a proton, or H^+ ion, donor. A base is a proton acceptor. When an acid donates its proton, what remains is called the conjugate base. When a base accepts a proton, it is converted into its conjugate acid.

$$\text{acid} \rightarrow H^+ + \text{conjugate base}$$

$$\text{base} + H^+ \rightarrow \text{conjugate acid}$$

■ A conjugate acid–base pair differs in formula by one hydrogen ion. The conjugate acid has the extra hydrogen. The conjugate acid has an ionic charge one unit greater than the conjugate base.

$$HA \rightarrow H^+ + A^-$$
$$\text{acid} \qquad \text{conjugate base}$$

■ An amphiprotic species can react as either an acid or a base. Water is the most common example of an amphiprotic species. When water reacts as a base, it becomes its conjugate acid, the hydronium ion (H_3O^+). When water reacts as an acid, it gives the conjugate base, the hydroxide ion (OH^-).

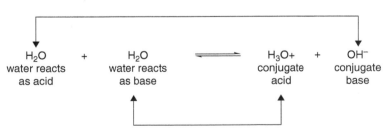

■ Strong acids are essentially 100% ionized in water. Strong acids are relatively rare. There are five common strong acids: sulfuric acid (H_2SO_4), hydroiodic acid (HI), hydrobromic acid (HBr), hydrochloric acid (HCl), and nitric acid (HNO_3). Virtually all other acids can be classified as weak acids.

■ Strong bases include the soluble ionic hydroxides. The most common strong bases are sodium hydroxide (NaOH), potassium hydroxide (KOH), and lithium hydroxide (LiOH). When a strong base dissolves in water, it essentially ionizes 100% to the cation and the hydroxide ion.

■ Weak acids (generically represented as HA) are not completely ionized in water, but instead establish an equilibrium between the un-ionized (molecular) form of the acid and its conjugate base.

$$HA \rightleftharpoons H^+ + A^-$$

■ Weak bases (generically represented as B) are not completely ionized in water, but instead establish an equilibrium between the un-ionized (molecular) form of the base and its conjugate acid.

$$B + H_2O \rightleftharpoons BH^+ + OH^-$$

■ The equilibrium constant (K_a) for an acid equilibrium reaction is a measure of the acid's strength.

$$K_a = \frac{[H^+][A^-]}{[HA]}$$

■ The equilibrium constant (K_b) for a base equilibrium reaction is a measure of the base's strength.

$$K_b = \frac{[BH^+][OH^-]}{[B]}$$

■ The p-function is used to condense a wide range of hydrogen ion concentration values down to a more manageable set. The p-function means "take the log, and then change the sign."

■ The pH of a solution means take the log of the hydrogen ion concentration and then change the sign.

$$pH = -\log[H^+]$$

■ To find the hydrogen ion concentration, the order of operations is reversed: Change the sign and take the antilog. On most calculators the antilog button is the 10^x.

$$[H^+] = \text{antilog}\,(-pH) = 10^{-pH}$$

■ The pOH of a solution is given by:

$$pH = -\log[OH^-]$$

■ To find the hydroxide ion concentration, change the sign of the pOH and take the antilog.

$$[OH^-] = 10^{-pOH}$$

■ Water is a weak acid and undergoes "self-ionization" into a hydrogen ion and a hydroxide ion.

$$H_2O \rightleftharpoons H^+ + OH^-$$

■ The self-ionization of water is governed by a constant called K_w, and this fixes a relationship between $[H^+]$ and $[OH^-]$. K_w has a value of 1.0×10^{14} at 25°C.

$$K_w = [H^+] \cdot [OH^-]$$

■ The pH and pOH values of a solution sum to 14.

$$pH + pOH = 14$$

■ The p-function can also be applied to K_a, K_b, and K_w:

$$pK_a = -\log(K_a)$$

$$pK_b = -\log(K_b)$$

$$pK_w = -\log(K_w) = -\log(1.0 \times 10^{-14}) = 14.00$$

■ Analogous to $[H^+]$ and $[OH^-]$, the product of K_a and K_b equals K_w.

$$K_a \cdot K_b = K_w = 1.0 \times 10^{-14}$$

■ Analogous to pH and pOH, pK_a and pK_b add to 14:

$$pK_a + pK_b = 14$$

■ To find the $[H^+]$ in a solution of a strong acid (e.g., HCl, HBr, HI, or HNO_3), you can assume the $[H^+]$ concentration is equal to the stated molarity of the strong acid.

- In finding [OH$^-$] in a solution of a strong base (e.g., LiOH, NaOH, or KOH), you can assume the [OH$^-$] concentration is equal to the stated molarity of the strong base.
- With a solution of a weak acid, the system is in a state of equilibrium, so you don't just use the stated concentration of the weak acid (c_{HA}) to find the pH. You have to begin with the ionization equilibrium reaction:

$$HA \rightleftharpoons H^+ + A^-$$

- When the system reaches equilibrium, the concentration of [H$^+$] will equal the concentration of [A$^-$], and the equilibrium concentration of [HA] will equal the stated concentration minus the [H$^+$]. Unless the solution is extremely dilute, [H$^+$] concentration is normally much less than the stated concentration of the weak acid, and we can assume the initial concentration of the weak acid remains constant. Putting these substitutions into the equilibrium constant gives:

$$K_a = \frac{[H^+][A^-]}{[HA]} = \frac{[H^+][H^+]}{C_{HA} - [H^+]} \approx \frac{[H^+]^2}{C_{HA}}$$

or

$$[H^+] = \sqrt{K_a \cdot C_{HA}}$$

- With a solution of a weak base, the system is in a state of equilibrium, so you don't just use the stated concentration of the weak base (c_B) to find the pOH. You have to begin with the ionization equilibrium reaction:

$$B + H_2O \rightleftharpoons BH^+ + OH^-$$

- When the system reaches equilibrium, the concentration of [OH$^-$] will equal the concentration of [BH$^+$], and the equilibrium concentration of [B] will equal the stated concentration minus the [OH$^-$]. Unless the solution is extremely dilute, the [OH$^-$] concentration is normally much less than the stated concentration of the weak base, and we can assume the initial concentration of the weak base remains constant. Putting these substitutions into the equilibrium constant gives:

$$K_b = \frac{[BH^+][OH^-]}{[B]} = \frac{[H^+][H^+]}{c_B - [OH^+]} \approx \frac{[OH^+]^2}{c_B}$$

or

$$[OH^-] = \sqrt{K_b \cdot c_B}$$

- The stronger the acid, the weaker the conjugate base. The stronger the base, the weaker the conjugate acid.
  - Na$^+$, Li$^+$, and K$^+$ do not affect the pH of a solution.

- The conjugate bases of strong acids (Cl^-, Br^-, and NO_3^-) are very weak bases and do not affect the pH of a solution.
- The conjugate bases of weak acids (also called the salts of weak acids) have base strength and raise the pH of a solution.
  - The K_b of the conjugate base of a weak acid is equal to K_w/K_a of the acid.
- The conjugate acids of weak bases (also called the salts of weak bases) have acid strength and lower the pH of a solution.
  - The pK_a of the conjugate acid of a weak base is equal to K_w/K_b of the base.
- Nonmetal oxides dissolve in water to give acids. The most common nonmetal oxide in physiology is carbon dioxide.
- A buffer is a solution that contains a weak acid and its conjugate base.
  - A buffer solution resists changes in pH.
  - The weak acid absorbs any added strong base.
  - The weak base absorbs any added strong acid.
- The pH of a buffer can be calculated using the Henderson–Hasselbalch equation:

$$pH = pK_a + \log\frac{[A^-]}{[HA]}$$

Review Questions for Acids and Bases

1. What is a dynamic equilibrium?

2. What is Le Châtelier's Principle?

3. What is an equilibrium constant, and what does its size indicate?

4. The ionization of water is an endothermic process. Do you expect the pH of water at 35°C to be above 7 or below 7? Explain.

5. What is an acid? A base? A conjugate acid?

6. What is the conjugate base of hydroxide? Is it a stronger or weaker base than hydroxide?

7. Is the pH of solutions containing these salts likely to be acidic, basic, or neutral? NaCl, NaI, NH_4Cl, cocaine hydrochloride, sodium acetate, sodium palmitate.

8. Calculate the pH of a 0.010 M solution of each of these compounds:
   a. Acetic acid
   b. Sodium acetate
   c. Ammonia
   d. Ammonium chloride

9. What role does the self-ionization of water play in acid–base chemistry?

10. What is the pH scale?

11. What is a p-function?

12. What is the pH of vomit, if the H^+ concentration is 0.1 mol/L?

13. What is an amprotic species?

14. What is a buffer? How do you prepare a buffer?

15. What is the pH of 0.15 M HCl?

16. If blood pH is 7.30, what is the concentration of the H^+ ion, the OH^- ion, and the pOH?

17. When acids and bases react, you normally get ___ and ___.

18. What is the pH of 0.10 M lactic acid ($K_a = 1.3 \times 10^{-4}$)?

19. What is the pK_a of lactic acid?

20. What is the pH of a 0.10 M solution of sodium lactate?

21. What is the pH of a solution that is equal parts sodium lactate and lactic acid?

22. Use the alpha plot for lidocaine (Figure 9.7) to estimate the percent of the material that is present as the hydrochloride (conjugate acid form) at a pH of:
   a. 6.5
   b. 7.2
   c. 8.0

23. How does a buffer prevent the pH from rising when a base is added?

24. Do you expect the equilibrium constant for binding carbon monoxide to hemoglobin (Hb) to be larger or smaller than the equilibrium constant for binding oxygen to hemoglobin? Explain.

$$Hb + 4\ CO = Hb(CO)_4$$

$$Hb + 4\ O_2 = Hb(O_2)_4$$

a. Larger
b. Smaller
c. Nearly zero
d. Nearly infinite

25. One treatment for carbon monoxide poisoning is administration of oxygen. The high concentration of oxygen binds to free hemoglobin, including the Hb from the equilibrium between Hb and CO. This is an application of:
a. Henry's Law
b. Raoult's Law
c. Le Châtelier's Principle
d. Trying really hard not to kill the patient

26. All of these compounds are considered to be acids by chemists. Which is NOT a Brønsted acid?
a. BF_3
b. H_3PO_4
c. HCl
d. All of these are Brønsted acids

27. What is the conjugate ACID of nitric acid (HNO_3)?
a. H^+
b. H_3O^+
c. HNO_2
d. $H_2NO_3^+$

28. To be an effective buffer, the acid must:
a. Have a pK_a nearly equal to the desired pH
b. Be a strong acid
c. Have a strong conjugate base
d. All of these

29. Given the pK_a values for phosphoric acid,
$H_3PO_4 = H^+ + H_2PO_4^- \ pK_1 = 2.2$
$H_2PO_4^- = H^+ + HPO_4^{-2} \ pK_2 = 7.2$
$HPO_4^{-2} = H^+ + PO_4^{-3} \ pK_3 = 12.3$
Which of these is the strongest base?
a. H_3PO_4
b. $H_2PO_4^-$
c. HPO_4^{-2}
d. PO_4^{-3}

30. Which conjugate acid–base pair is most important in buffering blood?
    a. H_3PO_4 and $H_2PO_4^-$
    b. $H_2PO_4^-$ and HPO_4^{-2}
    c. HPO_4^{-2} and PO_4^{-3}

31. What is the pH of a 0.001-M solution of HCl?
    a. 1
    b. 2
    c. 3
    d. 7

32. What is the hydrogen ion concentration in a solution whose pOH is 12.0?
    a. 2 M
    b. 1×10^{-12} M
    c. 1×10^{-2} M
    d. More information is needed.

33. What is the conjugate base of bicarbonate, HCO_3^-?
    a. $H_2CO_3^-$
    b. CO_3^{2-}
    c. OH^-

34. If the K_a of bicarbonate is 4.8×10^{-11}, what is the pK_a of bicarbonate?
    a. 10.30
    b. 3.70
    c. 7.00
    d. 14.00

35. What is the pH of a buffer that contains equal amounts of HCO_3^- and CO_3^{2-}?
    a. 10.30
    b. 7.00
    c. 3.70

36. You have 1 liter of water. You add 1.0×10^{-8} mol of HCl, so now the molarity of the HCl is 1.0×10^{-8} M. What is the pH of the solution? Be careful!
    a. 8.00
    b. 6.98
    c. 7.00

37. Write an equilibrium constant expression for each of these chemical reactions.
    a. $HCN \rightleftharpoons H^+ + CN^-$
    b. $C_5H_5N + H_2O \rightleftharpoons HC_5H_5N^+ + OH^-$
    c. $CO_2^- + H_2O \rightleftharpoons H^+ + HCO_3^-$
    d. $H_2PO_4^- \rightleftharpoons H^+ + HPO_4^{2-}$
    e. $HPO_4^{2-} + H_2O \rightleftharpoons H_2PO_4^- + OH^-$

38. Using Le Châtelier's Principle, say whether each of these changes will cause the solubility equilibrium of carbon dioxide in water to shift toward starting materials or products.

$$CO_2 + H_2O \rightleftharpoons H^+ + HCO_3^-$$

   a. Increasing the pressure of the carbon dioxide (think about Henry's Law)
   b. Increasing the [H⁺] concentration
   c. Lowering the pH
   d. Raising the pOH
   e. Removing bicarbonate

39. Using Le Châtelier's Principle, say whether each of these changes will cause the solubility equilibrium of calcium oxalate (one mineral found in kidney stones) to shift toward products or starting materials. The oxalate anion is the conjugate base of a weak acid and picks up a hydrogen ion to form bioxalate.

$$CaC_2O_4 \rightleftharpoons Ca^{2+} + C_2O_4^{2-}$$

$$C_2O_4^{2-} + H^+ \rightleftharpoons HC_2O_4^-$$

   a. Increasing [Ca²⁺]
   b. Decreasing [C₂O₄²⁻]
   c. Increasing the pH
   d. Decreasing the [H⁺]
   e. Adding more bioxalate

40. Give the formula of the conjugate base of each of these compounds.

a. HNO_2	b. HF	c. H_2O_2	d. $HCNO$
e. HCO_2H	f. $HBrO_3$	g. H_3PO_4	h. $H_2PO_4^-$
i. HSO_4^-	j. H_3O^+	k. $HN_2H_4^+$	l. $HC_5H_5N^+$

41. Give the formula of the conjugate acid of each of these compounds.

a. NH_3	b. OH^-	c. CH_3NH_2	d. N_2H_4
e. $C_{10}H_{10}N_2$	f. CO_3^{2-}	g. HCO_3^-	h. SO_4^{2-}
i. HSO_3^-	j. CH_3OH	k. NH_2^-	l. HNO_3

42. Each of these compounds can be amphiprotic. Write an equation for each behaving as an acid and then a second equation for each behaving as a base.

a. H_2O	b. $H_2PO_4^-$	c. HCO_3^-	d. $HC_2H_4NO_2$ (glycine)

43. Write an ionization equilibrium equation for each of these acids and the corresponding equilibrium constant expression.

a. HNO_2	b. HF	c. HCO_2H	d. HCNO

44. Write an ionization equilibrium equation for each of these bases and the corresponding equilibrium constant expression.

a. CH_3NH_2	b. N_2H_4	c. CO_3^{2-}	d. HSO_3^-

45. Provide the missing information for each of these solutions.

SOLUTION	pH	pOH	$[H^+]$	$[OH^-]$
1	6.52			
2		4.64		
3			5.2×10^{-02}	
4				5.6×10^{-05}
5	3.11			
6		5.81		
7			2.0×10^{-04}	
8				1.9×10^{-13}

46. Provide the missing information for each of these weak acids or bases.

WEAK ACID	CONJUGATE BASE	K_a FOR ACID	PK_a FOR ACID	K_b FOR BASE	PK_b FOR CONJUGATE BASE
$HC_2H_3O_2$ (acetic acid)	(acetate)	1.74×10^{-5}			
(ammonium)	NH_3 (ammonia)		9.25		
$HC_6H_5O_2$ (benzoic acid)	(benzoate)			1.60×10^{-10}	
(hydrofluoric acid)	F$^-$ (fluoride)				10.83
HCN (hydrocyanic acid)	(cyanide)	6.17×10^{-10}			
lactic acid	$C_3H_5O_2^-$ (lactate)		3.86		

47. Give the pH of solutions of these strong acids.
    a. 0.015 M HCl
    b. 0.0032 M HBr
    c. 0.163 M HNO$_3$
    d. 0.00478 M HI

48. Give the pH of solutions of these strong bases.
    a. 0.015 M NaOH
    b. 0.0032 M KOH
    c. 0.163 M LiOH

49. Give the pH of solutions of these weak acid solutions.

SOLUTION	ACID	CONCENTRATION	K_a
1	acetic acid	0.25	1.74×10^{-05}
2	acetic acid	0.025	1.74×10^{-05}
3	acetic acid	0.0025	1.74×10^{-05}
4	ammonium ion	0.15	5.62×10^{-10}
5	ammonium ion	0.015	5.62×10^{-10}
6	ammonium ion	0.0015	5.62×10^{-10}
7	benzoic acid	0.10	6.25×10^{-05}
8	benzoic acid	0.010	6.25×10^{-05}

9	benzoic acid	0.0010	6.25×10^{-05}
10	lactic acid	0.037	1.38×10^{-04}
11	lactic acid	0.050	1.38×10^{-04}
12	lactic acid	0.0056	1.38×10^{-04}

50. Give the pH of solutions of these weak base solutions. Do not give the pOH!

SOLUTION	WEAK BASE	CONCENTRATION	K_a OF CONJUGATE ACID	K_b
1	acetate	0.25	1.74×10^{-05}	5.75×10^{-10}
2	acetate	0.025	1.74×10^{-05}	
3	acetate	0.0025	1.74×10^{-05}	
4	ammonia	0.15	5.62×10^{-10}	
5	ammonia	0.015	5.62×10^{-10}	
6	ammonia	0.0015	5.62×10^{-10}	
7	benzoate	0.10	6.25×10^{-05}	
8	benzoate	0.010	6.25×10^{-05}	
9	benzoate	0.0010	6.25×10^{-05}	
10	lactate	0.037	1.38×10^{-04}	
11	lactate	0.050	1.38×10^{-04}	
12	lactate	0.0056	1.38×10^{-04}	

51. Will the pH of aqueous solutions for these compounds be acidic, basic, or neutral?
    a. NaCl
    b. $NaNO_3$
    c. HCl
    d. NaOH
    e. Na_2CO_3
    f. $LiC_2H_3O_2$
    g. NH_3
    h. NH_4Cl

52. Calculate the pH of each of these acetate buffer solutions. The pK_a for acetic acid is 4.76.
    a. $[HC_2H_3O_2] = 0.15$ M; $[C_2H_3O_2^-] = 0.75$ M
    b. $[HC_2H_3O_2] = 0.55$ M; $[C_2H_3O_2^-] = 0.45$ M
    c. $[HC_2H_3O_2] = 0.25$ M; $[C_2H_3O_2^-] = 0.25$ M
    d. $[HC_2H_3O_2] = 0.75$ M; $[C_2H_3O_2^-] = 0.75$ M

53. Calculate the pH of each of these phosphate buffer solutions. The pK_a for dihydrogen phosphate is 7.20.
    a. $[H_2PO_4^-] = 0.50$ M; $[HPO_4^{2-}] = 0.50$ M
    b. $[H_2PO_4^-] = 0.15$ M; $[HPO_4^{2-}] = 0.30$ M
    c. $[H_2PO_4^-] = 0.50$ M; $[HPO_4^{2-}] = 0.25$ M
    d. $[H_2PO_4^-] = 0.15$ M; $[HPO_4^{2-}] = 0.15$ M

54. Calculate the ratio of sodium dihydrogen phosphate (a weak acid) to sodium monohydrogen phosphate (its conjugate base) that would give a buffer solution having these pH values. The pK_a for dihydrogen phosphate is 7.20.
    a. 7.20
    b. 7.50
    c. 7.00

Electricity and Electrical Safety

Technological advances have made modern living safer, more convenient, and more efficient. A quick look around us reveals a plethora of devices and gadgets that we rely on every day. These devices wake us, make coffee, deliver IV fluid, and monitor fetal heart rates. Many of these items are electronic in nature and thus require some sort of energy supply such as a wall plug or batteries. Our dependence on electricity has become so heavy that a simple electrical outage can cause extreme hardships, both real and perceived. In this chapter we explore electrical circuits and the nature of their behavior.

ELECTRICITY AND ELECTRICAL CHARGE

We are all familiar with the word "electricity"; however, its definition can be elusive. Some might say that electricity is the "stuff" that comes out of wall outlets or from batteries, or perhaps it's what makes up lightning. The term *electricity* can be interpreted in many ways. In this text we define electricity as that which results from mobile charges.

Charge

Now that we have a simple definition for electricity, we need to understand the meaning of charge. The nice thing about charges is the fact that they come in only two varieties: positive and negative. Over the years, humans have observed that the two types of charges—which, by the way, were given arbitrary names by Benjamin Franklin—behave in a very unique, reproducible manner. Positive charges are attracted to negative charges and vice versa, while positive charges are repelled by other positive charges. In addition, negative charges are repelled by other negative charges. In short, like charges repel and unlike charges attract. Charges also exhibit behavior similar to other small entities in our universe. That is, they are quantized or come in discrete units. Finally, scientists have concluded that the total amount of electric charge in the universe is constant. This is known as the Law of Conservation of Electric Charge. To date, there are no known exceptions to this law.

Some of you might be thinking back to the days of your youth when you would walk around in your socks trying to shock other people. So if the total amount of electric charge is constant in the universe, how can you shock others by rubbing your socked feet against carpet? The answer is surprisingly simple. During a process like this, charge is neither created nor destroyed; it is simply transferred from one object to another. Thus, we see that charges must be able to move. Typically, the charges that move are negatively charged electrons.

The SI unit of electrical charge is called the coulomb (C), and one electron has a charge of -1.602×10^{-19} C. Atoms are neutral species in which the number of protons in an atom's nucleus exactly equals the number of electrons in the electronic cloud. Thus, the charge on one proton is equal to $+1.602 \times 10^{-19}$ C. It should be mentioned at this point that the other constituent in the nucleus, the neutron, has no electrical charge. The symbol e is used to represent the *magnitude* (without regard to sign) of the charge on any electron, thus:

$$e = 1.602 \times 10^{-19} \text{ C}$$

Coulomb's Law

The force of attraction or repulsion between two charges is called the electrostatic or coulombic force. The term *electrostatic* implies that the charges are stationary or fixed in position. The magnitude of the electrostatic force between two fixed point charges is given by Coulomb's Law:

$$F = k\frac{|q_1||q_2|}{r^2}$$

In this equation, $k = 8.99 \times 10^9$ N $\cdot$ m²/C² and is called Coulomb's Law Constant or the Coulomb Constant, and $|q_1|$ and $|q_2|$ represent the magnitude (not the sign) of the charge in coulombs on objects 1 and 2, respectively. The quantity r is the distance in meters between the two point charges. Note that Coulomb's Law provides only the magnitude of the electrostatic force between two point charges, not the direction. The directionality of this force is determined by the signs of the charges on the two bodies involved (i.e., like charges repel while opposite charges attract).

The force between charges is very large. For example, let's calculate the magnitude and direction of the force that exists on an electron due to the presence of another electron at a distance of 1.0×10^{-10} m (approximately the distance across an atom).

Let's start by making a diagram showing the physical layout of the problem. We will place the electron of interest (q_1) at the origin and the other electron (q_2) on the x-axis a distance of 1.0×10^{-10} m away, as shown in Figure 10.1.

Using Coulomb's Law, we plug in the desired quantities to get:

$$F_{1,2} = k\frac{|q_1||q_2|}{r^2} = \left(8.99 \times 10^9 \, \frac{\text{N} \cdot \text{m}^2}{\text{C}^2}\right)\frac{|-1.602 \times 10^{-19} \text{ C}||-1.602 \times 10^{-19} \text{ C}|}{(1.0 \times 10^{-10} \text{ m})^2}$$

Note the subscript used on the force ($F_{1,2}$). This signifies the force on electron 1 due to the presence of electron 2. Simplification of the above expression yields the magnitude of the force:

$$F = 2.3 \times 10^{-8} \text{ N}$$

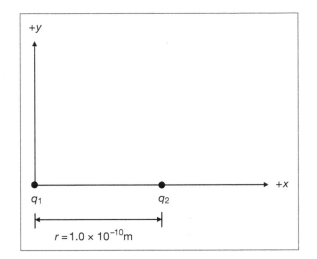

Figure 10.1 Coordinate System for Two Charged Particles

Before you say that's a tiny amount, remember this is for just two electrons. On a more macroscopic level, say two moles of electrons, this force would be multiplied by Avogadro's number.

$$2.3 \times 10^{-8} \, \text{N} \left(\frac{6.02 \times 10^{23}}{\text{mol}} \right) = 1.4 \times 10^{16} \, \text{N}$$

This is the weight equivalent to 3×10^{15} (3 million billion) pounds.

We must now determine the direction. Looking at Figure 10.1, we see that both charges are negative. Hence, the force between them will be repulsive. The force felt by the charge at the origin ($F_{1,2}$) will be directed toward the left (along the $-x$-axis). The other electron will feel a repulsive force ($F_{2,1}$) of the same magnitude but directed toward the right (along the $+x$-axis). These forces are illustrated in Figure 10.2.

Figure 10.2 Forces Between Two Charged Particles

The two forces labeled as $F_{1,2}$ and $F_{2,1}$ also constitute an action–reaction pair, in agreement with Newton's third law.

Electric Field

Contemporary physics uses the idea of a force field to explain the behavior of the electrostatic force between charged entities. Imagine a positive point charge $+q$ located and fixed at the origin of a coordinate axis system. A second positive charge, q_0, called a test charge, is located at point 1, as indicated in Figure 10.3. The test charge does not have a fixed position and is

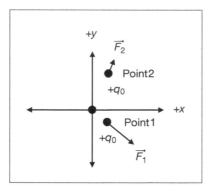

Figure 10.3 Electric Field

thus mobile. The repulsive electrostatic force felt by the test charge at point 1 is represented in the figure by the vector $\vec{F}_1$. Let's move the test charge to point 2. Upon doing so, the repulsive electrostatic force now felt by q_0 is represented by the vector $\vec{F}_2$. We can easily see that there will be a force at every point around our central charge q. This group of forces around q is known as the *force field*. All charges have an associated force field. We can thus say that the force field is what exerts the electric force on a test charge located at any point around a central charge.

It's obvious that the magnitude of the electrostatic force is directly proportional to the magnitude of the test charge q_0. We can remove this dependence by dividing the electrostatic force by q_0 and obtain a quantity known as the electric field, which is in reality a force per unit charge. Since the electrostatic force is really a vector and q_0 is a scalar, the electric field is also a vector:

$$\vec{E} = \frac{\vec{F}}{q_0}$$

The SI units of electric field are newtons per coulomb (N/C). We can use the above equation to calculate the force on a test charge if we know the magnitude of the test charge and the electric field strength.

Electric Potential Energy

Since the test charge in our previous example is free to move, once we place it at point 1, the electrostatic force will cause this charge to accelerate in accordance with Newton's second law. Thus, a charge initially at rest (no kinetic energy) will acquire kinetic energy. However, if we were to somehow hold the charge in place so that it could not move, the charge would have the potential to move and therefore have potential energy. This type of potential energy is called *electric potential energy* and is symbolized by U.

It can be shown, using mathematical techniques that are outside the scope of this text, that the electric potential energy U for two point charges q and q_0 separated by a distance r is given by:

$$U = k\frac{qq_0}{r}$$

The SI unit for electric potential energy is the joule. Note how similar the electric potential energy expression is to Coulomb's Law. One key difference is in the denominator, which has r versus r^2 in Coulomb's Law.

Electric Potential

It is convenient to define another quantity that is independent of q_0. We can divide the electric potential energy U by q_0 to obtain the new quantity called the *electric potential*. The electric potential V is thus defined to be:

$$V = \frac{U}{q_0} = k\frac{q}{r}$$

The SI units for electric potential are energy per unit charge, which is called a volt (V).

$$1\frac{\text{joule}}{\text{coulomb}} = 1\frac{\text{J}}{\text{C}} = 1 \text{ Volt} = 1 \text{ V}$$

Extreme caution must be used when dealing with the concepts of electric potential energy and electric potential. Even though they are related, they are two separate notions differentiated by a single word, "energy." Let the student beware!

Electric Current

Charges have the ability to move, and so we need a way to quantify the movement of charge. Electric current I is defined as the amount of charge flowing per unit time:

$$I = \frac{\Delta Q}{\Delta t}$$

The SI unit of current is the ampere (or amp; A) defined as:

$$1 \text{ A} = 1\frac{\text{coulomb}}{\text{second}} = 1\frac{\text{C}}{\text{s}}$$

OHM'S LAW AND ELECTRICAL CIRCUITS

Electrical Conductors and Insulators

Electrical conductors are materials in which charges can easily move. Most metals are decent electrical conductors. In contrast, electrical insulators are materials in which charges cannot freely move. Most insulators are nonmetallic in nature. One major exception is graphite, which conducts electricity fairly well. Just because a material is an insulator does not mean that it cannot be charged. On the contrary, good electrical insulators can hold a significant static charge. For example, what happens when you rub a balloon on your hair? Charge is transferred between the balloon and your hair, resulting in your hair standing up and the balloon being able to stick to objects like a wall. A similar phenomenon results when you are at a gas station on a cold day and decide to get back into your vehicle while the gas is being pumped. By sliding your pants or skirt on the car seat you can get a charge buildup. Discharge of this static charge can be an ignition source in the presence of flammable gases.

Good electrical conductors are made of atoms that do not hold onto electrons tightly. In other words, the electrons can freely move from one atom to the next. Metals are typically good conductors because they have outer electron shells that are mostly empty. For example, copper has its highest energy electron orbital only half full. In one approach to bonding known as Molecular Orbital (MO) Theory, atomic orbitals from individual atoms are added together to form molecular orbitals that extend over the entire molecule. Remember, orbitals are just mathematical wave functions that describe where an electron can be found in space. In that sense, MO theory is a natural extension of valence bond theory, which localizes the mixing of orbitals between two atoms. The mathematical outcome of the MO approach is typically a large collection (or band) of MOs that is only partially filled. Furthermore, the entire collection of metal atoms is now one large "molecule." Thus, there is room to accept new electrons into the "molecule," or for electrons to move over the entire "molecule" with very little energy. So, if a metal is placed into an electric field, the electrons on the metal develop a potential (and potential energy) and move in accordance with the applied electric field.

Chemists use MO diagrams to display the results of their very computer-intensive quantum mechanical calculations. An example diagram is Figure 10.4. In this diagram, the horizontal lines represent energies of the various allowed states of the molecule or contributing atom, while the arrows represent electrons. In such a diagram, each energy level can have, at most, two electrons of opposite spin. Thus the directions of the arrows represent their spin.

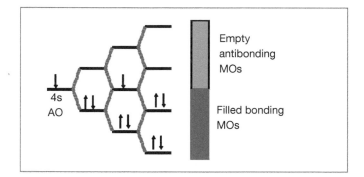

Figure 10.4 Formation of Conductance Bands

Resistance

Electrical conductors allow electrons to pass through their conduction band with very little effort. Some effort, however, is nonetheless required to get electrons to flow. Thus, under ordinary conditions, all materials offer some resistance to the flow of electrons. If this were not the case, a current would continue to flow in a loop of copper wire with no need to expend additional energy once the current was started. The energy required to push electrons through a material is a measure of the resistance of the material.

If we assume that a material such as a wire has a constant resistance R, the relationship between the potential difference V (also known as the voltage) and current I is given by Ohm's Law:

$$V = IR$$

By rearranging Ohm's Law, we find that resistance is simply the ratio of voltage to current, or the ratio of how hard some potential is pushing the electrons to the flow rate of electrons.

$$R = \frac{V}{I}$$

Focusing on the SI units, we find

$$\text{resistance} = R = \frac{V}{I} = \frac{\text{volt}}{\text{ampere}} = \frac{V}{A}$$

Note that an italicized "V" is the potential difference, whereas a nonitalicized "V" is the SI unit known as the volt. Additionally, a volt divided by an ampere is called an ohm, Ω, the SI unit of resistance:

$$1\ \Omega = 1\ \frac{V}{A}$$

Ohm's "law" is really misnamed because there are many materials that do not obey it. Hence, these materials are known as nonohmic, while materials that do obey Ohm's Law are termed ohmic.

Let's now look at an application of Ohm's Law. A 6.0-V battery is connected to a light bulb that allows 0.75 A of current to flow. What is the resistance of the light bulb? Since we are asked to calculate the resistance, we should solve Ohm's Law for resistance and then fill in the values for voltage and current:

$$R = \frac{V}{I} = \frac{6.0\ V}{0.75\ A} = 8.0\ \Omega$$

Conductance

The conductance G is the reciprocal of resistance:

$$G = \frac{1}{R}$$

Conductance is measured in siemens (S). You may also occasionally see the older unit which is the mho, which is ohm spelled backwards. Who says physicists don't have a sense of humor? Notice that the siemens and the mho are both equal to 1/ohm.

Electrical Diagrams

Scientists and engineers frequently use diagrams to schematically represent electric circuits. A circuit exists when charge is able to flow around a closed path. There are two types of circuits:

direct current (DC) circuits and alternating current (AC) circuits. In DC circuits the current flows in one direction only, while in AC circuits the current periodically changes direction. DC circuits commonly use batteries, whereas most AC circuits get their energy from wall outlets or AC generators. We will discuss only DC circuits in detail in this textbook.

Electrical diagrams contain symbols used to represent the various circuit elements. For example, Figure 10.5 is a simple DC circuit containing a battery, a switch, and a resistor.

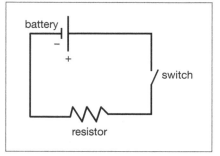

Figure 10.5 A Simple Circuit

This circuit could represent, for instance, a flashlight. Notice that there is only one path that can be taken by the current; thus we call this a series circuit. Parallel circuits offer the current more than one path and will have junctions where wires intersect. Of course, circuits, and hence circuit diagrams, can get very complicated. As such, rules have been developed to help simplify or reduce circuits to equivalent circuits containing fewer components.

Series Circuits

When batteries are connected in series, as in a multibattery flashlight where the positive terminal of one battery contacts the negative terminal of the next battery, the voltages of the batteries simply add together (Figure 10.6). For example, if we connect a 3-V battery and a 6-V battery in series, the equivalent voltage, V_{eq}, would be equal to 9.0 V. For the purposes of a circuit diagram, the two batteries could be replaced by a single 9.0-V battery. Your 12-V car battery is comprised of six 2-V electrical cells, connected in series.

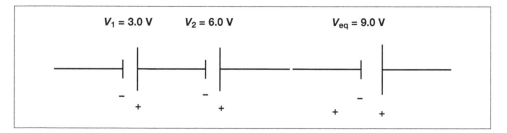

Figure 10.6 Batteries Connected in Series

Resistors in series behave in the same manner as batteries in series. The equivalent resistance is simply the sum of the individual resistances. Figure 10.7 illustrates this point.

$$R_{eq}=R_1+R_2+\cdots+R_n = \sum R_{individual}$$

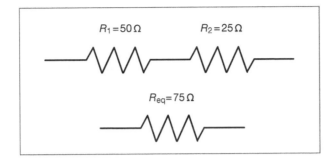

Figure 10.7 Resistors Connected in Series

Parallel Circuits

When resistors are connected in parallel (Figure 10.8), the current has more than one path that it can take. Thus there will be a unique current that flows through each resistor independently. Household circuits are parallel circuits, so that electricity can run through any individual appliance without the requirement that every appliance be turned on in order to complete the electrical circuit.

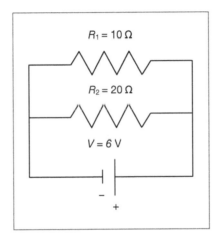

Figure 10.8 Parallel Resistors

The equivalent resistance for a parallel circuit having a battery and more than one resistor is given by:

$$\frac{1}{R_{eq}} = \sum \frac{1}{R_{individual}}$$

So resistors in parallel add inversely. Let's take a look at an example to illustrate this concept. Let's find the equivalent resistance in the parallel circuit pictured in Figure 10.8.

To get the equivalent resistance, we must add the two resistances inversely.

$$\frac{1}{R_{eq}} = \frac{1}{10\ \Omega} + \frac{1}{20\ \Omega} = 0.15\ \Omega^{-1}$$

$$R_{eq} = \frac{1}{0.15\ \Omega^{-1}} = 6.7\ \Omega$$

So, Figure 10.9 shows the two parallel resistors as a reduced circuit with the equivalent resistance. Notice the equivalent resistance is less than the resistance of either resistor.

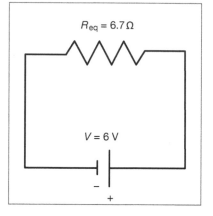

Figure 10.9 Parallel Resistors (Reduced Circuit)

Now, let's calculate the total current in this equivalent circuit using Ohm's Law:

$$I = \frac{V}{R_{eq}} = \frac{6.0 \text{ V}}{6.67 \text{ }\Omega} = 0.8996 = 0.90 \text{ A}$$

Real circuits can have a combination of both series and parallel arrangements, which may or may not be readily reduced to simpler circuits. Let's try one of these more complicated situations. Imagine the circuit shown in Figure 10.10 in which $R_1 = R_2 = R_3 = R = 10 \text{ }\Omega$ and $V = 10 \text{ V}$. Let's find the total current that flows through this circuit.

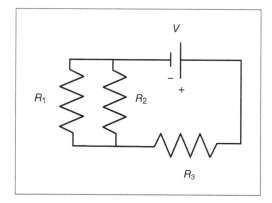

Figure 10.10 Parallel and Series Circuit

Inspection of the circuit reveals that resistors R_1 and R_2 are in a parallel arrangement, while R_3 is in series with these two. Our strategy will be to reduce R_1 and R_2 to an equivalent resistance, which we'll call $R_{eq,1}$. After this, we'll reduce $R_{eq,1}$ and R_3 to another equivalent resistance known as $R_{eq,2}$. Since R_1 and R_2 are in a parallel arrangement, their resistances add inversely:

$$\frac{1}{R_{eq,1}} = \frac{1}{10 \text{ }\Omega} + \frac{1}{10 \text{ }\Omega} = 0.20 \text{ }\Omega^{-1}$$

$$R_{eq,1} = 5.0 \text{ }\Omega$$

We can now redraw this circuit, replacing R_1 and R_2 with the equivalent resistance $R_{eq,1}$. This is shown in Figure 10.11.

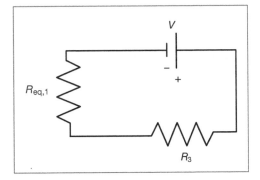

Figure 10.11 Parallel and Series Circuit (2)

We'll now add $R_{eq,1}$ to R_3 to get $R_{eq,2}$:

$$R_{eq,2} = R_{eq,1} + R_3 = 5.0\ \Omega + 10\ \Omega = 15\ \Omega$$

Our reduced circuit now has only a single resistor with an equivalent resistance of 15 Ω as shown in Figure 10.12.

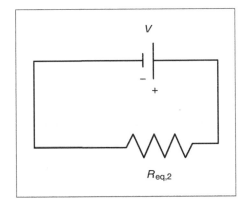

Figure 10.12 Parallel and Series Circuit (3)

We can finally calculate the total current I through the circuit using Ohm's Law:

$$I = \frac{V}{R} = \frac{10\ V}{15\ \Omega} = 0.67\ A$$

As you can imagine, very complicated circuits can be reduced in a similar stepwise fashion. The only thing required to perform such tasks is a little patience.

Electrical Power

Since ordinary circuits have finite resistances, work must be done to move the charges through the circuit. Therefore, power will be consumed. Another way of looking at this is from the point of electric potential energy U. A change in electric potential energy results when a charge ΔQ moves across a potential difference V. This change in electric potential energy is given by:

$$\Delta U = \Delta Q V$$

Since the power P is the rate at which energy is expended or consumed, we can say:

$$P = \frac{\Delta U}{\Delta t} = \frac{\Delta Q V}{\Delta t} = \left(\frac{\Delta Q}{\Delta t}\right) V = IV$$

Note in the above expression we recognized that the current is the rate of flowing charges, $I = \Delta Q / \Delta t$. So, in general, the power consumed or dissipated by an electrical circuit is given by $P = IV$. The SI unit of power is the watt, so we see that a watt is related to an ampere and a volt by:

$$1 \text{ watt} = 1 \text{ W} = 1 \text{ A} \cdot \text{V}$$

For the special case involving a resistor, where $I = V/R$, the power can be expressed as:

$$P = IV = \left(\frac{V}{R}\right) V = \frac{V^2}{R}$$

or as:

$$P = IV = I(IR) = I^2 R$$

Electrical Energy

In the hot summer months, most of us rely on air conditioning to cool our homes. While we enjoy the comfortable temperatures, we do not necessarily like the associated energy bill that accompanies it. Some folks mistakenly refer to their electric energy bill as the "power" bill. To see why this is so, all we have to do is look at the definition of power and rearrange it to get:

$$\text{energy} = \text{power} \times \text{time}$$

So a power unit (e.g., watts or kilowatts) multiplied by a time unit (e.g., hours or seconds) gives an energy unit. Electric companies charge customers for how many units of energy they consume each month. The energy unit used by electric companies is the kilowatt-hour (kWh). To calculate how much you owe, companies take the number of kilowatt-hours used times the cost per kilowatt-hour. For example, if the going rate is 6 cents per kilowatt-hour and you used 1466 kWh, then your energy charge would be:

$$\text{energy charge} = 1466 \text{ kWh} \times \frac{\$0.06}{\text{kWh}} = \$87.96$$

Let's look at an example that is a bit more involved. If electricity costs 10 cents per kilowatt-hour, how much does it cost to operate a 220-V air conditioner that draws 30 A of current for 12 hours per day for 30 days? To solve this problem, we will need to know how much energy (the number of kilowatt-hours) is involved. This in turn is composed of power and time. Thus, a good place to start is power:

$$P = IV = (10 \text{ A})(220 \text{ V}) = 2200 \text{ W} \times \frac{1 \text{kW}}{1000 \text{ W}} = 2.2 \text{ kW}$$

We can calculate the total number of hours involved by taking the number of hours per day times the number of days:

$$\text{total hours} = \left(\frac{12 \text{ hours}}{\text{day}}\right) \times (30 \text{ days}) = 360 \text{ hours}$$

The number of kilowatt-hours is then simply:

$$\text{\# kilowatts} \times \text{hours} = (2.2 \text{ kW}) \times (360 \text{ hr}) = 792 \text{ kWh}$$

The last step involves multiplying the number of kilowatt-hours by the cost per kWh:

$$\text{energy charge} = 792 \text{ kWh} \times \frac{\$0.10}{\text{kWh}} = \$79.20$$

SEMICONDUCTORS

Semiconductors are materials with electrical conducting properties somewhere between those of insulators and conductors. Semiconductors are prepared from semimetals, most commonly silicon. Semiconductors are used in many electronic devices, including computers. What makes these materials so popular is the ability to control the conductivity by the addition of small amounts of impurities called *doping agents*.

p-Type Semiconductors

A p-type (for positive type) semiconductor is an electron-poor material comprised of silicon (a Group IV element) doped with something like boron (Group III). Boron atoms have one less valence electron than silicon atoms, so the crystalline lattice has fewer electrons—or more "positive holes"—in it compared to pure silicon. This is illustrated in Figure 10.13.

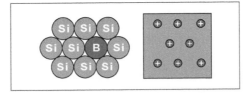

Figure 10.13 p-Type Semiconductor

n-Type Semiconductors

An n-type (for negative type) semiconductor is an electron-rich material comprised of silicon (Group IV) doped with something like arsenic (Group V). Arsenic atoms have one additional valence electron compared to silicon atoms, so the crystalline lattice has more electrons compared to pure silicon. This is shown in Figure 10.14.

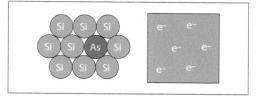

Figure 10.14 n-Type Semiconductor

Diodes

Diodes are circuit elements that have a large conductance in one direction and a smaller conductance in the reverse direction. Thus, they can be used to control the direction of current. Effective diodes can be made by having p-type and n-type regions next to each other within a single crystal of silicon. Such a junction between the two regions is called a *pn junction*.

A diode is said to be forward biased if a voltage is applied to the diode so that the n-region is biased negative compared to the p-region. Under forward biasing, the holes and electrons will move (because of the electric field) toward the pn junction. This is illustrated in Figure 10.15. If the diode is kept biased, the DC power supply will continue to provide electrons to the n-region and produce holes in the p-region. These electrons and holes will move toward the junction and annihilate. Under forward bias, a diode will thus continue to have current flow. This annihilation is exothermic, and the energy is released as a photon of light. Because these diodes emit light, they are called light-emitting diodes, or LEDs.

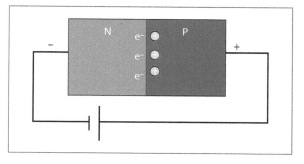

Figure 10.15 Forward-Biased Diode

Something interesting happens if we reverse the polarity of the DC power supply. The diode is now said to be under reverse bias. In this situation, the holes and electrons will migrate away from the pn junction, creating a volume called a *depletion region*, with very few charges. The resistance to current flow under these circumstances is very high compared to forward bias. A reverse-biased diode is shown in Figure 10.16.

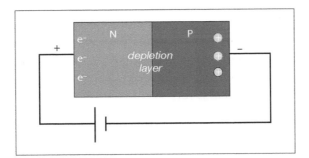

Figure 10.16 Reverse-Biased Diode

Because the current versus voltage relationship is nonlinear, diodes are non-ohmic. Diodes are very popular circuit elements in today's world. For instance, diodes can be used to charge a battery and simultaneously prevent the battery from discharging. Another very popular type of diode is the LED. These devices typically serve as signal lights on many electronic devices such as computers and cell phones because they are small, last a long time, and use very little energy. In the future, the use of LEDs will greatly expand as worldwide energy requirements increase.

Triodes

A triode is made from two closely situated pn junctions. These devices are also known as *transistors*, which is a circuit element that is useful for amplifying or switching currents. Solid-state transistors have revolutionized the electronics industry. Today, very complicated integrated circuits can be made of thousands of semiconductor elements, all contained in a very small package. Such integrated circuitry is vital to the computer, as well as scientific, telecommunication, and many other businesses. Our age is truly the age of the semiconductor!

SPECTROSCOPY

Many analyses are done quickly and efficiently using spectroscopy. The SMA 20 chemical profiles and oxygen pulse oximeters are just two applications of spectroscopy in medicine. The only requirement is that the analyte (the substance being analyzed) must interact with light. When white light passes through a colored substance, certain wavelengths are absorbed. The rest are transmitted or reflected, giving the material its characteristic color. A "color wheel" of the primary and secondary colors is an effective first approximation of which color of light a sample will absorb. A substance absorbs its complementary color. For example, red things absorb green light. An example color wheel is shown in Figure 10.17.

One way to think of light is as "particles" called photons. Photons have a characteristic wavelength. Different colors of light have different wavelengths. Furthermore, different wavelengths of light have different energies. Purple photons are more energetic than red photons. Light interacts with matter when the energy of a photon exactly matches the difference in energy between two electron energy levels. Visible and ultraviolet light cause valence electrons to jump to higher energy levels. By conservation of energy, the energy of the photon is stored as potential energy by the excited electrons. Thus, when white light passes through a colored

Figure 10.17 Color Wheel

species, fewer photons emerge than enter. The percentage of light that passes through a sample is called the percent transmittance, %T. The detector in a spectrometer measures the intensity, or number of photons per second per unit area, of the light before it encounters the sample (I_0) and after (I). The ratio of I to I_0 gives the transmittance:

$$T = \frac{I}{I_0}$$

The amount of light absorbed by the sample, the absorbance or A, is related to T by the following logarithmic relationship.

$$A = -\log(T) = 2 - \log(\%T)$$

Absorbance, like most other logarithmic parameters, has no units.

Instrumentation

Video 10.1

Regardless of the instrument, most spectrometers have the same basic design. Furthermore, all instruments are based on some physical or chemical phenomenon generating an electric current. The source emits the desired wavelengths of electromagnetic radiation. For visible spectroscopy, a tungsten light bulb is used. A pulse oximeter uses LEDs. A slit provides a narrow beam of light that passes through a monochromator and is spread into a spectrum. You can think of a monochromator as a prism, although most instruments use diffraction gratings—not prisms—as monochromators. Manipulation of the monochromator directs the desired wavelength of light through a second slit and then through the sample. A monochromator is not necessary with LED sources because LEDs emit highly monochromatic light. Finally, a detector measures the intensity of the light. The detector exploits the photoelectric effect. A photon of light strikes the detector and ejects an electron. This signal may be amplified into many more electrons by a device such as a photomultiplier tube (PMT). The electrons move through an electric circuit. The electrical current thus generated is directly proportional to the number of ejected electrons and, consequently, the number of photons striking the detector. A spectrophotometer schematic is shown in Figure 10.18.

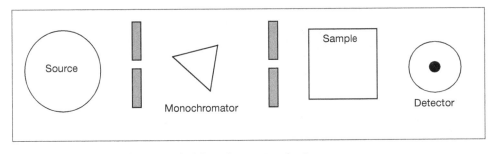

Figure 10.18 Schematic of a Spectrometer

Solid-state detectors are based on diodes. When a reverse-biased diode is struck by a photon, an electron can be ejected from the depletion layer. This creates a mobile hole that moves toward the negative potential in the diode. This flow of current is registered by external circuitry, translating the incident photon as an electric current.

Beer's Law

Three factors affect how much light is absorbed by the sample. A probability factor called the *absorbtivity* (*a*) describes the likelihood that a photon will excite the molecule, and is a constant for each substance at a given wavelength. A greater concentration (*c*) of the analyte increases the chance that a photon will encounter a molecule with which to interact. The same is true of increasing the distance (*b*) the light beam travels through the sample. Beer's Law gives the simple relationship between these variables:

$$A = abc$$

Now, one crucially important feature of Beer's Law is that it is strictly true only when the analyte concentration is zero. So, as concentration increases, the linear relationship between absorbance and concentration begins to break down. You are generally okay as long as the absorbance is between 0.1 and 1.

By measuring the absorbance of several solutions containing an analyte at various known concentrations, you can discern this relationship. Graphing the absorbance versus the concentration yields a "calibration curve," as shown in Figure 10.19. If you measure the absorbance

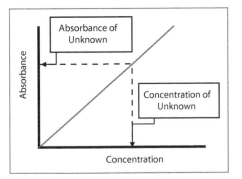

Figure 10.19 Standard Beer's Law Curve

of an analyte solution of unknown concentration, the calibration curve gives the concentration of analyte that corresponds to that particular absorbance.

Electrical Safety

Video 10.2

Electrical devices are an important part of our every day life, both at home and at work. Because of the ubiquitous nature of energized electrical devices and circuits, exposure to potential harm is a real concern. Although modern electrical devices and circuits have a multitude of built-in safety features, we must still treat them with a healthy respect.

Electrical Shock

One of the most misunderstood words involving electricity is the word "electrocute." According to the *American Heritage College Dictionary* this word has two meanings: (1) "to kill with electricity" and (2) "to execute (a condemned prisoner) by means of electricity." The word "shock," on the other hand, according to *American Heritage* means "the sensation and muscular spasm caused by an electric current passing through the body." With this distinction in mind, our modern world is full of electrical shock hazards. Every energized device or circuit has the potential to shock us. Electrocution is even possible in some situations. In the absence of any safety devices, a person can readily experience a prolonged electrical shock by simply completing an electrical circuit. In other words, if one touches an energized circuit, a path can be established where electricity will flow through the body. Imagine a classic blender with an old-style two-prong plug that is inserted into a live 120-volt receptacle. If an internal wire comes loose and subsequently touches the metal casing of the blender, a shock hazard develops whenever the blender is turned on. One can readily receive an electrical shock by simply touching the casing of the blender. The outcome of this event will depend on the path of the current, the magnitude of the current, and the time of contact. In addition, electrical shock can result from stray capacitance and inductance in electrical equipment that is properly operating.

Macroshock Versus Microshock

At one extreme, macroshock occurs when a relatively large amount of current flows through an individual potentially resulting in injury or death. For example, macroshock can occur if one were to touch an energized electrical transformer or stick a metal object into an electrical outlet. At the other extreme, microshock occurs when a relatively minor current is delivered directly to the heart. Microshock can occur when an individual has an external, low-resistance pathway that directly contacts the heart. Such a person would be classified as electrically susceptible. An example of a low-resistance pathway would be a pacemaker wire or a pulmonary catheter carrying a saline solution. Ventricular fibrillation can be caused by very small currents in electrically susceptible patients. Table 10.1 summarizes the bodily effects of macroshock and microshock for 1-second contact times with 60-Hz AC.

Table 10.1 EFFECTS OF MACROSHOCK AND MICROSHOCK ON THE HUMAN BODY

MACROSHOCK, mA	EFFECT
1	Lower limit of perception
5	Maximum harmless current magnitude
10–20	"Let go" current without sustained muscle contraction
50	Pain with possible loss of consciousness
100–300	Start of ventricular fibrillation
6000	Extensive physiological damage
MICROSHOCK, µA	EFFECT
10	Maximum recommended leakage current
100	Ventricular fibrillation

Fuses and Circuit Breakers

Over the years, electrical safety standards have changed, making the possibility of electrical shock progressively smaller. Modern electrical circuits and equipment have a variety of built-in safety devices to minimize the possibility of electrical shock. First and foremost, electrical circuits have fuses or circuit breakers typically installed in a panel. These devices are designed to prevent too much current from flowing because of a short circuit or circuit overload. A fuse contains a thin metal strip that heats up due to the resistance of the metal to the electrical current. If the current exceeds the rated value of the fuse, the metal strip will become too hot and melt, thus interrupting the current to the circuit. When the fuse has melted, it cannot be reused and therefore must be replaced. There are several different types of circuit breakers. One type is made with a switch and a strip composed of two metals. This bimetallic strip will warp when heated because of its resistance to the current. If the current is above the rated value of the circuit breaker, the bimetallic strip will bend enough to open, or trip, the switch and stop the flow of current. Once the strip cools, it will return to its original shape, thus allowing the circuit breaker to be easily reset with a flip of the switch.

Polarized Plugs

Polarized plugs have one wide and one narrow prong, as shown in Figure 10.20. Likewise, polarized receptacles have a wide opening and a narrow opening, as illustrated in Figure 10.21.

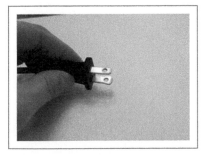

Figure 10.20 A Polarized Plug

Figure 10.21 A Polarized Receptacle

By convention, the narrow opening of the receptacle is at high potential while the wide opening is at low potential. This geometry forces the plug to be oriented in a specific direction when inserted into an outlet. Manufacturers design electrical appliances so that the case is connected to the low potential, wide prong of the plug. Thus, when the device is plugged in, the casing is at low potential. In addition, when the electrical appliance is turned off, the polarized plug ensures high potential only exists from the receptacle to the power switch on the appliance while the remainder of the appliance is at low potential.

Three-Pronged Grounded Plugs

Figure 10.22 illustrates a three-pronged grounded plug that has two equal-sized flat prongs and a third, round dedicated ground prong that is wired directly to the case of the electrical device. This style of plug also forces the user to insert it into an outlet using a specific orientation. If, for some reason, the high potential wire happens to come into contact with the casing of the appliance, the third prong provides a path for the current directly to ground, thus protecting someone who might touch the case of the appliance.

Figure 10.22 A Three-Pronged Grounded Plug

Ground Fault Circuit Interrupter

A ground fault circuit interrupter, or GFCI, is used in circuits that are near sources of water such as sinks, tubs, and swimming pools. If an electrical device is operating as designed, the current flowing to the device from the high-potential wire should equal the current returning from the device via the low-potential (neutral) wire. The GFCI is designed to immediately disrupt the current in a circuit if an imbalance between the two currents is detected. GFCIs are designed to detect very small current differences on the order of 5 mA or less. A GFCI-equipped outlet looks similar to an ordinary outlet but typically has two additional buttons on its face, as shown in Figure 10.23. One of the buttons is for testing the functionality of the outlet while the second button is used to reset the circuit. For circuits containing more than one outlet, a GFCI-equipped outlet can be installed as the first receptacle in the circuit and the remaining outlets can be the non-GFCI style. In this case, the GFCI-equipped outlet will protect an individual from accidental shock if a current surge is experienced through the GFCI outlet or through any of the other outlets located after it in the circuit. Alternatively, a GFCI may be installed in the form of a circuit breaker inside of a breaker box. In such a case, all of the outlets in the circuit may appear to be ordinary non-GFCI outlets, but are in fact protected by the GFCI circuit.

Figure 10.23 A GFCI With Three-Pronged Receptacle

Ungrounded Systems

Although grounded electrical systems are designed to minimize shock hazards, these systems can, in certain situations, still present a reasonable possibility of a shock. A primary example is the modern operating room. With an abundance of electrical devices and various conducting fluids, the risk of macroshock can be unacceptably high. The potential for operating room personnel or patients to unintentionally complete a circuit by accidentally touching something is a real possibility. To further reduce the likelihood of electric shock in this type of environment, most operating rooms use isolated ungrounded electrical circuits. Many times these types of circuits are simply referred to as isolated circuits or ungrounded circuits. Figure 10.24 shows such a circuit.

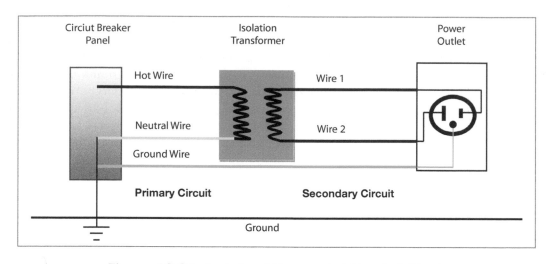

Figure 10.24 An Isolated Ungrounded Electrical Circuit

To make an ungrounded circuit, one needs an isolation transformer. The operating principle behind isolation transformers is electromagnetic induction. Faraday's Law of Induction is the quantitative relationship that mathematically describes electromagnetic induction. Simply put, Faraday's Law of Induction states that a changing magnetic flux in one coil produces a changing voltage, and hence a changing current, in a second coil. One way to change the magnetic flux is to change the magnetic field. This is easily accomplished by using AC in the first coil. A typical isolation transformer consists of a primary circuit connected to an AC power source and a secondary circuit, each with coils wrapped around a common iron core. Note that there is no direct contact between the primary and secondary circuits. In addition, the secondary circuit is not directly connected to ground and thus consists of only two wires, with a 120-V potential between them. The benefit of an ungrounded system is that one of the wires can be touched and no current will flow. This is because a complete circuit would not exist in this situation due to the lack of a direct ground connection, in stark contrast to a grounded system. A person would therefore have to touch both wires of the ungrounded system simultaneously before experiencing an electrical shock. Electrical outlets connected to ungrounded circuits still contain dedicated ground wires. These dedicated ground wires; however, are also not in direct contact with the secondary circuit. If the plug of an electrical device with a short is inserted into an ungrounded circuit, the excessive current simply flows through the dedicated ground wire, thus converting the circuit into a grounded circuit. In this case, there is no significant electrical hazard to the user of the equipment. A second fault, such as touching a live wire, can, however, lead to an electrical shock in this situation.

Line Isolation Monitors

All real AC electrical circuits experience a phenomenon known as leakage current where small amounts of current (on the order of mA) flow to ground, even though there is no direct connection to the ground wire. The reason behind this is beyond the scope of this book. Suffice to say, the presence of leakage current results in an ungrounded circuit becoming grounded, although to a very minor extent. Contact with such small leakage currents presents no real threat to operating room personnel. However, this can become an issue when multiple devices are plugged into an ungrounded circuit resulting in a potentially large amount of leakage current.

Thus, it is necessary to continually check the grounding status of an isolated circuit. A line isolation monitor (LIM) is designed to measure the impedance to ground of both lines in an isolated ungrounded circuit. Impedance is a generalization of the concept of resistance, which is simply the opposition of a circuit to the flow of current. High impedance would result in very little current while low impedance can result in large currents. As an example, suppose a current of 5 mA (5×10^{-3} A) was to flow to ground from one of the lines in an isolated circuit. Applying Ohm's Law and using impedance, Z, in place of resistance, R, yields:

$$V = IZ$$

Rearrangement of this expression gives:

$$Z = \frac{V}{I} = \frac{120V}{5 \times 10^{-3}A} = 24,000 \ \Omega$$

Thus, 5 mA of current corresponds to 24,000 Ω of impedance. LIMs display the current that corresponds to the measured impedance to ground in an isolated circuit. It must be emphasized that the displayed value is the current that would result if a fault were to occur, not necessarily the current that is actually flowing at that moment. A fault could result, for example, if a piece of equipment with a short is plugged into an isolated circuit, thus reducing the impedance to ground. Upon connecting this faulty piece of equipment to the circuit, the LIM would alarm, indicating the presence of the first fault. Electrical shock could occur if a second fault were to develop. Depending upon the specific LIM model used, the alarm is set to trigger at either 2 or 5 mA. An alarm indicates that the impedance to ground of one line has fallen below the preset limit and thus a current greater than either 2 or 5 mA could flow. In short, a LIM alarm indicates that the ungrounded state of an isolated circuit has been compromised. An important aspect of LIMs is that they allow the circuit to remain on, so that vital equipment can continue to be used.

False LIM alarms can be annoying, but they can also lead to complacency. All LIM alarms should thus be taken seriously, no matter how many false alarms one has experienced. The validity of an alarm should first be investigated by examining the current gauge. For example, if the LIM has a 2 mA alarm limit and the gauge currently reads somewhere between 2 and 5 mA, there is a distinct possibility that too many electrical devices are connected to the isolated circuit, resulting in too much leakage current. To remedy the situation, any unnecessary electrical devices should be unplugged from the isolated circuit. Readings of >5 mA typically indicate that a real fault has occurred. The faulty device can be identified by sequentially unplugging one piece of equipment at a time until the indicated current on the LIMs drops below the preset limit and the alarm is silenced. Once the faulty piece of equipment has been identified, an evaluation of the necessity of the equipment needs to be performed. If the equipment can be easily replaced, it should be replaced. If the equipment is not necessary then it should be removed. If the piece of equipment is absolutely necessary, and cannot be replaced, its operational status should be verified. If the offending piece of equipment is still operational, it may still be used, with the recognition that the isolated circuit is no longer ungrounded, thus posing an increased risk of electrical shock in the event of a second fault. Established facility procedures for evaluating LIM alarms and potentially faulty equipment should always be followed.

Electrosurgery

Electrosurgery (ESU) is a methodology that employs high frequency current in the approximate range of 0.3 – 2 MHz. This corresponds to the radiofrequency region of the electromagnetic spectrum. If a high frequency current passes through a large resistance, the instantaneous power dissipated by the resistor is:

$$P = I^2 R$$

where I is the instantaneous current in amperes and R is the resistance in ohms. However, since ESU uses alternating current, we must deal with the average value of the power. Thus, the average power, P_{av}, dissipated by the resistor is:

$$P_{av} = I_{rms}^2 R$$

where I_{rms} is a special type of average current known as the root mean square current. The resistance is provided by the tissue to which the ESU electrode is applied. The power will be dissipated as heat. If this power is distributed over a small area, very large intensities result, thus giving the surgeon the ability to cut or coagulate tissue at a desired location. Another benefit of the high frequency current is poor tissue penetration and its inability to excite contractile cells. Thus, these high frequency currents have the ability to pass right across the precordium and not cause ventricular fibrillation.

In order for current to flow in ESU, a conducting path or circuit must be present. In unipolar ESU, the current is introduced into the body via a small electrode that produces a very high current density at its tip. A large-area-return electrode is used to provide a safe return path through the body. The large area of the return electrode ensures that the current disperses as it makes its way through the body. By keeping the current density sufficiently low at locations away from the small electrode, little or no tissue damage will result as the current travels through the body and into the return electrode. It is vitally important that the return electrode makes proper physical and electrical contact with the patient's body. If the electrode does not make proper contact, severe burns can result. Bipolar ESU uses two closely spaced small electrodes. For example, the two electrodes can be incorporated into a device that looks like a pair of forceps. As the high frequency current flows between the two electrodes, the current only flows through a small volume of tissue before exiting the body. Bipolar ESU therefore does not need a large-return electrode connected to the patient.

Summary

- Electricity results from mobile charges.

- Magnitude of the charge on an electron:

$$e = 1.602 \times 10^{-19} C$$

- Coulomb's Law:

$$F = k\frac{|q_1||q_2|}{r^2}$$

- The electric field is defined as:

$$\vec{E} = \frac{\vec{F}}{q_0}$$

- Electric potential energy:

$$U = k\frac{qq_0}{r}$$

- Electric potential (voltage):

$$V = \frac{U}{q_0} = k\frac{q}{r}$$

- Electric current:

$$I = \frac{\Delta Q}{\Delta t}$$

- Ohm's Law:

$$V = IR$$

- Conductance:

$$G = \frac{1}{R}$$

- Series circuits have only a single pathway through which the current can flow.

- Resistors in series:

$$R_{eq} = R_1 + R_2 + \cdots + R_n = \sum R_{individual}$$

- Parallel circuits have two or more pathways through which the current can flow.

- Resistors in parallel:

$$\frac{1}{R_{eq}} = \frac{1}{R_1} + \frac{1}{R_2} + \dots + \frac{1}{R_n} = \sum \frac{1}{R_{individual}}$$

- Electrical power:

$$P = IV = \frac{V^2}{R} = I^2 R$$

- Semiconductors are materials with electrical-conducting properties between those of insulators and conductors.
- Diodes are electric circuit elements that have a large conductance in one direction and a smaller conductance in the reverse direction. Diodes can be used to control the direction of current in an electrical circuit.
- SMA 20 chemical profiles and oxygen pulse oximeters rely on spectroscopy for fast efficient analyses.
- The transmittance is defined as:

$$T = \frac{I}{I_0}$$

- Absorbance:

$$A = -\log(T) = 2 - \log(\%T)$$

- Beer's Law:

$$A = abc$$

- Electrical shock can occur if one comes into contact with an external source of electricity, including stray capacitances and inductances, and completes a path through which the current can flow.
- Macroshock occurs when a relatively large amount of current flows through an individual, potentially resulting in injury or death.
- Microshock occurs when a relatively minor current is delivered directly to the heart. Microshock can occur when a person has an external, low-resistance pathway that directly contacts the heart. Such a person would be classified as electrically susceptible.
- The longer an individual is in contact with a current, the greater the damage will be.
- Modern electrical appliances and circuits can have one of many built-in safety devices including polarized plugs, three-pronged grounded plugs, and ground fault circuit interrupters.
- Modern operating rooms have isolated circuits to prevent shock hazards.
- Isolation transformers use magnetic inductance to produce an ungrounded secondary current that is isolated from the primary, grounded system.

- Isolation monitors warn operating room personnel if a fault in an ungrounded system is present.
- Electrosurgery is a technique that uses high-frequency current to cut or coagulate tissue. Based on the configuration of the electrodes, there are two categories of electrosurgery: unipolar and bipolar.

Review Questions for Electricity

1. Describe and give units for charge, electrical potential, current, resistance, and power.

2. Describe Coulomb's Law and use it to calculate forces between charges.

3. Describe electrical conductors and insulators.

4. Describe Ohm's Law.

5. What voltage is needed to drive a current of 3.0 A through a $1\bar{0}0$ kΩ resistor?

6. What is the current that flows through a toaster if its resistance is $5\bar{0}$ Ω and operates at 120 V?

7. A 1.5-V battery drives a $1\bar{0}0$-mA current through a flashlight bulb. What is the resistance?

8. Calculate the electrical power dissipated by the lightbulb (a resistor) that draws a $1\bar{0}0$-mA current from a 1.5-V battery.

9. How much current does a $1\bar{0}00$-W hairdryer draw? Assume $V=120V$.

10. A string of lights draws 1.25 A. What is the power usage? Assume $V=120V$.

11. A 10.0-Ω, 5.00-Ω, and 2.50-Ω resistor are wired in series. What is the total resistance?

12. A 10.0-Ω, 5.00-Ω, and 2.50-Ω resistor are wired in parallel. What is the total resistance?

13. Describe semiconductors, p-type semiconductors, and n-type semiconductors.

14. Describe diodes and triodes and give an application of each.

15. A LIM gauge indicates 3.9 mA of leakage current. Calculate the impedance corresponding to this leakage current.

16. An OR LIM is set to alarm if the impedance to ground in an isolated circuit drops below $6\bar{0},000$ Ω. If the LIM alarms, at what current will the alarm be triggered?

17. What is the purpose of the return electrode in unipolar electrosurgery (ESU)?

Classes of Organic Compounds

FUNCTIONAL GROUPS

Organic compounds are based on carbon. You have no doubt had an introduction to organic chemistry during your undergraduate studies, and you know that organic chemistry is the foundation of biochemistry. Carbon merits its own branch of chemistry because it is nearly unique in its ability to form bonds to other carbon atoms, including straight-chained, branched-chained, and cyclic structures. Carbon also forms stable covalent bonds with many other elements, including hydrogen, oxygen, nitrogen, phosphorus, and sulfur. Because of this versatility in bonding, carbon-based compounds exhibit isomerism. Isomers are different compounds that have the same molecular formula. Isomers can differ in either the atom connectivity (i.e., which atom is bonded to which) or in stereochemistry (i.e., how the atoms are arranged in three-dimensional space).

So, there are millions of organic compounds. Even though organic chemistry presents a bewildering number of substances, it is a well-understood branch of chemistry because this vast array of compounds fit into about a dozen or so functional groups, and the chemical properties of these functional classes can be described in about 10 or so unique mechanistic (bond-forming/bond-breaking) processes. Of course, as any organic chemistry student can tell you, the devil is indeed in the details, so a thorough exploration of the chemistry of organic compounds is beyond the scope of our purposes. The purpose of this chapter is to review the kinds and structures of the common classes of organic compounds, and briefly to explore their chemical and physical properties and their nomenclature.

Organic molecules have two parts: a carbon backbone that is relatively inert and one or more functional groups. Because the carbon–carbon bond is relatively strong, a carbon backbone provides a stable template for the functional groups. A functional group is a set of atoms bonded together in a specific way, and it is the functional groups that largely define the chemical and physical properties of the compound. Put another way, you can expect molecules with similar functional groups to have similar physical and chemical properties. New drug discovery often exploits this observation. Chemists prepare new potential medications by synthesizing derivatives of known compounds by manipulating the functional groups, attempting to fine-tune their physiological properties.

HYDROCARBON FUNCTIONAL GROUPS

Hydrocarbons contain only hydrogen and carbon. The hydrocarbon functional groups include alkanes, alkenes, alkynes, and arenes (aromatic compounds). Simple hydrocarbons have few medicinal applications, but are the feedstock of the petrochemical industry to produce plastics, dyes, solvents, detergents, and adhesives (to name just a few). Therefore, hydrocarbons are essential to the medical field. Additionally, all hydrocarbons are flammable and, therefore, find application as fuels. For example, gasoline is a mixture of hydrocarbons.

Alkanes

Also known as paraffins, alkanes are the simplest of the organic compounds. Alkanes are characterized by carbon–carbon single bonds. Thus, alkanes are the carbon backbone for other functional groups. Methane (CH_4) is the simplest alkane. Ethane, CH_3–CH_3 is the next member of the alkane family. There are several ways we can represent ethane.

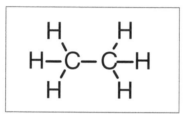

Figure 11.1 Lewis Structure of Ethane

Figure 11.1 is, of course, a Lewis structure. In a Lewis structure, the atoms are represented by their atomic symbol and the covalent bonds are represented by lines. Figure 11.2 is a ball-and-stick model of ethane. Here, the gray spheres represent the carbon atoms and the blue spheres are the hydrogen atoms. Later, we will encounter molecules with oxygen or nitrogen atoms, which will be represented by red and light blue spheres, respectively. A ball-and-stick model better represents the three-dimensional structure of ethane. Notice that each carbon has

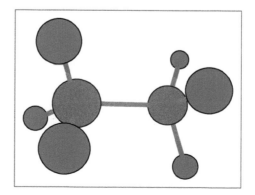

Figure 11.2 Ball-and-Stick Model of Ethane

a tetrahedral geometry, which is what you would expect for an atom with four groups bonded to it. Finally, Figure 11.3 is a space-filling model of ethane. In a space-filling model, the spheres represent the boundaries of the orbitals of the electrons that bind the atoms into a molecule. The space-filling model appears the most congested, and many people find this to be the most confusing representation. However, a space-filling model is more realistic because there are no empty spaces between the atoms in a molecule, as suggested by the ball-and-stick model.

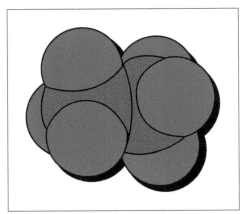

Figure 11.3 Space-Filling Model of Ethane

Table 11.1 lists the normal alkanes. Normal alkanes have a linear chain of carbon atoms with no branching. Notice that the stem names for the C5–C10 alkanes are the Latin equivalent of the number: pent = 5; hex = 6, and so on.

Table 11.1 THE NORMAL ALKANES

CH_4 methane	$CH_3-(CH_2)_4-CH_3$ hexane
CH_3-CH_3 ethane	$CH_3-(CH_2)_5-CH_3$ heptane
$CH_3-CH_2-CH_3$ propane	$CH_3-(CH_2)_6-CH_3$ octane
$CH_3-(CH_2)_2-CH_3$ butane	$CH_3-(CH_2)_7-CH_3$ nonane
$CH_3-(CH_2)_3-CH_3$ pentane	$CH_3-(CH_2)_8-CH_3$ decane

Bond-Line Structures (Molecular Graphs)

As we explore increasingly complex structures, drawing complete Lewis structures that show every atom becomes increasingly tedious. Furthermore, the underlying geometry of a structure can be obfuscated by drawing in every atom. So, organic chemists usually use bond-line structures (or molecular graphs) to represent organic molecules. In a bond-line structure, we draw lines to represent the carbon–carbon bonds. A carbon atom is located wherever the molecular graph terminates or changes direction. It is also understood that every carbon is bonded to enough hydrogen atoms to satisfy the octet rule. For carbon, that (almost) always means that every carbon atom has a total of four covalent bonds. For simplicity and clarity, though, the hydrogen atoms are rarely included in the structure. For example, consider the two representations of 3-methylpentane shown in Figure 11.4. The compound has five carbon atoms in a chain, with a one-carbon methyl group (shown in blue) appended to the middle carbon.

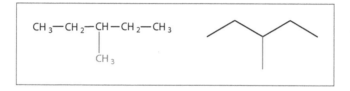

Figure 11.4 Bond-Line Structure of 3-Methylpentane

There is free rotation around carbon–carbon single bonds. Therefore, alkane chains are quite flexible and can adopt a large number of conformations. For example, two conformations of butane are shown in Figure 11.5. These two structures are not isomers of each other because it is not possible to separate them.

Figure 11.5 Free Rotation Around Single Bonds

Alkyl Groups

If a hydrogen atom is removed from an alkane, it leaves an alkyl group. You cannot isolate an alkyl group and keep it on the lab bench in a test tube. An alkyl group is a stick-on group that is always connected to a molecular parent or to another group of atoms. Alkyl groups are named by replacing "ane" on the alkane name with "yl". Table 11.2 lists some common alkyl groups. The dash that is all the way to the right represents the bond that connects the alkyl group to the rest of the molecule. The isopropyl group is not a systematic name but has been accepted by the International Union for Pure and Applied Chemistry, or IUPAC. The isopropyl group is an *isomer* of the propyl group. While the propyl group connects to another group of atoms at the first carbon in the three-carbon chain, the isopropyl group is attached at the center atom to another group. The isobutyl group is less commonly encountered in the chemical literature, but is more common in the medical

Table 11.2 SOME COMMON ALKYL GROUPS

CH_3- methyl	CH_3-CH_2- ethyl
$CH_3-CH_2-CH_2-$ propyl	$CH_3-CH_2-CH_2-CH_2-$ butyl
$(CH_3)_2CH-$ isopropyl	$(CH_3)_2CH-CH_2-$ isopropyl

literature. For example, 2-[4-*isobutyl*phenyl]propanoic acid is known in the medical community as *ibu*profen.

Alkyl groups are generally symbolized by R. So, when you see R in an organic structure, it means some generic alkyl group. It could be ethyl or methyl, and it really doesn't matter. The R group is the carbon backbone of the molecule, and we can largely ignore it and focus on the functional groups. This sort of generalization works in organic chemistry because the principal chemical and physical properties of a compound are determined by its functional groups, not the carbon backbone.

This is not to say, however, that the identity of the R group is absolutely irrelevant. As the R group becomes larger and bulkier, it does have an impact on the chemical and physical properties of the functional groups present. For example, since C–H and C–C bonds are essentially nonpolar, larger R groups increase the hydrophobic nature of the molecule. Also, as the R groups become bulkier and more congested (e.g., isopropyl compared to methyl), the R groups require more space and can prevent incoming reagent molecules from approaching or require incoming molecules to adopt a specific orientation. This is the key behind the remarkable specificity of enzymes.

Organic Nomenclature: IUPAC Rules

In order to communicate with each other, chemists must be able to assign unambiguous names to every compound. A valid chemical name points to a single substance. Unfortunately, in organic chemistry especially, the reverse is not always true. Organic compounds sometimes have more than one commonly accepted name. For example, 2-propanol, isopropanol, and isopropyl alcohol all refer to the main ingredient in rubbing alcohol.

The IUPAC publishes the definitive rules for chemical nomenclature. The IUPAC name is the official name for an organic compound. All other names are called trivial or common names. While the details can sometimes be very challenging, the basic rules are really quite simple.

- Name the longest chain as the parent alkane.
- Name any substituent alkyl groups in alphabetical order.
- Use multipliers to indicate the number of identical substituents
  (di = 2, tri = 3, tetra = 4, penta = 5, etc.).

■ Number the parent chain (begin at one end, not in the middle), and indicate substituent position(s) with lowest possible numbers (called *locants*).

■ Numbers are separated from letters by hyphens. Numbers are separated from numbers by commas.

In the examples in Figure 11.6, the parent chain is shown in black, and the alkyl substituents are shown in blue.

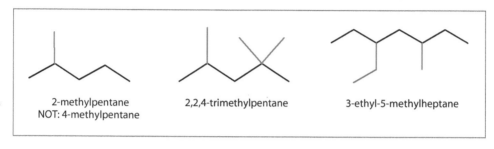

2-methylpentane
NOT: 4-methylpentane

2,2,4-trimethylpentane

3-ethyl-5-methylheptane

Figure 11.6 Alkane Nomenclature Examples

Notice in 2,2,4-trimethylpentane that each methyl group gets a locant number, even though two of the locants are the same. With 3-ethyl-5-methylheptane, either the ethyl group or the methyl group could be assigned the lower number, and we would still have the same set of locants. In this case, the ethyl group gets the lower number because it is listed first in the name.

Cycloalkanes

A chain of carbon atoms can form ring structures. To name a cycloalkane, we simply add the prefix "cyclo" to the parent name. Some examples of cycloalkanes are given in Figure 11.7. Cyclopropane has been used as an anesthetic gas.

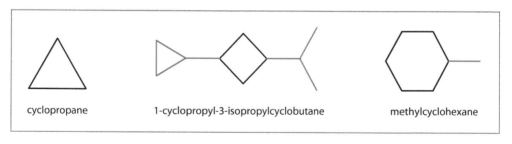

cyclopropane

1-cyclopropyl-3-isopropylcyclobutane

methylcyclohexane

Figure 11.7 Cycloalkane Nomenclature Examples

It is acceptable to omit locant numbers if there is no ambiguity in the location of the substituent groups. For example, the locant number 1 is unnecessary in methylcyclohexane. However, it is *never* wrong to include locant numbers.

Saturated Compounds

The generic formula of a *saturated* alkane is C_nH_{2n+2}. Saturated molecules contain the maximum number of hydrogen atoms for a given number of carbon atoms. Forming rings or double bonds "costs" hydrogen atoms. Therefore, alkenes and cycloalkanes have fewer hydrogen atoms per carbon atom than alkanes. For this reason, compounds that contain double bonds or rings are called *unsaturated*. We will discuss unsaturation in greater detail in the context of fatty acids and triglycerides.

ALKENES

Alkenes, also known as olefins, have a carbon–carbon double bond functional group. The simplest alkene is ethene (aka ethylene in industrial chemistry), and some representations of ethene are given in Figure 11.8. Notice that the geometry around alkene carbons is trigonal planar.

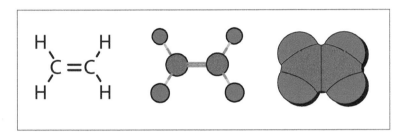

Figure 11.8 Some Models of Ethene

To name alkenes, we follow the basic IUPAC rules, but add the suffix "ene" to indicate the main functional group is an alkene. The location of the double bond is given by a locant number, which identifies the first carbon in the double bond. The main functional group always gets the lowest number. Figure 11.9 illustrates some example alkene structures and names.

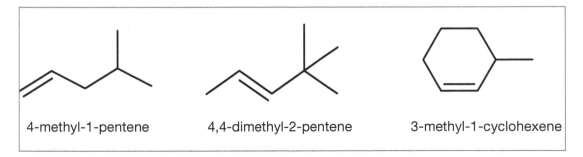

4-methyl-1-pentene   4,4-dimethyl-2-pentene   3-methyl-1-cyclohexene

Figure 11.9 Alkene Nomenclature Examples

Cis/Trans Isomerism

There is normally free rotation around the carbon–carbon single bonds in alkanes. The alkene functional group has two carbon–carbon bonds. The introduction of the second bond freezes rotation around the carbon–carbon double bond, so alkenes are less flexible than alkanes.

Alkenes, therefore, exhibit a kind of stereoisomerism called *cis/trans isomerism*. We will restrict our consideration to alkenes in which each alkene carbon has a bond to a hydrogen atom.[1]

The cis isomer has both of the hydrogens on the same side of the double bond, whereas the trans isomer has each hydrogen on opposite sides of the double bond. Figure 11.10 shows the cis and trans isomers of 2-butene.

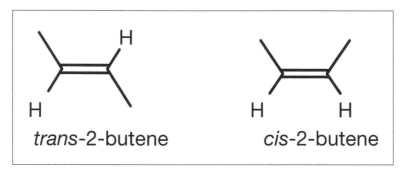

Figure 11.10 Cis and Trans Isomers

Alkenes are also much more reactive than alkanes. This is because the electrons that form the second bond occupy an orbital that lies relatively far away from the center of the molecule. The red and blue lobes in Figure 11.11 describe the area occupied by the "double-bond" electrons.

Figure 11.11 The Alkene Pi Bond

For any purists, these are called pi electrons. Therefore, reagents that are attracted to electrons (i.e., reagents that bear a partial positive charge) are very responsive to alkenes.

Reactions of Alkenes

The chemistry of alkenes is quite extensive, but we will consider only two reactions. The first is called *hydrogenation*. In the presence of an appropriate catalyst, such as finely divided platinum

---

[1] To name alkene diastereomers in general, organic chemists use the *E* (entgegen; against) and *Z* (zusammen; together) stereochemical descriptors. If you are interested in learning more, please refer to any organic chemistry text.

metal, hydrogen gas reacts with an alkene to form an alkane. Each of the alkene carbons gets a bond to a hydrogen atom (shown in blue in Figure 11.12). During this process, the double-bond electrons are used to form one of these new bonds, and the bonding electrons from the hydrogen molecule are used to form the other.

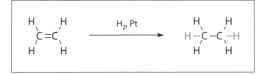

Figure 11.12 Hydrogenation of an Alkene

Hydrogenation converts liquid vegetable oils into solid vegetable shortening (e.g., Crisco®). If you add yellow food coloring and a few artificial flavors to vegetable shortening, you get margarine. The fatty acids in vegetable oils frequently contain double bonds and are, therefore, said to be unsaturated. The double bonds in naturally occurring vegetable oils have a cis configuration. The cis configuration forces a kink into the carbon chains, so the vegetable oil molecules have difficulty in settling into the solid state. Therefore, they have low melting points and are liquids at room temperature. Solid fats, on the other hand, have more saturated carbon chains, and the greater flexibility of the alkane chains facilitates fat molecules settling into the solid state. So when the double bonds in a vegetable oil are hydrogenated, the liquid is converted into a solid. Unfortunately, the hydrogenation process is somewhat reversible. During the hydrogenation of a vegetable oil, pairs of hydrogen atoms can be removed by the platinum catalyst, reintroducing a double bond into the carbon chain. Trans double bonds are more stable than cis double bonds, so the double bonds introduced into the chain are largely trans. This is the source of trans fats that have been implicated in heart disease.

Alkenes are also one main feedstock for preparation of polymers (i.e., plastics). Again, the double-bond electrons play a critical role. In the presence of a suitable catalyst, the double-bond electrons can be used to stitch together a large number of alkene molecules into immensely long alkanes (as seen in Figure 11.13). For example, polymerization of ethene gives polyethylene.

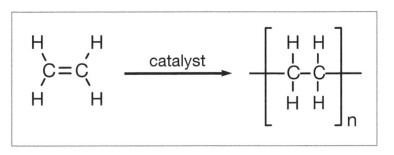

Figure 11.13 Polymerization of an Alkene

The length of the chain depends on the reaction conditions. The longer the chain, the denser and stronger the polymer. Cross-linking agents, such as 1,3-butadiene ($CH_2{=}CH{-}CH{=}CH_2$), also influence the density and strength of polyethylene. High-density polyethylene is used for beverage bottles, whereas low-density polyethylene is used in plastic films. Ultra-high-molecular-mass polyethylene finds applications in surgical prosthetic devices.

Alkynes

Alkynes, also known as acetylenes, have a carbon–carbon triple bond functional group (see Figure 11.14). Alkynes are relatively rare in medical settings, so we present them only for the sake of completeness. The simplest alkyne is ethyne, which is more commonly known as acetylene. Alkynes are named by adding the suffix "yne" and following the normal IUPAC rules.

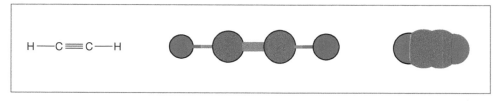

Figure 11.14 Models of Ethyne

Aromatics

Aromatic compounds, sometimes called arenes, have a functional group called a *benzene ring*. Actually, this is too narrow a definition, because these are nonbenzenoid aromatic compounds. However, this definition is a good starting point. Aromatic compounds are very common in nature because they are especially stable. Aromatic rings are frequently a structural element in medications. Benzene is the archetype aromatic compound and has the molecular formula C_6H_6. The carbon atoms form a six-membered ring, and the entire molecule is flat. Models of benzene are shown in Figure 11.15.

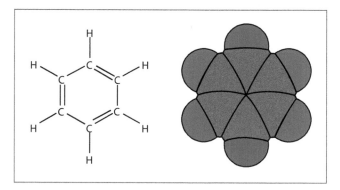

Figure 11.15 Models of Benzene

The ring appears to have alternating double and single bonds, but this is not technically correct. Benzene exhibits resonance, which means the double bonds are actually delocalized around the entire six carbon atoms in the ring. It's not that the double bonds are moving between different pairs of carbon atoms; it's that the double bonds were never localized between any two carbon atoms in the first place. You will sometimes see the double bonds replaced by a circle, which emphasizes the delocalization of the pi electrons. These concepts are illustrated in Figure 11.16.

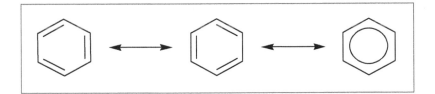

Figure 11.16 Resonance Structures of Benzene

In naming aromatic compounds, you may name the benzene ring as the parent or as a substituent, depending on which strategy is more convenient. When a benzene ring is a stick-on group, it is called a phenyl group. Aromatic compounds very frequently have common or trivial names. In the examples given in Figure 11.17, the IUPAC name is given first, and the common name follows in italics.

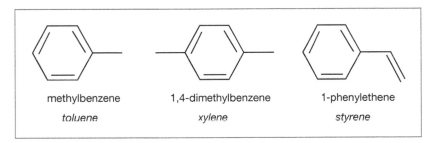

Figure 11.17 Aromatic Compound Nomenclature Examples

ORGANOHALOGEN COMPOUNDS

Organohalogen compounds are organic compounds that contain one or more halogen atoms. The halogens are found in Group 7A of the periodic table and include F, Cl, Br, and I. If the parent compound is an alkane, introduction of a halogen affords a haloalkane. Haloalkanes are also called *alkyl halides*. If the parent compound is benzene, then the halogenated derivatives are called *aryl halides*. The generic formula of an alkyl halide is R–X, where R is a generic alkyl group (i.e., methyl or ethyl, etc.) and X is a generic halide. Haloalkanes, especially if ether functional groups are present, are common anesthetic agents. We will discuss ethers later. Figure 11.18 shows some example organohalides with their IUPAC name listed first and their common name listed in italics.

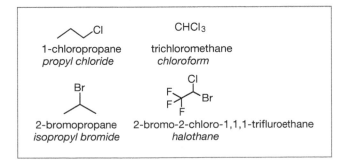

Figure 11.18 Organohalide Nomenclature Examples

The IUPAC rules name organohalides as alkanes with halogen substituents. The halogens are named as halo groups: fluoro, bromo, chloro, and iodo. The position of each halo group is indicated with a locant number. The common way to name an alkyl halide is to simply name the alkyl group and then the halide. Some alkyl halides have other trivial names. For example, 1,1,1-trichloromethane is commonly called chloroform. Of course, chloroform was once used as an anesthetic reagent but is no longer used because of its toxicity.

FUNCTIONAL GROUPS BASED ON WATER

The next two functional groups are alcohols and ethers. Both are derivatives of water. If we replace one hydrogen on the water molecule with an alkyl group, we have an alcohol. If we replace both hydrogens with alkyl groups, we have an ether.

Alcohols

Alcohols have a generic formula of R–O–H, as shown in Figure 11.19. The functional group of an alcohol is the hydroxyl group, O–H. Since the hydroxyl group is covalently bonded to the carbon chain, alcohols do not contain a hydroxide ion. Therefore, alcohols are not strong bases like the ionic compound sodium hydroxide (NaOH).

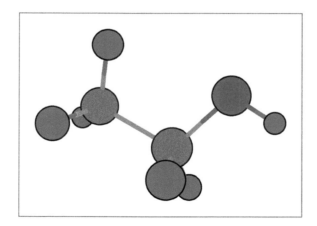

Figure 11.19 Ball-and-Stick Model of an Alcohol (Ethanol)

Alcohols have a hydrogen atom directly bonded to an oxygen atom. Therefore, alcohols can form hydrogen bonds. If the R group contains three or fewer carbon atoms, the alcohol is perfectly soluble in water. As the R group becomes larger, the solubility in water decreases. Introduction of additional hydroxyl groups increases the solubility of the compound in water. The hydrogen-bonding capacity of alcohols finds applications in living systems as well. Molecular systems, such as enzyme–substrate complexes and nucleic acid transcription/translation, "recognize" each other through hydrogen bonding, and alcohol functional groups are often involved.

To name alcohols by the IUPAC rules, add the suffix "ol" to the name of the parent and indicate the position of the hydroxyl group with a locant number (Figure 11.20). Simple alcohols are also commonly named by naming the alkyl group and then adding "alcohol."

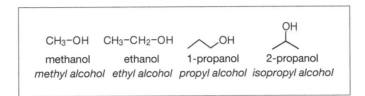

Figure 11.20 Alcohol Nomenclature Examples

The simplest alcohol, methanol, is commonly known as wood alcohol, because it was once obtained by heating wood in the absence of air, a process that also produced charcoal. Now methanol is synthesized from methane in natural gas. Methanol itself has a relatively low toxicity. However, methanol is oxidized (metabolized) in the liver to formaldehyde and then to formic acid, both of which are much more toxic.

If the compound contains more than one hydroxyl group, you need to include a prefix multiplier to indicate the number of hydroxyl groups. Two examples are shown in Figure 11.21.

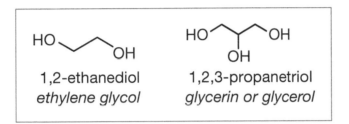

Figure 11.21 Poly-Alcohol Nomenclature Examples

Ethylene glycol is used as a freezing-point depressant in automotive antifreeze. It is highly toxic because the enzyme alcohol dehydrogenase and the coenzyme nicotinamide adenine dinucleotide (NAD) oxidize ethylene glycol to much more liver-toxic compounds like glyoxal, hydroxyacetaldehyde, glyoxylic acid, and oxalic acid, all of which are structurally similar to either formaldehyde or formic acid. Ethanol is one treatment for ethylene glycol poisoning (as well as methanol poisoning). Ethanol competes with ethylene glycol for the enzyme and coenzyme, and gives the body time to simply excrete the water-soluble ethylene glycol through urination.

Glycerin is essentially nontoxic. It serves as a backbone that holds fatty acids into triglycerides and is one of the main ingredients in K-Y Jelly. Treating glycerin with nitric and sulfuric acids introduces three nitro groups ($-NO_2$) into the hydroxyl groups, to afford nitroglycerin (Figure 11.22). Nitroglycerin is used in explosives and to treat angina. The nitration of glycerin is an extremely temperamental process that usually leads to an explosion in the hands of the inexperienced. So, kids, don't try this at home!

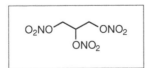

Figure 11.22 Nitroglycerin

Aromatic alcohols are called *phenols*. The word "phenol" comes from phenyl alcohol. Phenol itself, formerly known as *carbolic acid*, finds application as a preservative. Three example phenols are illustrated in Figure 11.23. Hexachlorophene, a phenolic aryl halide, was once a common active ingredient in over-the-counter antibacterial soaps.

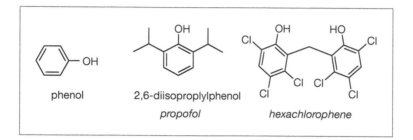

phenol   2,6-diisoproplylphenol

propofol   *hexachlorophene*

Figure 11.23 Some Examples of Phenols

Propofol is properly named 2,6-diisoproplylphenol. Because there are 12 carbon atoms, this compound is not very soluble in water, and it is much more soluble in hydrophobic solvents. Therefore propofol is formulated as an emulsion[2] of the 2,6-diisoproplylphenol suspended in a mixture of soybean oil and water. An egg-based phospholipid (e.g., lecithin) is added to stabilize the emulsion. By the way, you can buy flavored water–soybean oil–lecithin mixtures, but you call it *mayonnaise*.

Ethers

Ethers have a generic formula of R–O–R′, so the functional group of an ether is an oxygen bridge between two alkyl groups, as shown in Figure 11.24.

The prime mark (R′) means the two alkyl groups may be the same or different. Ethers are rather unreactive in organic chemistry, so they are often used as solvents and protecting groups that hide the much more reactive hydroxyl group of alcohols. However, in medical settings,

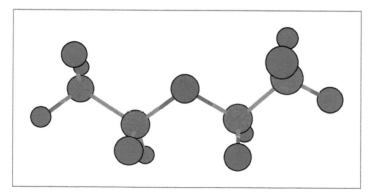

Figure 11.24 A Ball-and-Stick Model of an Ether (Diethyl Ether)

---

[2] An emulsion is a mixture comprised of small droplets of one liquid suspended in a second, immicisble liquid.

ethers continue to find use as anesthetic reagents. Diethyl ether, also known as *ethyl ether* or simply *ether*, is one of the most common solvents in the organic chemistry laboratory. Diethyl ether was once used as an anesthetic. However, because of its extreme flammability, diethyl ether is rarely used as an anesthetic today. Introduction of halogens onto the organic backbone partially oxidizes the molecule, rendering it much less flammable and much safer to work with.

The IUPAC rules name ethers as *alkoxy alkanes*. The larger R group is named as an alkane, and the smaller R group, along with the oxygen atom, is named as an alkoxy substituent. Simple ethers can also be named by citing the two alkyl groups and then adding "ether." Three examples are given in Figure 11.25.

CH₃–O–CH₃   CH₃–CH₂–O–CH₂–CH₃   CH₃–CH₂–O–CH₃

methoxymethane   ethoxyethane   methoxyethane
dimethyl ether   *diethyl ether*   *ethyl methyl ether*

Figure 11.25 Ether Nomenclature Examples

The most common anesthetic gases are halogenated ethers. Notice the "theme-and-variation" relationship between these compounds. Researchers moved and exchanged halogen atoms around on the same basic skeleton in a systematic attempt to fine-tune the physiological properties of the substance. Figure 11.26 lists some common anesthetic ethers.

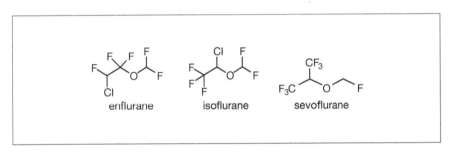

enflurane   isoflurane   sevoflurane

Figure 11.26 Common Anesthetic Ethers

AMINES

Amines are functional groups derived from ammonia (NH₃). So, amines are nitrogen analogs of alcohols, and the chemistry of the two functional groups is similar. The generic formula of an amine is R₃N, where R can either be alkyl groups or hydrogen atoms, in any combination. So, R₃N, or R₂NH, or RNH₂ are all generic formulas of amines. An example is shown in Figure 11.27.

Like alcohols, amines can form hydrogen bonds and tend to be more soluble in water than many other functional classes. The hydrogen bonds formed by amines in nucleic acids play a critical role in the genetic code. The hydrogen bonds allow the genetic bases to recognize each other, and this controls DNA synthesis and transcription, as well as protein synthesis.

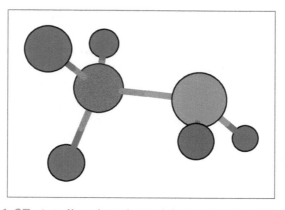

Figure 11.27 A Ball-and-Stick Model of an Amine (Methylamine)

While the IUPAC rule for naming amines calls for naming the alkane parent and adding "amine," most organic chemists still adhere to the older system of naming the groups on the nitrogen as alkyl groups and adding "amine." In the IUPAC system, substituent groups on a nitrogen atom are given a locant N. Figure 11.28 gives some examples of amine nomenclature.

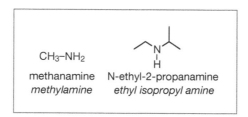

Figure 11.28 Amine Nomenclature Examples

A large number of medications are amines, and a few examples are shown in Figure 11.29. It is interesting to note that mechlorethamine, one of the first anticancer chemotherapy drugs, belongs to a class of compounds known as *nitrogen mustards*, the same class of compounds used in chemical warfare during World War I.

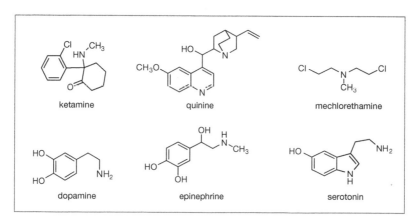

Figure 11.29 Some Medically Important Amines

Nitrogenous heteroaromatic compounds comprise an important class of nonbezenoid aromatic compounds. Some of these have a chemistry similar to that of amines. This mouthful translates as follows. Hetero means other. So heteroaromatic compounds have one or more atoms other than carbon in the ring. Nitrogenous means that the other atom is a nitrogen atom. As you can probably guess from Figure 11.30, there is a staggering variety of heteroaromatic compounds. Pyridine is the most common heteroaromatic compound, but purine and pyrimidine are the templates upon which the genetic bases are built. Pyrrole and imidazole are both based on a five-membered ring and occur frequently in many biologically important molecules.

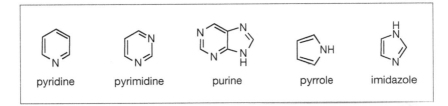

Figure 11.30 Some Nitrogenous Heteroaromatic Compounds

Probably the most important chemical reaction of amines, at least in medical applications, is the basicity of amines. Most amines have a noticeable base strength and will accept a proton from a strong acid to form its conjugate acid. The conjugate acid of an amine is called an *ammonium salt*. The ionic salt is much more soluble in water than the electrically neutral free-base form. So, by controlling the pH of the medium, you can control the solubility of many medications. Because amines are weak bases, naturally occurring amines are called alkaloids. When the alkaloid dissolves in water, the solution becomes very slightly alkaline.

$$\text{Free-base form} \qquad \text{Ammonium salt}$$

CARBONYL FUNCTIONAL GROUPS

A carbonyl group is a carbon doubly bonded to an oxygen (Figure 11.31). This bonding arrangement is very strong and occurs frequently throughout the chemical world. The carbonyl carbon atom needs two more bonds to complete its octet of electrons. The next set of functional groups

Figure 11.31 A Carbonyl Group

is derived by supplying various hydrogen-, carbon-, oxygen-, and nitrogen-based substituents to the carbonyl carbon.

Aldehydes

Aldehydes have an alkyl group and a hydrogen atom bonded to a carbonyl group. The generic formula of an aldehyde is shown in Figure 11.32. Figure 11.33 illustrates a ball and stick model of an aldehyde. Aldehydes are relatively rare among naturally occurring compounds because most aldehydes are easily oxidized to carboxylic acids, even by the oxygen in air. The aldehyde functional group is represented in medicine largely in the carbohydrates, and we will discuss that group in the next chapter.

Figure 11.32 A Generic Aldehyde

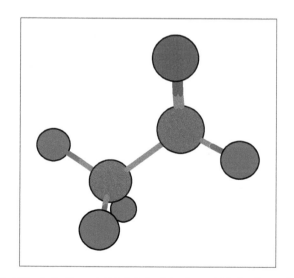

Figure 11.33 A Ball-and-Stick Model of an Aldehyde (Ethanol)

The IUPAC rules for naming aldehydes append the suffix "al" to the parent name. The aldehyde carbon is always the first carbon in the chain, so a locant is not necessary. The simplest aldehyde, methanal, is commonly known as formaldehyde, a highly toxic gas used to preserve biological specimens. Ethanal is the next aldehyde, although most organic chemists call it acetaldehyde. 2,2,2-trichloroethanal, more commonly known as chloral, reacts with water to form the sedative chloral hydrate. Phenylmethanal, more commonly known as benzaldehyde, is used as artificial cherry or almond flavoring. Nomenclature examples of aldehydes are shown in Figure 11.34.

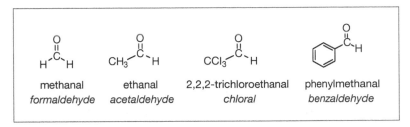

Figure 11.34 Aldehyde Nomenclature Examples

Ketones

Ketones have two alkyl groups bonded to a carbonyl group. The generic formula of a ketone is shown in Figure 11.35. Figure 11.36 illustrates a ball-and-stick model of a ketone. Ketones are very common in nature and are represented in a large variety of natural compounds and medicines.

Figure 11.35 A Generic Ketone

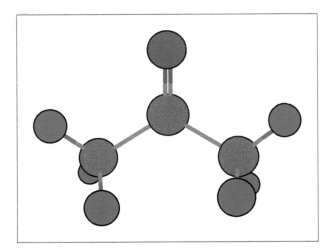

Figure 11.36 A Ball-and-Stick Model of a Ketone (Propanone)

The IUPAC rules for naming ketones append the suffix "one" to the parent name. The position of the ketone carbon is indicated by a locant number. Figure 11.37 gives three nomenclature examples of ketones. The simplest ketone, propanone, is commonly known as *acetone*. Acetone is an excellent solvent for most organic compounds and is the main ingredient in fingernail polish remover. Acetone is one of the ketone bodies that build up in the bloodstream

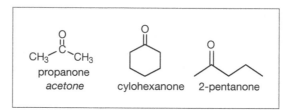

Figure 11.37 Ketone Nomenclature Examples

from excessive metabolism of fats. Because ketones typically have a sweet taste and odor, this can give a patient with ketosis a characteristic "acetone breath."

Some biologically and medically important ketones are shown in Figure 11.38.

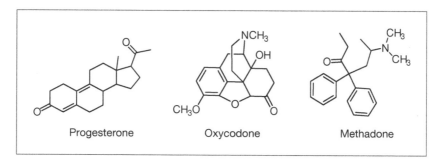

Figure 11.38 Some Medically Important Ketones

Reactions of Aldehydes and Ketones

Aldehydes and ketones are central players in organic synthesis because of the wide variety of reactions they undergo. Likewise, their chemistry is important to biological systems. We will consider just two of the reactions involving these compounds: oxidation/reduction and formation of acetals and ketals.

Oxidation and reduction are chemical reactions that transfer electrons between reacting species. The general definition of oxidation and reduction states oxidation is as follows: *The loss of electrons and reduction is the gain of electrons.* Oxidation may be manifest by an increase in ionic charge (e.g., $Fe^{2+} \rightarrow Fe^{3+}$). In organic chemistry, we find it easier to recognize oxidation as increasing the number of bonds to oxygen. Usually, increasing the number of bonds to oxygen accompanies decreasing the number of bonds to hydrogen, so this is also a way to recognize oxidation. Reduction is the opposite of oxidation. So when the number of bonds to oxygen decreases or the number of bonds to hydrogen increases, the species has undergone reduction. We live in an oxidizing environment because there is so much oxygen in the atmosphere. The oxidizing power of oxygen makes many oxidation reactions exothermic and thermodynamically favorable processes. All our foods contain organic compounds, and most organic compounds are relatively easy to oxidize. So oxidation represents a chemical pathway to extract the chemical potential energy stored in the carbon–hydrogen and carbon–carbon bonds in an organic molecule. Carbohydrates have a generic formula of $C_x(H_2O)_y$. So the carbon backbone in carbohydrates is already partially oxidized. Fats have a much higher carbon to oxygen ratio. This is why carbohydrates contain fewer calories per gram than fats.

Alcohols are oxidized to give ketones or aldehydes, depending on the substitution pattern on the alcohol. Aldehydes and ketones are reduced to alcohols. Notice that, when going from an alcohol to an aldehyde or ketone, two hydrogen atoms (shown in blue in Figure 11.39) are removed, while, at the same time, the carbon–oxygen bond order increases from a single bond to a double bond. In reduction, the reverse is true.

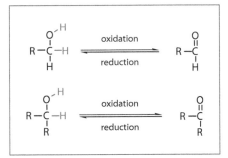

Figure 11.39 Organic Redox Reactions

In the laboratory, there are a variety of oxidizing and reducing agents. In living systems, these transformations are facilitated by enzymes and driven by various coenzymes, such as NAD. Ultimately, oxygen recycles these coenzymes, so oxygen is the ultimate (in both senses of the word) oxidizing agent in biological systems.

Aldehydes react with alcohols to form hemiacetals and then acetals. In the first reaction, the alcohol adds across the carbonyl double bond, reminiscent of how hydrogen adds to the carbon–carbon double bond of alkenes. This product is called a *hemiacetal*. The functional group of a hemiacetal consists of an alcohol functional group and an ether functional group on the same carbon atom. The hemiacetal carbon atom also has one bond to a hydrogen atom. The process of forming a hemiacetal is reversible, and normally the position of the equilibrium favors the aldehyde and alcohol starting materials. This is shown in Figure 11.40.

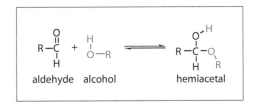

Figure 11.40 Formation of a Hemiacetal

In the second step, the hydroxyl group of the hemiacetal is replaced by a second alcohol molecule, as shown in Figure 11.41. This is an example of a *condensation reaction*, because two molecules are joined together, with the concomitant expulsion of a water molecule. This is also an example of a substitution reaction, because a hydroxyl group is replaced by an alkoxy group. The functional group of an acetal is two ether functional groups on the same carbon which also has one bond to a hydrogen atom. This process is an equilibrium reaction. The position of the equilibrium is controlled by the concentration of water. If we rigorously exclude water from the system, then the reaction proceeds in the forward direction toward formation

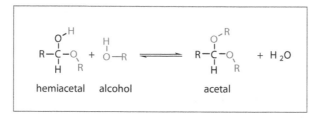

Figure 11.41   Formation of an Acetal

of the acetal. If there is a high concentration of water, then the equilibrium favors the starting materials. Since water tears the acetal back into a hemiacetal and an alcohol, this is an example of a hydrolysis reaction.

Ketones react with alcohols to form hemiketals and then ketals. This pair of reactions is exactly analogous to the formation of hemiacetals and acetals. In the first reaction (Figure 11.42), the alcohol adds across the carbonyl double bond to give a hemiaketal. This equilibrium also generally favors the starting materials. The functional group of a hemiketal also consists of an alcohol functional group and an ether functional group on the same carbon atom.

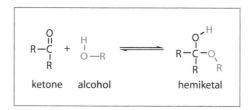

Figure 11.42   Formation of a Hemiketal

The hydroxyl group of the hemiketal can be replaced by a second alcohol molecule, as can be seen in Figure 11.43. As with acetals, a ketal can be hydrolyzed by water back into a hemiketal and alcohol.

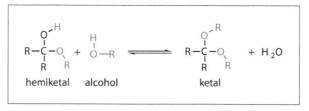

Figure 11.43   Formation of a Ketal

Acetal/ketal formation and hydrolysis occur under very mild chemical conditions, and they do not require large amounts of energy to drive either the forward process or the reverse process. In carbohydrate chemistry, acetal and ketal functional groups play the crucial role of connecting monosaccharides into polysaccharides. This is important, since carbohydrates are our principal energy source. Linking simple sugars together as acetals or ketals is an efficient strategy for storing these molecules in a compact form. Facile hydrolysis of the acetal/ketal linkages allows easy access to the chemical potential energy. We will explore this chemistry in greater detail in the next chapter.

Carboxylic Acids

The functional group of a carboxylic acid is the carboxyl group, which is a carbonyl group with a hydroxyl group bonded to the carbonyl carbon. The generic formula of a carboxylic acid is shown in Figure 11.44. Figure 11.45 shows a ball-and-stick model of a carboxylic acid. Carboxylic acids are important intermediates in metabolism, and fatty acids are long-chained carboxylic acids that we use as our principal long-term energy storage molecules.

Figure 11.44 A Generic Carboxylic Acid

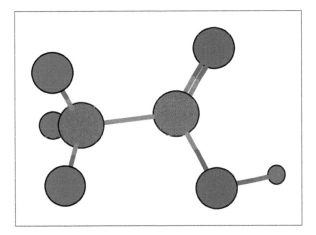

Figure 11.45 A Ball-and-Stick Model of a Carboxylic Acid (Ethancic Acid)

The IUPAC rules for naming carboxylic acids append the suffix "oic acid" to the parent name. The simplest carboxylic acid, methanoic acid, is commonly known as formic acid. The next carboxylic acid is ethanoic acid, much more commonly known as acetic acid. The acetyl group or the carbon backbone of acetic acid is the principal "chemical currency" that runs the Krebs Cycle. Carbohydrate and lipid metabolism both eventually chew the molecules up into two-carbon acetate fragments, which are esterified to form acetyl coenzyme A, which feeds the Krebs Cycle. Benzoic acid, or more correctly its conjugate base (benzoate), is frequently used as a food preservative. Long-chain carboxylic acids with R groups having 11 to 17 carbon atoms are called fatty acids. We will explore the details of fatty acids in the next chapter. Four example carboxylic acids, along with their names, are shown in Figure 11.46.

One important reaction of carboxylic acids is their ability to act as acids. Carboxylic acids are medium-strong acids, with pK_a values typically between 4 and 5. The conjugate bases of

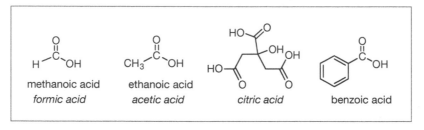

Figure 11.46 Carboxylic Acid Nomenclature Examples

carboxylic acids are named by adding the suffix "oate" to the parent name (Figure 11.47). One application of carboxylate anions is to serve as biologically inert anions for positively charged medications. For example, the citrate ion is commonly used as a counterion for positively charged medications, for example, sildenafil citrate (Viagra).

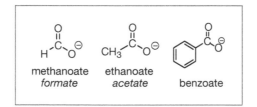

Figure 11.47 Carboxylate Anion Nomenclature Examples

Esters

Esters are condensation products between acids and alcohols. In Figure 11.48, we see the generic structure of an ester. In Figure 11.49, a ball-and-stick model of an ester is illustrated. The portion in black is derived from a carboxylic acid, and the portion in blue is derived from an alcohol.

Figure 11.48 A Generic Ester

Almost uniquely among the organic functional groups, many esters have a pleasant, fruity odor. Naturally occurring esters give fruits and flowers their odors. Synthetic (and some naturally occurring esters) are used in the fragrance and flavor industry. An ester functional group binds three fatty acid molecules to a glycerin backbone, forming a triglyceride. Figure 11.50 shows a generic triglyceride. The R groups have 11 to 17 carbon atoms. The glycerin backbone is shown in blue.

The IUPAC rules name esters as alkyl alkanoates. That is, the portion of the ester derived from the alcohol is named as an alkyl group. The portion of the ester that is derived from the

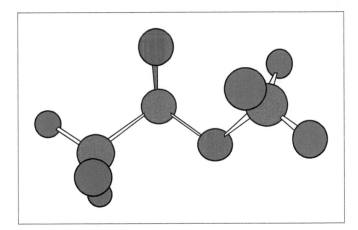

Figure 11.49 A Ball-and-Stick Model of an Ester (Methyl Ethanoate)

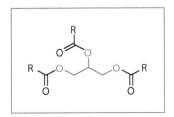

Figure 11.50 A Generic Triglyceride

carboxylic acid is named as the conjugate base of that acid. It is easy to distinguish these parts. The half derived from the carboxylic acid has the carbonyl group. Pentyl ethanoate, or pentyl acetate, is one ester used as artificial banana flavoring. Figure 11.51 shows three ester nomenclature examples.

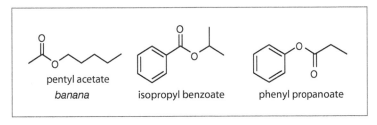

Figure 11.51 Ester Nomenclature Examples

Carboxylic acids react with alcohols to form esters. This reaction requires special reactions conditions, but let's not worry about that now. The equation shown in Figure 11.52 is phenomenonologically correct, if not mechanistically accurate. The box shown in the equation captures the elements of water, H–OH, from the reaction, and water appears on the product side of the equation. This reaction is a reversible process, and so an ester can be torn apart back into a carboxylic acid and an alcohol. Since water is responsible for tearing the ester molecule apart, the reverse reaction is called *hydrolysis*.

Figure 11.52 Formation of an Ester

The position of this equilibrium in the laboratory is controlled by water. If the reaction is conducted under conditions in which water is scarce or absent, then the position of the equilibrium favors formation of the ester. If there is a lot of water present, then the equilibrium favors formation of the carboxylic acid and alcohol. This is just one more example of Le Châtelier's Principle in action.

We said esters are condensation products between carboxylic acids and alcohols, but this definition is really too restrictive. In the most general sense, an ester is the condensation product between an alcohol and an acid, not just a carboxylic acid. We saw an example of a nitrate ester earlier in the structure of nitroglycerin. In biochemistry, phosphate esters are one of the most important classes of esters. A phosphate ester is derived from an alcohol and phosphoric acid (H_3PO_4). Again, the acid and alcohol are condensed (under special reaction conditions) to give the ester and a molecule of water. This is shown in Figure 11.53. Phosphate esters form the backbone of DNA and RNA.

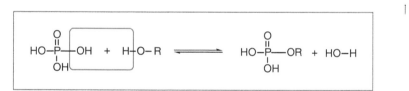

Figure 11.53 Formation of a Phosphate Ester

Amides

Amides are condensation products between carboxylic acids and amines. Figure 11.54 shows the structure of a generic amide, while Figure 11.55 shows a ball-and-stick model of an amide.

Figure 11.54 A Generic Amide

The portion of the molecule derived from the amine is shown in blue, and the portion of the molecule derived from the carboxylic acid is shown in gray. The R groups on the nitrogen can either be alkyl groups or hydrogens. Proteins are the most biologically important example of amides, and we will discuss proteins in greater detail in the next chapter.

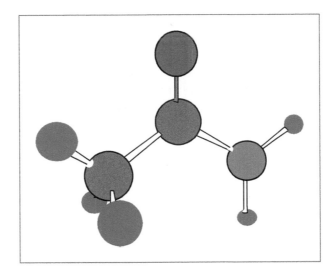

Figure 11.55 A Ball-and-Stick Model of an Amide (Ethanamide)

The IUPAC rules name amides by appending the suffix "amide" to the parent name of the carboxylic acid from which the amide is derived. If there are substituents on the amide nitrogen, they are given the locant N. Three examples are given in Figure 11.56.

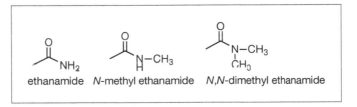

ethanamide   *N*-methyl ethanamide   *N,N*-dimethyl ethanamide

Figure 11.56 Amide Nomenclature Examples

Carboxylic acids react with amines to form amides. Okay, this reaction does not occur as written in Figure 11.57, but it does illustrate the overall process. The box shown in the equation captures the elements of water, H–OH, from the reaction, and water appears on the product side of the equation. This reaction is also a reversible process, and so an amide can be torn apart by water back into a carboxylic acid and an amine.

Figure 11.57 Formation of an Amide

The hydrolysis of amides requires harsher reaction conditions than the hydrolysis of esters, and this is a good thing. Esters in triglycerides bind fatty acids to glycerin as an energy-dense storage molecule called a *triglyceride*. We need to be able to tear the triglycerides back apart to access the chemical energy stored in the fatty acids. Proteins, on the other hand, are polymers held together by amide bonds. We need our structural proteins to be very durable to withstand the presence of all the water in our bodies. While proteolytic enzymes can tear down the amide bonds in proteins, amides are essentially impervious to 0.1 M HCl at 37°C. These are approximately the conditions in our stomachs, the harshest chemical conditions in our bodies.

Summary

- Organic compounds are compounds that contain carbon.
- Organic compounds consist of a carbon backbone and one or more functional groups.
- Hydrocarbon functional groups consist of only carbon and hydrogen and include alkanes, alkenes, alkynes, and aromatic compounds.
  - Alkanes are hydrocarbons that contain all single bonds.
  - Alkyl groups are alkanes minus one hydrogen atom. Alkyl groups are "stick on" groups appended to larger organic molecules. A generic alkyl group is represented as R.
  - The alkene functional group is the carbon–carbon double bond.
  - The alkyne functional group is the carbon–carbon triple bond.
  - The aromatic functional group (to a first approximation) is a benzene ring.
- Alcohols and ethers are derived from water.
  - The alcohol functional group is the hydroxyl (OH) group, so a generic alcohol may be represented as R-OH.
  - The ether functional group is an oxygen bonded to two alkyl groups. A generic ether may be represented as R-O-R′.
- The functional group of an alkyl halide is a halogen atom (F, Cl, Br or I). Generically, halogens are represented as X, so a generic alkyl halide may be represented as R-X. Many inhaled anesthetic reagents are alkyl halides that often contain ether functional groups.
- Amines are derived from ammonia, in which one or more of the hydrogens on ammonia have been replaced by alkyl groups. RNH_2, R_2NH, and R_3N all represent generic amines.
- The carbonyl group is a carbon doubly bonded to an oxygen. The carbonyl-based functional groups include aldehydes, ketones, carboxylic acids, esters, and amides.
  - Aldehydes have a hydrogen atom bonded to a carbonyl group. A generic aldehyde may be represented as $R(C = O)H$.
  - Ketones have two alkyl groups bonded to a carbonyl group. A generic ketone may be represented as $R_2C = O$.
  - Carboxylic acids have a hydroxyl group bonded to a carbonyl group. A generic carboxylic acid may be represented as $R(C = O)OH$. Sometimes you'll see this written as RCO_2H or $RCOOH$.
  - Esters have an alkoxy (OR) group bonded to a carbonyl group. A generic ester may be represented as $R(C = O))OR′$. Sometimes you'll see this written as $RCO_2R′$ or $RCOOR′$. Esters are condensation products between a carboxylic acid and an alcohol:

$$RCO_2H + R′OH \rightarrow RCO_2R′ + H_2O$$

- Amides have a nitrogen atom bonded to a carbonyl group. A generic amide may be represented as $R(C = O)NR_2$. The R groups on nitrogen may be either alkyl groups or hydrogens. Amides are condensation products between a carboxylic acid and an alcohol:

$$RCO_2H + R_2NH \rightarrow R(C = O)NR_2 + H_2O$$

- Aldehydes and ketones undergo addition reactions with alcohols to form hemiacetals or hemiketals. The functional group of a hemiacetal/hemiketal is characterized by an ether functional group and an alcohol functional group on the same carbon.

$$R_2C=O + R'OH \rightarrow R_2C(OH)OR'$$

- In the presence of excess alcohol (and an acid catalyst), aldehydes and ketones condense with two alcohol molecules to form acetals or ketals. The acetal/ketal functional group is characterized by two ether functional groups on the same carbon.

$$R_2C=O + 2 R'OH \rightarrow R_2C(OR')_2$$

Review Question for Organic Compounds

Identify all of the functional groups in these medications and drugs. All of the drug names are registered trademarks of their respective manufacturers (except LSD, of course).

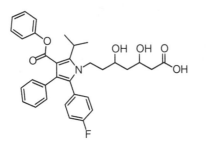

- Lipitor

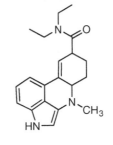

- LSD

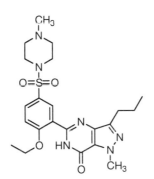

- Viagra

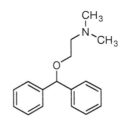

- Benadryl

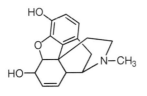

- Morphine

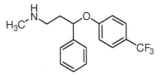

- Prozac

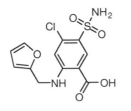

- Lasix

- Phenergan

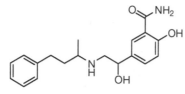

- Labelalol

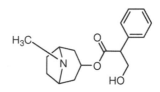

- Atropine

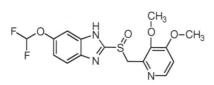

- Protonix

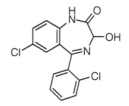

- Ativan

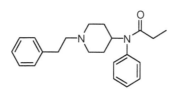

- Fentanyl

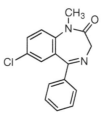

- Valium

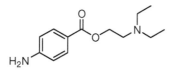

- Procaine

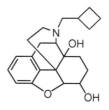

- Nubain

Biochemistry

BIOMOLECULES

Biochemistry is the study of the chemistry that occurs in living systems, and it focuses on the biomolecules that are the building blocks of living organisms. Biomolecules are organic molecules that (roughly) fall into four categories. Each class serves one or more physiological purposes and is categorized largely on the basis of its organic functional groups. First are the carbohydrates or sugars. Carbohydrates are ketones or aldehydes that also contain multiple alcohol groups. Carbohydrates are used primarily for energy storage and structure (at least in plants and insects). Proteins are polymers of amino acids, which contain, not surprisingly, an amine functional group and a carboxylic acid functional group. These two functional groups are condensed into amide functional groups that knit the amino acids into a polypeptide chain (aka a protein). Proteins serve as structural elements in animals and, more critically, as chemical catalysts called enzymes that make life possible. Nucleic acids are polymers of sugars, joined together by phosphate ester linkages, which contain aromatic amines whose shape forms the chemical genetic code through hydrogen bonding. Nucleic acids are the architects and construction contractors of proteins. Finally, lipids are the very diverse group of biomolecules that are characterized by a physical property rather than a characteristic organic functional group. Lipids are biomolecules that are more soluble in organic solvents, such as ether, than in water. Lipids are largely comprised of nonpolar hydrocarbon functional groups, in addition to a small proportion of polar functional groups (primarily alcohols, ethers, esters, and ketones) that establish the form and function of the molecule.

CARBOHYDRATES

Carbohydrates (or sugars) were originally believed to be "hydrates of carbon," because they have the general formula $C_x(H_2O)_y$. We now know that carbohydrates are polyalcohol aldehydes or ketones. Carbohydrate names are largely based on trivial names, rather than systematic names generated by IUPAC rules. However, most carbohydrate names end in "ose."

There are several useful ways to categorize carbohydrates. Monosaccharides are sugars that cannot easily be broken down into simpler sugars. Glucose is by far the most common monosaccharide. The alcohol and aldehyde or ketone functional groups in monosaccharides can be used to join simple sugars into more complex structures by forming acetal or ketal functional groups. Disaccharides are composed of two monosaccharides joined by an acetal or ketal linkage. Sucrose, or table sugar, is a very common disaccharide comprised of glucose and fructose. Polysaccharides are chains comprised of many (hundreds or thousands) monosaccharide units. Polysaccharides include starches and cellulose.

We can also classify monosaccharides on the basis of the number of carbon atoms they contain. Trioses have three carbon atoms, tetroses have four, pentoses have five, and hexoses have six. Sugars that contain an aldehyde functional group are called aldoses, and sugars that contain a ketone functional group are classified as ketoses. Aldehyde functional groups are comprised of a carbon atom with a double bond to an oxygen atom and a single bond to a hydrogen atom. To simplify drawing the structures of sugars, the aldehyde functional group is often shown as CHO. For example, ribose is an aldopentose and fructose is a ketohexose (Figure 12.1).

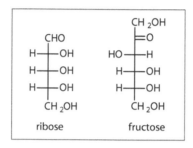

Figure 12.1 Aldoses and Ketoses

FISCHER PROJECTIONS

Most of the carbon atoms in sugars have a tetrahedral geometry. Therefore, sugar molecules are not flat but have a three-dimensional structure. The three-dimensional structure of carbohydrates is commonly depicted using Fischer projections, named in honor of Emile Fischer. Fischer worked out the structures of most of the carbohydrates in the first part of the twentieth century with little technology but a great deal of genius. In a Fischer projection, you imagine the main carbon skeleton to be vertical with the carbon–carbon single bonds rotated so that the carbon chain is curved out toward you (Figure 12.2).

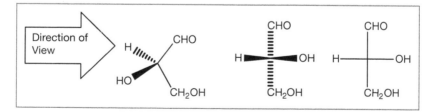

Figure 12.2 Fischer Projection of Glyceraldehyde

Let's take a look at glyceraldehyde. From the viewer's perspective, the aldehyde carbonyl group (CHO) points up, and the CH_2OH group points down. The OH group points to the viewer's right, and the H points to the viewer's left. Therefore, the Fischer projection puts each of these groups in that orientation. It is important to recognize that, although a Fischer projection is drawn in the plane of the page, it represents a three-dimensional molecule. The horizontal bonds are projected toward you, and the vertical bonds are projected away from you.

Chirality

Chirality is a topological property of an object, and an object is chiral if it is different from its mirror image. The word "chiral" comes from *chiros*, the Greek word for "handed." Since chiral has a Greek origin, it is pronounced with a hard *ch*, such as "*kiral*," not "*shiral*." For example, your right and left hands are mirror images of each other, but they are not identical. So, your hands are chiral. Socks are not chiral, so there are not left-handed socks and right-handed socks. Molecules can also be chiral, and this subtle difference is crucially important in the world of biochemistry.

Many molecules—especially biomolecules—are chiral. A molecule is chiral if there is a "right-handed" form of the molecule that is different from a "left-handed" form. The right-handed and left-handed forms[1] are called enantiomers. Enantiomers are mirror images of each other, but they are different compounds, even though they have identical formulas and identical atom connectivity. Enantiomers have identical chemical and physical properties, except for how they interact with other chiral molecules and how they interact with plane polarized light. You already have experience with how chiral objects interact with other chiral objects. Your right hand interacts differently with right-handed gloves and left-handed gloves. The same is true for chiral molecules. We explore the interaction of chiral molecules with polarized light in the "Optical Activity" section.

Carbohydrates belong to the right-handed family, and amino acids belong to the left-handed family. In general, our bodies can use only one of the handed isomers of a chiral molecule. Introducing the wrong-handed isomer can have effects ranging from amusing to a noneffect to disastrous. For example, the right-handed isomer of carvone smells like spearmint, while the left-handed isomer smells like caraway. Underground chemists have learned that the pure left-handed isomer of methamphetamine is much more potent than a mixture of the two isomers. Presumably, our bodies simply ignore the right-handed isomer. The antinausea drug phthalidomide was once prescribed to fight morning sickness during pregnancy. The correct-handed isomer did indeed reduce morning sickness. However, the wrong-handed isomer caused horrific birth defects known as *phthalidomide babies*.

In order for a molecule to be chiral, a molecule must contain one or more stereocenters.[2] A carbon is a stereocenter when it has four different groups. Notice we didn't say "chiral carbon," because chirality is a property of the molecule as a whole and is not localized on any single atom or group of atoms.

Let's take a look at glucose and decide which carbons are stereocenters (Figure 12.3).

---

[1] For the sake of simplicity, we admit to being somewhat cavalier in our use of the terms *right-handed* and *left-handed*. Interested readers are directed to any modern organic chemistry textbook.
[2] This is not strictly true. There are some special (and rare) classes of molecules that are chiral because of restricted rotation around single or double bonds that impose a nonplanar and chiral conformation. Substituted allenes are the classic example of axially dissymmetric chiral molecules.

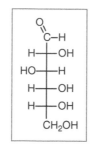

Figure 12.3 Stereocenters in Glucose

The C1 carbon (the carbonyl carbon) in glucose is not a stereocenter. Yes, it does have a total of four bonds, but it has bonds to only three different groups: the oxygen, the hydrogen, and the next carbon in the chain. The bottom carbon is also not a stereocenter, because the carbon is bonded to a carbon group, an OH group, and two *identical* H atoms. In order to be a stereocenter, the carbon must be bonded to four different groups. All of the other carbons in glucose are stereocenters. For example, C2 has bonds to an OH group, an H, the aldehyde functional group, and the rest of the chain. C3 has bonds to an OH group, an H, a two-carbon group above, and a three-carbon group below. Notice we said stereocenters have bonds to four different groups, not four different kinds of atoms.

There are several conventions to indicate the configuration of a stereocenter. The most precise is the IUPAC system that names the configuration around a stereocenter as *R* (for right handed; *rectus*) or *S* (for left handed; *sinister*). You have probably seen the *R/S* descriptors on drug information sheets. The details of the *R/S* system are beyond our purposes here, but any standard organic chemistry text will discuss how to assign these descriptors.

Optical Activity

One cool property of chiral molecules is optical activity. Enantiomers of a chiral molecule interact with polarized light, but one enantiomer tilts the polarization in one direction, while the other enantiomer tilts it in the opposite direction. This phenomenon is called *optical activity*.

Light is called electromagnetic radiation because it can be described by an electric field and a magnetic field (Figure 12.4). These two fields oscillate like waves, at right angles to each other and to the direction in which the light is traveling. These fields can oscillate in any direction as long as they are at right angles to each other. If you imagine the light is heading away from you, you could observe the electric field to be oscillating up and down, or left to right, or at any intermediate orientation.

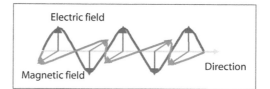

Figure 12.4 Electric and Magnetic Fields of Light

You get polarized light by passing regular (unpolarized) light through a polarizing filter, which allows oscillation of the electric field only in one plane. When polarized light passes through a sample of a chiral (nonracemic) material, the orientation of the polarized light is rotated (Figure 12.5).

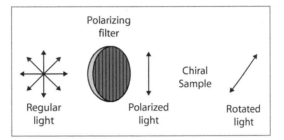

Figure 12.5 Rotation of Plane-Polarized Light by a Chiral Sample

Enantiomers that rotate polarized light in a clockwise direction (to the right) are called the *dextrorotatory* isomer and labeled as the (+) enantiomer. The enantiomer that rotates polarized light in a counterclockwise direction (to the left) is called the *levorotatory* isomer and is labeled as the (−) enantiomer. A solution of glucose rotates polarized light to the right, and this is why glucose still has a trivial name of dextrose. Fructose rotates polarized light to the left and was formerly known as levulose.

A racemic mixture contains equal amounts of the (+) enantiomer and the (−) enantiomer and has the designation (+/−). Racemic mixtures are not optically active because the rotation of the dextrorotatory enantiomer cancels out the rotation of the levorotatory enantiomer. Synthesis or isolation of a single enantiomer in the laboratory is a challenging task, and most syntheses of chiral molecules result in a racemic mixture containing both enantiomers. The Food and Drug Administration (FDA) policy statement drafted in 1992 and updated in 2005[3] requires pharmaceutical companies to characterize the properties of single enantiomers, and this adds difficulty, time, and expense to the development of new medicines.

D-Sugars and L-Sugars
When glyceraldehyde is drawn as a Fischer projection, the OH group on the stereocenter points to the right (Figure 12.6).

$$
\begin{array}{c}
CHO \\
H \!-\!\!|\!-\! OH \\
CH_2OH
\end{array}
$$

Figure 12.6 D-Glyceraldehyde

Emil Fischer called this the D-isomer. It turns out that virtually all naturally occurring monosaccharides have the hydroxyl group on the next-to-last carbon atom pointed to the right (when drawn as a Fischer projection). So, all naturally occurring sugars belong to the D-family.

---

[3] http://www.fda.gov/cder/guidance/stereo.htm, accessed February 1, 2013.

To decide which family a sugar belongs to, first draw the molecule as a Fischer projection with the aldehyde or ketene functionality at the top of the structure. Then look at the OH group on the next-to-bottom carbon. If that OH points to the right, the sugar is a D-sugar. If the OH points to the left, it is an L-sugar.

Figure 12.7 shows the D- and L-isomers of glucose. Notice the D- and L-isomers are mirror images of each other. Therefore, D- and L-isomers are enantiomers. Both compounds have some OH groups that point to the left and other OH groups that point to the right. However, in determining which structure is the D-sugar, you only consider the next-to-bottom carbon (shown in blue).

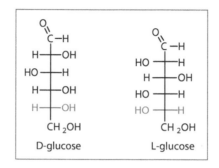

Figure 12.7 D-Glucose and L-Glucose

The D-Aldohexoses

Figure 12.8 shows the structures and names of the D-aldohexoses. These compounds all have six carbon atoms, an aldehyde functional group, and the OH group on C5 pointed to the right. Of these eight compounds, glucose is by far the most common.

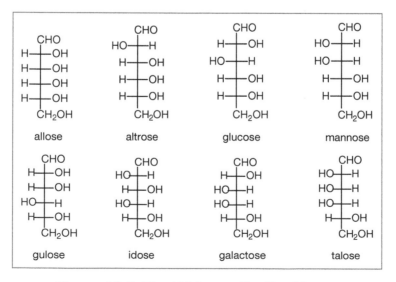

Figure 12.8 The Aldohexose Family of Sugars

Fructose

Fructose is the most common ketohexose. The structure of fructose is very similar to that of glucose. The functional groups on C1 and C2 are simply reversed (Figure 12.9).

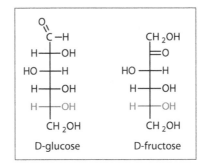

Figure 12.9 D-Glucose and D-Fructose

D-Aldopentoses

The D-aldopentoses are five-carbon sugars. The most important of these are ribose and deoxyribose because they are used to construct the nucleic acids RNA and DNA. Figure 12.10 gives the names and structures of the D-aldopentoses.

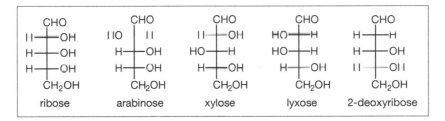

Figure 12.10 The Aldopentose Family of Sugars

A Mnemonic to Learn the Structures of the Aldose Sugars

Many students use the mnemonic device *"All Altruists Gladly Make Gum In Gallon Tubs"* to learn the names of the aldohexoses, coupled with drawing eight "blank" Fischer projections and then following a pattern of putting all hydroxyl groups on C5 on the right, then alternate by fours on C4, by twos on C3, and by ones on C2. This technique works, but it is not particularly convenient or efficient to identify the structure of a single aldohexose without drawing all eight isomers.

In a system consisting of multiple components, each of which is restricted to one of two states, an effective method for enumerating all of the possible configurations is to count the possibilities using the binary number system. Starkey reported a strategy for using the binary number system to describe the structures of the aldohexoses.[4] Hydroxyl groups on stereocenters that

---

[4] Starkey, R. (2000). "SOS: A mnemonic for the stereochemistry of glucose." *Journal of Chemical Education, 77, 734.*

point to the right are a "zero," and those that point to the left are a "one." If we assign C2 as the 2^0 (or 1), C3 as 2^1 (or 2), and C4 as 2^2 (or 4), we can count off the sugars in the mnemonic in the order shown in Table 12.1.

Table 12.1 BINARY CODE FOR THE D-ALDOHEXOSES

MNEMONIC	SUGAR	POSITION OF HYDROXYL GROUPS			BINARY CODE	DECIMAL NUMBER
		C4	C3	C2		
All	Allose	Right	Right	Right	000	0
Altruists	Altrose	Right	Right	Left	001	1
Gladly	Glucose	Right	Left	Right	010	2
Make	Mannose	Right	Left	Left	011	3
Gum	Gulose	Left	Right	Right	100	4
In	Idose	Left	Right	Left	101	5
Gallon	Galactose	Left	Left	Right	110	6
Tubs	Talose	Left	Left	Left	111	7

Although this approach has an interesting theoretical foundation, its application lacks a fundamental simplicity. However, we note that aldohexoses have four stereocenters, and the human hand has four fingers (and a thumb, of course). We can map each finger to the positions of the hydroxyl groups. Hold your open right hand in front of you, with your thumb pointing up and your fingers projected horizontally out to the left. Each finger can either be extended (pointing to the left) or closed (pointing to the right). Your index finger corresponds to the hydroxyl group on C2, your middle finger to the hydroxyl group on C3, your ring finger to the hydroxyl group on C4, and your pinky to the hydroxyl group on C5. Since C6 is not a stereocenter, it makes no difference where it is projected, so we don't need a finger to account for its spatial orientation.

Now, let's learn to count from 0 to 7 in base 2. The index finger corresponds to 2^0 (or 1), the middle finger to 2^1 (or 2), the ring finger to 2^2 (or 4), and the pinky to 2^3 (or 8). It might be helpful to write the value on the pad of each finger (where the "fingerprint" is). When the finger is extended, you can see the number, and it is counted. When the finger is closed, you can't see the number and the value is ignored. For example, when all the fingers on your hand are closed, you see no numbers, and the value is zero. If you extend your index and middle fingers, you can see the "1" and the "2," for a total value of three. A value of six requires extending the middle and ring fingers, while keeping the index and pinky fingers in the closed position (Figure 12.11).

The order of the mnemonic is fundamentally based on a binary strategy. The first sugar, allose, corresponds to a value of zero, altrose to a value of one, and so forth. Holding your right hand in front of you, with your thumb pointing upward, and all your fingers in the closed position, gives the binary value of zero. Where are all of your fingers pointed? They

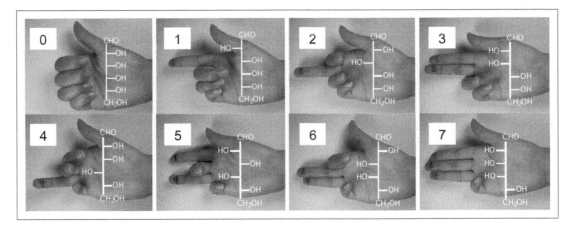

Figure 12.11 Binary Mnemonic for D-Aldohexoses

are all pointed to the right, and every hydroxyl group in the Fischer projection of allose is pointed to the right. Altrose is the next sugar in the mnemonic, corresponding to one, which means extending your index finger. Now your fingers, from index to pinky, point left, right, right, right, respectively, as do the hydroxyl groups in altrose. Glucose is the next sugar in the mnemonic and corresponds to the binary value of two. Glucose, in the form of cellulose and other polysaccharides, accounts for the majority of the biomass on this planet and is arguably the most important of all organic compounds. It is amusing to note that glucose requires a finger combination best not displayed in public. Figure 12.11 gives the structures of the aldohexoses next to the corresponding binary-finger values. To access the structure of any aldohexose, one simply says *"All (0); Altruists (1); Gladly (2); Make (3); Gum (4); In (5); Gallon (6); Tubs (7)"* while counting through the binary-finger values. When the desired sugar name is reached, the positions of the fingers indicate in which direction the hydroxyl groups point for that particular sugar.

A nice feature of this method is that it is easily applied to other aldose systems and to the L-family of sugars. The aldopentoses are given by the mnemonic *"Ribs Are eXtra Lean."* While not factually correct, this phrase corresponds to the sugars ribose, arabinose, xylose, and lyxose, and the binary values of zero, one, two, and three, respectively. To access the structures of the L-family of sugars, just use your left hand.

Cyclic Structures of Sugars

Monosaccharides have multiple alcohol functional groups and either a ketone or an aldehyde functional group. As we saw in the previous chapter, aldehydes react with alcohols to form hemiacetals, and ketones react with alcohols to form hemiketals. When this *intra*molecular reaction occurs, the linear monosaccharide forms a cyclic structure. Generally, cyclization reactions favor formation of five-membered rings or six-membered rings because these ring sizes are the most stable. Formation of a hemiacetal or a hemiketal is especially favorable when the reacting functional groups are contained in the same molecule. Therefore, monosaccharides exist almost exclusively as cyclic hemiacetals and hemiketals.

The cyclic structures are called Haworth projections. In a Haworth projection, the lower horizontal bond in the ring is understood to be projected toward you, above the plane of the page. This bond is usually in boldface. The upper horizontal bond is understood to be projected away from you, behind the plane of the page.

Let's take a look at glucose. The Fischer projection of glucose is shown in Figure 12.12. Remember, in a Fischer projection the carbon skeleton is curved out toward you. Now, let's rotate the Fischer projection clockwise 90°. Carbon 1 and carbon 6 are labeled to help you keep track of which carbon is which.

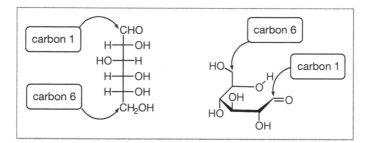

Figure 12.12   Relating Fischer Projection to Haworth Projection

Now, if the hydroxyl group on carbon 5 attacks the carbonyl carbon, we get a six-membered ring, as shown in Figure 12.13. The hemiacetal functional group is shown in blue.

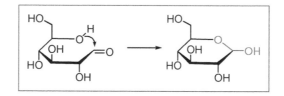

Figure 12.13   Cyclization of Glucose to Glucopyranose

When a monosaccharide forms a cyclic hemiacetal (or hemiketal), the carbonyl carbon becomes a *stereocenter*. Thus, cyclization leads to formation of two possible stereoisomers. These isomers are called *anomers*, and the former carbonyl carbon is called the *anomeric carbon*. The isomer with the anomeric OH (shown in blue) pointed down is the alpha anomer. The isomer with the anomeric OH (shown in blue) pointed up is the beta anomer (Figure 12.14).

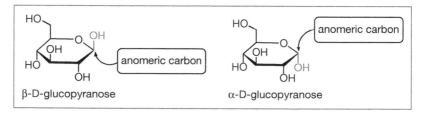

Figure 12.14   Anomers of Glucose

A six-membered cyclic sugar is called a *pyranose*. Glucose is almost always found as a pyranose ring system and is formally named *glucopyranose*.

A five-membered cyclic sugar ring is called a *furanose*. Fructose prefers a furanose ring system and is formally named *fructofuranose*. Like glucose, fructose can cyclize and can form either an alpha anomer or a beta anomer. Notice that, with fructose, the anomeric identity is determined by the direction the hydroxyl group points. The anomers of fructose are shown in Figure 12.15.

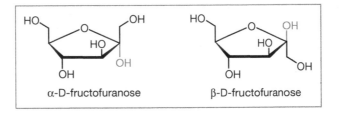

Figure 12.15 Anomers of Fructose

Glycoside Bonds

We saw in the previous chapter that aldehydes react with alcohols to form hemiacetals, and hemiacetals react with another alcohol to afford acetals. Well, the *intra*molecular reaction of the aldehyde carbon of an aldose monosaccharide with one of the hydroxyl groups in the same molecule affords a cyclic hemiacetal. An *inter*molecular reaction of this hemiacetal with the hydroxyl group of another sugar molecule provides an acetal functional group. This kind of acetal linkage connects monosaccharides into polysaccharides. Biochemists refer to these acetal (and ketal) bonds as *glycoside bonds*.

Glycoside bonds can be alpha or beta. Generally, we can digest only sugars that are connected by alpha glycoside bonds. Figure 12.16 illustrates the formation of an alpha glycoside bond. Notice that, in the product disaccharide, the linking oxygen atom (shown in blue) is pointed down, relative to the glucose residue on the left. Therefore, this linking oxygen atom is in the alpha position, and this is an alpha glycoside bond. The residue on the right still has a free anomeric hydroxyl group that can be either alpha or beta. The squiggly line means that either we are not specifying which anomer is present or that both anomers are present.

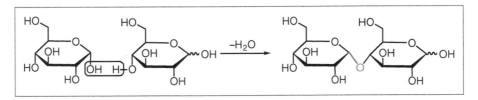

Figure 12.16 Formation of an Alpha Glycoside Bond

Figure 12.17 illustrates the formation of a beta glycoside bond. Again, the anomeric hydroxyl group on the glucose residue on the right can be either alpha or beta. In the product disaccharide, the linking oxygen atom (shown in blue) is pointed up, relative to the glucose residue on the left. Therefore, this linking oxygen atom is in the beta position. It is irrelevant that this oxygen is pointed down, relative to the glucose on the right, because glucose on the right contributed a regular alcohol functional group in forming the acetal link, not the hemiacetal functional group.

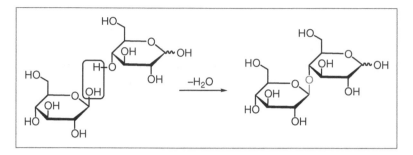

Figure 12.17 Formation of a Beta Glycoside Bond

Some Common Disaccharides

Disaccharides are comprised of two monosaccharides joined by a glycoside bond. Figure 12.18 gives the structures of some common disaccharides. Maltose, or malt sugar, is formally named α-D-*glucopyranosyl* [1↔ 4] β-D-*glucopyranose*. This name makes more sense if we break it into its components. α-D-glucopyranosyl means that the first residue is a glucose molecule cyclized into a six-membered ring, and the oxygen on the anomeric carbon is in the alpha configuration. [1↔ 4] means the glycoside bond is between carbon 1 on the first residue and carbon 4 on the second residue. β-D-*glucopyranose* means the second residue is a glucose molecule cyclized into a six-membered ring, and the oxygen on the anomeric carbon is in the beta configuration.

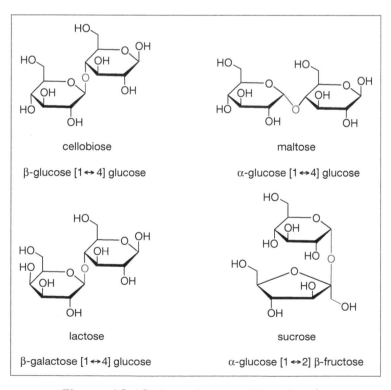

cellobiose

β-glucose [1↔ 4] glucose

maltose

α-glucose [1↔4] glucose

lactose

β-galactose [1↔4] glucose

sucrose

α-glucose [1↔2] β-fructose

Figure 12.18 Some Common Disaccharides

Cellobiose contains two glucose molecules joined by a beta glycoside bond and is formally named β-D-glucopyranosyl [1↔4] β-D-glucopyranose. Lactose contains a galactose residue joined by a beta glycoside bond to a glucose residue. Sucrose contains an alpha glucose joined to a beta fructose. Sucrose differs from the other examples because the glycoside bond connects the anomeric carbon of both sugar residues. The formal name of sucrose is α-D-glucopyranosyl [1↔2] β-D-fructofuranoside.

Polysaccharides

Two common polysaccharides are starch and cellulose. Both are polymers of glucose. Cellulose (Figure 12.19) is the structural material in plant matter and the major component of dietary fiber. Cellulose is poly [1↔ 4] β-glucose. The glycoside bonds in cellulose are beta. This configuration allows cellulose to adopt a fairly linear conformation, with a great deal of hydrogen bonding. This is part of the reason that cellulose is such a strong material.

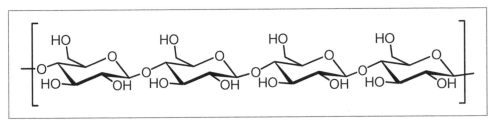

Figure 12.19 Cellulose

Starch (Figure 12.20) is a polymer of glucose that we can digest. Starch is poly [1↔ 4] α-glucose. The glycoside bonds in starch are alpha. This configuration puts a bend in the chain. In fact, starch adopts a helical conformation.

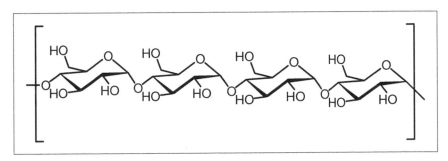

Figure 12.20 Starch

Chains of starch molecules can branch by forming [1↔6] glycoside bonds, as shown in Figure 12.21. This branching leads to more compact forms of starch, such as glycogen and amylopectin.

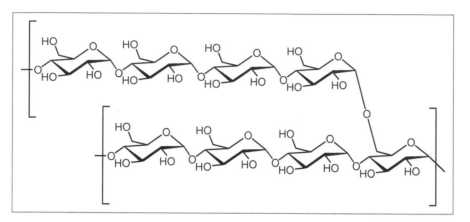

Figure 12.21 [1↔6] Branching in Starch

LIPIDS

While the other classes of biological molecules are characterized by structural features, lipids share a common physical property. Lipids are hydrophobic substances that are more soluble in organic solvents such as ether than in water. Lipids serve a myriad of biological purposes ranging from energy storage to sending chemical signals (both within an individual and between individuals), as well as serving as the main structural component of cell membranes for all living organisms. Lipids can be divided into two broad categories. The saponifiable lipids include triglycerides, waxes, and phospholipids. The nonsaponifiable lipids include steroids, prostaglandins, and fat-soluble vitamins.

Glycerides

Glycerides are esters composed of glycerin (1,2,3-propantriol) and fatty acids. Triglycerides have three fatty acid residues esterified to the glycerin backbone. Triglycerides are also known as triacyl glycerols. Mono- and diglycerides are mono- and di-esters of one or two fatty acids and glycerin.

Fatty Acids

Fatty acids are long-chained carboxylic acids. Although their name includes the word "fatty," fatty acids are not the same as fats. Fats are triglycerides, and fatty acids are one component of triglycerides. Naturally occurring fatty acids always have an even number of carbon atoms. Saturated fatty acids have carbon chains that contain only carbon–carbon single bonds. Unsaturated fatty acids have carbon chains that contain at least one carbon–carbon double bond. Monounsaturated fatty acids have one carbon–carbon double bond in the chain, and polyunsaturated fatty acids have two or more double bonds in the carbon chain. Naturally occurring unsaturated fatty acids have the cis double bond configuration.

Figure 12.22 gives the most common saturated fatty acids. The IUPAC names are given first. For dodecanoic acid, $C_{12}H_{24}O_2$, the parent chain has 12 carbons (do = 2 + deca = 10). The common names of the fatty acids are roughly derived from fats and oils that are rich in that fatty acid. For example, palm and palm kernel oils are rich in palmitic acid. Beef tallow is rich in stearic acid. Although stearic sounds a little bit like "steers," it is derived from the Greek word for tallow.

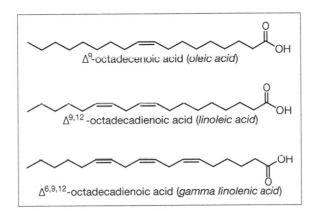

CH$_3$(CH$_2$)$_{10}$CO$_2$H	CH$_3$(CH$_2$)$_{12}$CO$_2$H
dodecanoic acid (*lauric acid*)	tetradecanoic acid (*myristic acid*)
CH$_3$(CH$_2$)$_{14}$CO$_2$H	CH$_3$(CH$_2$)$_{16}$CO$_2$H
hexadecanoic acid (*palmitic acid*)	octadecanoic acid (*stearic acid*)

Figure 12.22 Saturated Fatty Acids

Figure 12.23 presents three common unsaturated fatty acids. Notice that the configuration of the double bonds is cis. The delta nomenclature indicates the position of the double bond relative to the carboxyl group. The carbonyl carbon is carbon number one. For linoleic acid, $\Delta^{9,12}$-octadecadienoic acid, there are 18 carbons, and the double bonds begin on carbons 9 and 12. Linoleic and gamma linolenic acids are also classified as omega-6 fatty acids. The omega carbon is the last carbon in the chain. Omega-6 indicates there is a double bond six carbons from the end of the chain.

Δ^9-octadecenoic acid (*oleic acid*)

$\Delta^{9,12}$-octadecadienoic acid (*linoleic acid*)

$\Delta^{6,9,12}$-octadecadienoic acid (*gamma linolenic acid*)

Figure 12.23 Unsaturated Fatty Acids

Triglycerides

Structurally, triglycerides are tri-esters of glycerol (aka glycerin or 1,2,3-propantriol) and three fatty acids. Figure 12.24 gives the structure of a generic triglyceride. The glycerol backbone is shown in blue. Notice the three fatty acid chains (R$_1$, R$_2$, and R$_3$) are not necessarily the same.

Triglycerides are more commonly called fats or oils. A fat is a triglyceride that comes from animal sources, has a higher percentage of saturated fatty acids, and is a solid at room temperature.

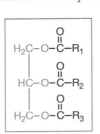

Figure 12.24 A Generic Triglyceride

An oil generally comes from plant sources, contains a larger percentage of unsaturated fatty acids, and is a liquid at room temperature. Fats and oils are not pure substances, because in a given sample of a fat or oil the identity of the R groups is not the same in all molecules. For example, most of the fatty acid residues in a sample of olive oil are oleic acid (ca. 70%) and palmitic acid (ca. 10%), along with smaller amounts of linoleic acid, stearic acid, and linolenic acid. However, the fatty acid composition on a given olive oil molecule is essentially a random selection from these choices of fatty acids.

Figure 12.25 shows space-filling models of tristearin and triolein. Tristearin is a triglyceride containing three stearic acid residues and represents a generic fully saturated triglyceride. Triolein is a triglyceride containing three oleic acid residues and represents a generic unsaturated triglyceride. The geometry of these models was optimized using a molecular mechanics program that treats each chemical bond as a spring and each atom as a ball, and jiggles the whole model until the energy of the system is minimized. The hydrogen atoms in each model are omitted for clarity. The three carbons in glycerol are colored blue, and the oxygen atoms are red. Notice that the cis double bonds in triolein impose an awkward kink in the molecule, and this hinders triolein in settling into the solid state. The more symmetrical tristearin fits more easily into a crystalline lattice and has greater access to intramolecular London forces, and, therefore, is a solid at room temperature.

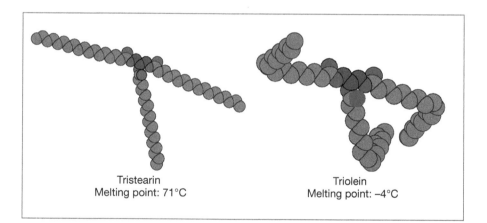

Tristearin
Melting point: 71°C

Triolein
Melting point: −4°C

Figure 12.25 Three-Dimensional Structures of Tristearin and Triolien

Saponification

Since triglycerides are esters, they undergo the same kinds of reactions as other esters. One important reaction is hydrolysis by aqueous sodium hydroxide, which is a reaction called *saponification* (see Figure 12.26). Saponification means "soap forming," and this is how

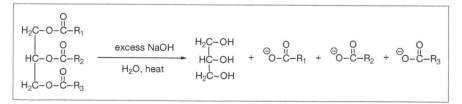

Figure 12.26 3-Saponification of a Triglyceride

soaps are synthesized. Hydrolysis of a triglyceride (with hot, aqueous NaOH) gives glycerin and three fatty acids. Since the reaction medium is strongly basic, the fatty acids are present in their conjugate base form. We have already discussed the function of soaps and other surfactants.

Detergents are synthetic soaps. Like soaps, detergents have a long, nonpolar tail and a polar head. The polar head can be anionic, cationic, or neutral. Most detergents are derived from triglyceride sources. Sodium dodecylsulfate (SDS) is a typical anionic detergent. SDS is also known as sodium laurylsulfate, because the carbon chain is synthesized from lauric acid. SDS is the active ingredient in laundry detergents such as Tide. SDS is shown in Figure 12.27.

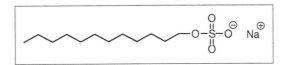

Figure 12.27 Sodium Dodecylsulfate

SDS has several advantages over soap. Unlike the carboxylate anions in soap, the dodecylsulfate anion does not form insoluble precipitates with hard-water ions (e.g., Ca^{2+}), leading to the formation of soap scum. Also, the dodecylsulfate anion is the conjugate base of a very strong acid (H_2SO_4). Therefore, SDS is a weak base, and solutions of SDS are pH neutral.

Phospholipids

Phospholipids are similar in structure to the glycerides, except that one of the fatty acid residues has been replaced by a phosphate ester group. Phospholipids are common in cellular membranes because of their surfactant properties. They have long, greasy tails that are hydrophobic. The polar phosphate group, or groups bonded to the phosphate group, are hydrophilic. Thus, phospholipids can form barriers to separate the aqueous cytoplasm from the aqueous extracellular fluid while facilitating the transport of ionic and polar materials through the membrane. Phosphatidylcholine, also known as lecithin, is a typical phospholipid (Figure 12.28).

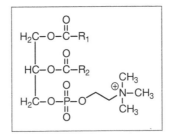

Figure 12.28 Phosphatidylchloline

Steroids

Steroids are characterized by the cyclopentanoperhydrophenanthrene ring system, which consists of three six-membered rings fused to a five-membered ring (Figure 12.29). Steroids play a variety of roles in living systems. Some common steroids include cholesterol, estrogen, and testosterone.

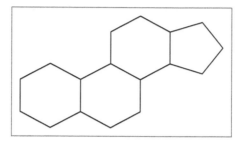

Figure 12.29 Steroid Ring System

Cholesterol is an essential component of cellular membranes. In addition to dietary sources, we can also synthesize cholesterol. Cholesterol is transported in the blood as a lipoprotein, which is an aggregate of water-soluble proteins, cholesterol, and other lipids, including triglycerides. Proteins are denser than lipids, so the ratio of protein to lipid determines the classification of the lipoprotein. High-density lipoproteins (HDL) have a greater protein-to-lipid ratio than low-density lipoproteins (LDL). Thus, from a chemical standpoint, there really aren't "good cholesterol" and "bad cholesterol." High levels of LDL are indicative of potential heart problems, not because of a different form of cholesterol, but because LDL particles are more likely to dump their load of cholesterol into the arteries, leading to plaque formation and atherosclerosis.

Other steroids include the sex hormones testosterone, progesterone, and the estrogen family (estrone, estradiol, estratriol, etc.; Figure 12.30).

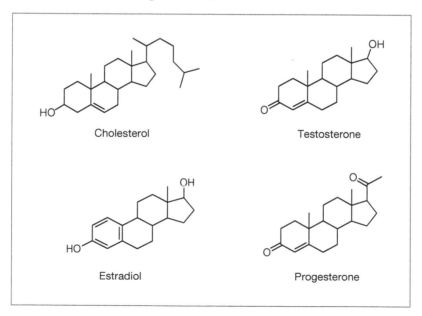

Figure 12.30 Some Common Steroids

Prostaglandins

Prostaglandins are powerful, but short-lived, hormones in mammalian systems that were first isolated from seminal fluid. The name prostaglandins is derived from "prostate gland." Prostaglandins are synthesized in vivo from an unsaturated fatty acid called arachidonic acid. The identifying structural feature of a prostaglandin is a five-membered ring with a 7-carbon side chain, R_7 (often ending in a carboxylic acid group) adjacent to an 8-carbon chain, R_8.

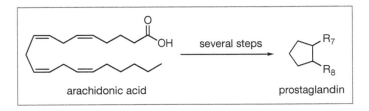

arachidonic acid   several steps   prostaglandin

Figure 12.31 shows some common prostaglandins. The letters PG stand for prostaglandin. The next letter indicates the functional groups in the ring. Sorry, there is no system to the letters. The number that follows indicates the number of double bonds in the side chains. In $PGF_{2\alpha}$, the alpha indicates the hydroxyl group on the lower side chain is pointed down (just as with the anomeric hydroxyl group in sugars).

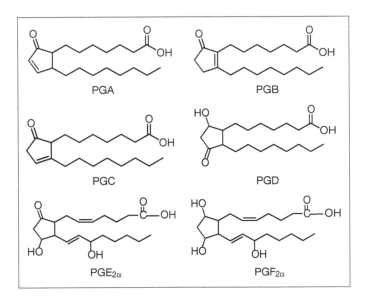

Figure 12.31 The Prostaglandins

METABOLISM OF CARBOHYDRATES AND FATTY ACIDS

Although carbohydrates and lipids play other roles in the body, one of our principal aims in eating fat and sugar is for the energy content. We need to access the energy stored in the chemical bonds of these compounds in order to move our muscles, keep our hearts beating, synthesize proteins and nucleic acids, and so on. Carbohydrates and triglycerides are metabolized through a series of chemical reactions, many of which are oxidation reactions. These pathways convert food molecules into high-energy biomolecules. The main high-energy molecule is adenosine triphosphate, more commonly known as ATP (Figure 12.32). ATP is the compound that our bodies most easily use for the processes of staying alive.

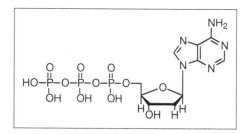

Figure 12.32 ATP

However, there are a number of other high-energy molecules that are produced in the metabolism of carbohydrates and fats. The first one we'll consider is acetyl coenzyme A (acetyl CoA). The structure of acetyl coenzyme A is given in Figure 12.33. The acetyl portion of the molecule is shown in blue, and, although this is a bit of an oversimplification, you can think of this acetyl group as being derived from acetic acid. The carbons in the acetyl group are derived by breaking the carbon–carbon bonds in fatty acids and glucose into two carbon units.

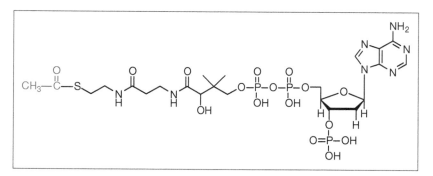

Figure 12.33 Acetyl Coenzyme A

The acetyl portion is bonded to coenzyme A through a thioester functional group. In organic chemistry, "thio" means sulfur. So a thioester is an ester in which one of the oxygen atoms has been replaced by a sulfur atom (see Figure 12.34). Because the carbon–sulfur bond is not

Figure 12.34 Formation of a Thioester

as strong as the carbon–oxygen bond in a regular ester, acetyl CoA readily releases the acetyl group to other organic molecules. So, the role of coenzyme A is to shuttle the carbons from metabolism of glucose and fatty acids into the Krebs Cycle.

Several reactions in metabolism are oxidation–reduction (or redox) reactions. Two of the principal redox carriers are nicotinamide adenine dinucleotide (NAD^+) and coenzyme Q. Remember that we live in an oxidizing world, so species that are in the reduced form are frequently high-energy compounds that react exothermically with oxygen. Also recall that organic molecules are reduced by adding bonds to hydrogen.

Figure 12.35 shows the reduction reaction of NAD^+ into the reduced form, NADH. The added hydrogen on NADH is shown in blue. NADH is the energized form that can feed into the electron transport chain to synthesize ATP.

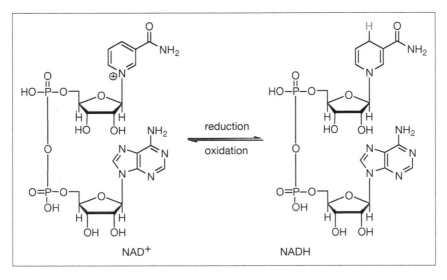

Figure 12.35 Nicotinamide Adenine Dinucleotide

Figure 12.36 shows the reduction reaction of coenzyme Q into the reduced form, QH_2. The added hydrogens are shown in blue. QH_2 is the energized form that can feed into the electron transport chain to synthesize ATP.

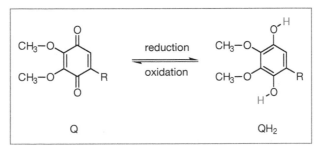

Figure 12.36 Coenzyme Q

Figure 12.37 gives the big picture for the metabolism of carbohydrates and triglycerides.

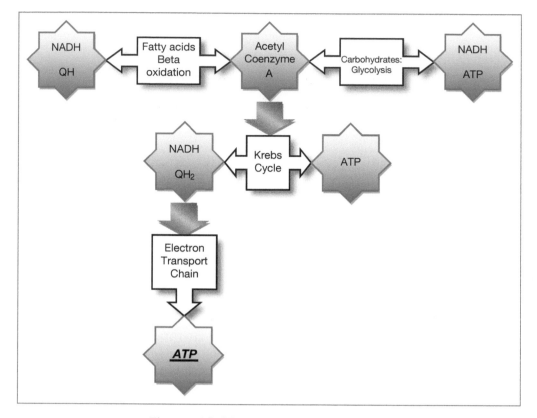

Figure 12.37 Overview of Metabolism

Carbohydrates are generally converted into glucose, and glucose feeds into the glycolysis pathway. The six carbons in glucose are broken down into two 3-carbon molecules of pyruvic acid, along with two molecules each of ATP and NADH for each glucose molecule metabolized. Glycolysis is shown in Figure 12.38.

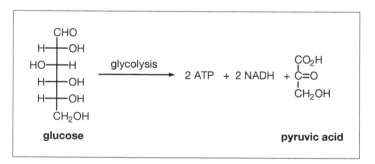

Figure 12.38 Glycolysis

Under anaerobic conditions, the pyruvic acid is reduced to lactic acid, which consumes the NADH produced (see Figure 12.39). Thus, anaerobic metabolism is highly inefficient because the more energetically valuable of the two high-energy products (NADH) is consumed.

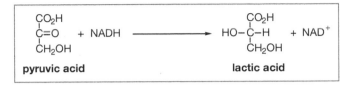

Figure 12.39 Reduction of Pyruvic Acid Into Lactic Acid

Under aerobic conditions, the pyruvic acid is converted into acetyl CoA. This pathway also provides an NADH, along with carbon dioxide, as shown in Figure 12.40.

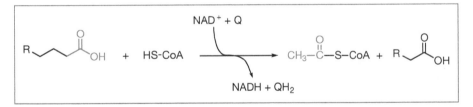

Figure 12.40 Formation of Acetyl Coenzyme A

Fatty acids are cleaved from triglycerides and metabolized through beta oxidation. This process is shown in Figure 12.41. Beta oxidation clips the fatty acids into two-carbon fragments. The 2-carbon fragments emerge from beta oxidation as acetyl CoA. Each acetyl CoA molecule is accompanied by the production of an NADH and a QH_2. Of course, when the chain is finally cut down to just two carbons, they are as an acetyl CoA. Since there are no carbon–carbon bonds to oxidatively cleave, no NADH or QH_2 are produced.

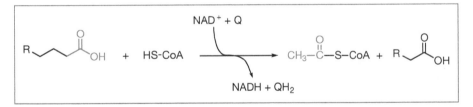

Figure 12.41 Beta Oxidation of Fatty Acids

The acetyl CoA molecules feed the Krebs Cycle, which is also known as the citric acid cycle or the tricarboxylic acid (TCA) cycle. Thus, one molecule of glucose can drive two rounds of the Krebs Cycle, whereas one molecule of stearic acid can drive nine rounds. The Krebs Cycle is the principal aerobic energy-producing pathway. Each turn of the Krebs Cycle affords one ATP[5] molecule, three NADH molecules, and one QH_2 molecule. Most of the energy produced directly from the Krebs Cycle is in NADH and QH_2, both of which require oxygen to convert their stored energy into ATP.

All of the reduced coenzymes, NADH and QH_2, feed into the electron transport chain, and this is where the real energy payoff occurs. Through a series of oxidation–reduction reactions, where oxygen gas is the final oxidant, the energy stored in the reduced coenzymes

---

[5] Actually, the product is guanosine triphosphate (GTP), but GTP is energetically equivalent to ATP.

drives the endothermic synthesis of ATP. On average, each NADH affords 2.5 ATP, and each QH_2 furnishes 1.5 ATP. Thus, each acetyl Co A is worth 10 ATP.

Table 12.2 summarizes the high-energy compounds derived from the metabolism of glucose and stearic acid. As you can see, fats offer much more energy than carbohydrates.

Table 12.2 NET ATP FROM GLUCOSE AND STEARIC ACID

FOOD SOURCE	CARBON ATOMS	HIGH-ENERGY PRODUCTS	NET ATP PRODUCED
Glucose	6	2 ATP 2 Acetyl CoA × 10 4 NADH × 2.5	32
Stearic acid	18	9 Acetyl CoA × 10 8 NAHD × 2.5 8 QH_2 × 1.5 −2 ATP to initiate reaction	120

PROTEINS AND AMINO ACIDS

Proteins are one of the most remarkable of the biomolecules. Proteins serve many roles in living systems, from transport molecules, such as hemoglobin, to structural and locomotion tissues, to enzymes that are necessary to catalyze virtually every chemical process that living organisms carry out. Proteins are polymers of relatively simple organic compounds called amino acids joined together by peptide bonds. These "polypeptide chains" may contain just a few amino acids or more than 1000 amino acids. When two amino acids are joined together, the protein is a dipeptide. When three amino acids are joined together, the protein is a tripeptide (etc.). When many amino acids are joined together, the protein is a polypeptide. So, while there are only about 20 common amino acids, there are nearly an infinite number of proteins. So, let's take a closer look at amino acids and how they combine to give proteins.

Amino Acids

Amino acids contain two organic functional groups: an amine group and a carboxylic acid group (aka a carboxyl group), as shown in Figure 12.42. These compounds are sometimes called

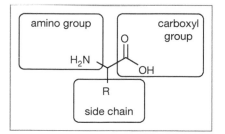

Figure 12.42 Generic Amino Acid

alpha amino acids because the amine functional group is located on the first carbon out from the main functional group (the carboxylic acid). Each amino acid has a unique side chain that gives the amino acid its characteristic physical and chemical properties. The combined influences of all of the amino acid side chains confer both structure and function upon the protein.

Side Chains

The identity of an amino acid is determined by the side chain. It is easiest to learn the side chains as organic groups (which is what they are). Each amino acid has a one-letter and a three-letter abbreviation. Although the three-letter abbreviations are more intuitive, most biochemists now use the one-letter abbreviations. Figure 12.43 gives the names, structures, and

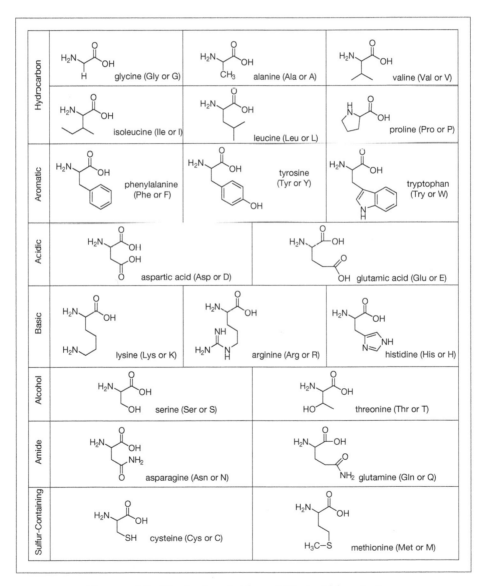

Figure 12.43 Amino Acids and Their Abbreviations

abbreviations of the amino acids, grouped according to the main functional group in the side chain. Figure 12.44 groups the amino acids according to their relative polarities. The least polar of the side chains are the hydrocarbon groups. The polar side chains include hydrogen-bonding

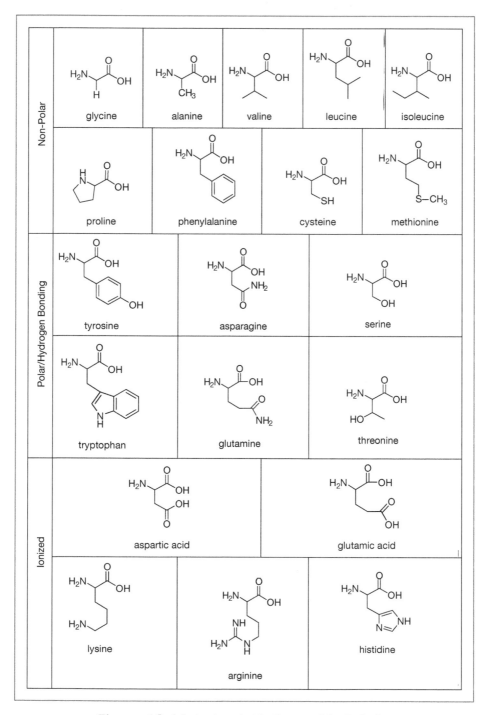

Figure 12.44 Amino Acids Grouped by Polarity

functional groups such as alcohols and amides. Remember, hydrogen bonds are possible when a hydrogen atom is directly bonded to an oxygen atom or a nitrogen atom. The acidic and basic side chains are the most polar groups because they are largely in their ionized form (conjugate acid or conjugate base) at physiological pH.

Stereochemistry of Amino Acids

Naturally occurring amino acids belong to the L-family. That is, when the amino acid is drawn as a Fischer projection, with the carboxyl group at the top and the side chain at the bottom, the amino group is on the left, as shown in Figure 12.45.

Figure 12.45 L-Amino Acid

Since all amino acids (except glycine) are chiral, proteins are also chiral. Thus, the inherent handedness of proteins enables proteins to differentiate between two enantiomeric isomers. Recall that enantiomers have identical chemical and physical properties, except for how they interact with polarized light and other chiral species. For example, this is why we can digest D-sugars but not L-sugars.

Acid–Base Properties of Amino Acids

For the sake of clarity, we have been drawing amino acids incorrectly, and now it is time to recognize this error. Amino acids have both an acidic functional group and a basic functional group. Remember that acids donate hydrogen ions, and bases accept hydrogen ions. The basic amino group takes a proton from the acidic carboxyl group. This intramolecular acid–base reaction results in formation of a zwitterion. A zwitterion is a species that is electrically neutral but has separated positive and negative ionic charges (see Figure 12.46). The word "zwitterion" comes from *zwei*, the German word for two.

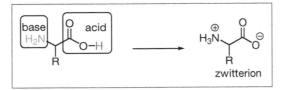

Figure 12.46 Formation of a Zwitterion

The carboxylate-ion-end of the zwitterion is the conjugate base of a weak acid, and the ammonium-ion-end of the zwitterion is the conjugate acid of a weak base. Thus, the zwitterion is a buffer, and solutions of amino acids (and proteins) resist changes in pH.

Zwitterions are an amphiprotic species and can behave as either an acid or a base. For example, if a strong acid, such as HCl, is added to a solution of an amino acid, the carboxylate-ion-end of the zwitterion accepts the proton from the stronger acid, and the zwitterion is converted into a cationic species. For convenience, let's represent the zwitterionic form, which has one H^+ to donate, as HA. The cationic form, which has two H^+ ions to donate and has a net positive charge, is represented as H_2A^+, as can be seen in Figure 12.47.

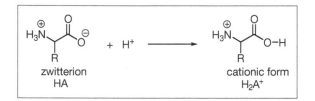

Figure 12.47 Formation of the Cationic Form of an Amino Acid

The zwitterion can also behave as an acid and donate a hydrogen ion from the ammonium group. When the ammonium-ion-end of the zwitterion surrenders an H^+ ion, the zwitterion is converted into an anionic form, A^- as can be seen in Figure 12.48.

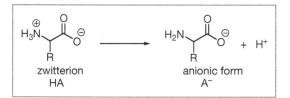

Figure 12.48 Formation of the Anionic Form of an Amino Acid

Thus, the net electrical charge on an amino acid is a function of pH, and this has both critical and useful implications. At very low pH, an amino acid is primarily in its fully protonated form (H_2A^+), and the net charge on an amino acid is positive. As pH increases, the cationic form is gradually converted to the zwitterionic form (HA), and the net charge approaches zero. As we continue to increase the pH to strongly basic conditions, the ammonium group surrenders the second proton, converting the amino acid into its fully deprotonated form (A^-), and the net charge on an amino acid is negative.

$$H_2A^+ \; \underset{\text{lower pH}}{\overset{\text{raise pH}}{\rightleftharpoons}} \; HA \; \underset{\text{lower pH}}{\overset{\text{raise pH}}{\rightleftharpoons}} \; A^-$$

An alpha plot is a convenient way to visualize the composition of an amino acid solution. Recall that alpha represents the percentage of each component in a chemical system. Let's consider the alpha plot of glycine shown in Figure 12.49. In strongly acidic solutions (pH < 1), nearly 100% of the glycine molecules are present as the cationic form (H_2A^+). As we increase the pH, the concentration of the cation decreases, and the concentration of the zwitterion

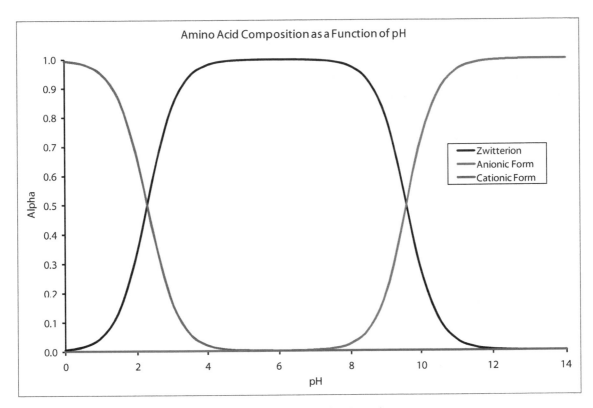

Figure 12.49 Alpha Plot for Glycine

increases. Notice that at $\alpha = 50\%$, the concentration of the weak acid (H_2A^+) is equal to the concentration of its conjugate base (HA). Since H_2A^+ is a weak acid and HA is its conjugate base, we have a buffer system. Recall that when the concentrations of the weak acid and its conjugate base are equal, the pH of the buffer is equal to the pK_a of the weak acid. So, we see the pK_a for H_2A^+ is equal to 2.3.

As we continue to increase pH, the zwitterion concentration reaches a maximum value at pH 6.0. Above this pH, the zwitterion is gradually converted into the anionic form (A^-). Once again, we have a buffer system comprised of a weak acid (HA) and its conjugate base (A^-). The point where the alpha values for these two species equal 50% corresponds to a pH of 9.6, which is the pK_a for the zwitterion.

The Isoelectric Point

The isoelectric point (pI) or isoelectric pH is the pH where the net charge on the amino acid is zero. At pI, the zwitterion is the dominant species, with an alpha value very close to 100%. The tiny equilibrium amounts of the cationic form and the anionic form are equal, so their opposite charges cancel each other out. The net charge on a collection of amino acid molecules at a pH less than pI is positive. The net charge on a collection of amino acid molecules at a pH greater than pI is negative.

The isoelectric point is equal to the average of the two pK_a values for the amino acid. For example, for a generic amino acid (HA), we can write two ionization equilibria and the corresponding equilibrium constant expressions.

$$H_2A^+ \rightleftharpoons HA + H^+ \quad K_{a1} = \frac{[H^+][HA]}{[H_2A^+]}$$

$$HA \rightleftharpoons A^- + H^+ \quad K_{a2} = \frac{[H^+][A^-]}{[HA]}$$

If we add these two equilibrium equations together, we have:

$$H_2A^+ \rightleftharpoons A^- + 2H^+ \quad K_a = \frac{[H^+]^2[A^-]}{[H_2A^+]}$$

Notice the equilibrium constant for these combined reactions equals the product of the individual equilibrium constant expressions, and the [HA] terms cancel out.

$$K_{a1} \cdot K_{a2} = \frac{[H^+][HA]}{[H_2A^+]} \cdot \frac{[H^+][A^-]}{[HA]} = \frac{[H^+]^2[A^-]}{[H_2A^+]}$$

If we solve this expression for the hydrogen ion concentration, we have:

$$[H^+]^2 = K_{a1} \cdot K_{a2} \cdot \frac{[H_2A^+]}{[A^-]}$$

Now let's specify that the hydrogen ion concentration is at the isoelectric point. Thus, $[H_2A^+]$ = $[A^-]$, and our expression simplifies to:

$$[H^+]^2 = K_{a1} \cdot K_{a2}$$

Now, let's take the negative logarithm of both sides of the equation:

$$-\log([H^+]^2) = -\log(K_{a1} \cdot K_{a2})$$

Recalling that a log of a power is equal to the exponent times the log of the base, and the log of a product is equal to the sum of the logs, we can simplify the equation to:

$$2 \cdot (-\log([H^+])) = [-\log(K_{a1})] + [-\log(K_{a2})]$$

Of course, the negative log of a quantity is the p-function, so we have:

$$2 \cdot pI = pK_{a1} + pK_{a2}$$

Solving for pI, we have:

$$pI = \frac{1}{2}[pK_{a1} + pK_{a2}]$$

So, for glycine, if $pK_{a1} = 2.3$ and $pK_{a2} = 9.6$, what is pI for glycine?

$$pI = \tfrac{1}{2}[2.3 + 9.6] = \tfrac{1}{2}[11.9] = 6.0$$

So, at a pH of 6.0, the net charge on glycine is zero. At a pH less than 6.0, the average charge on the collection of glycine molecules is positive, and as the pH decreases, the average charge on the collection of glycine molecules approaches a maximum value of $+1$. At a pH greater than 6.0, the average charge on the collection of glycine molecules is negative, and as the pH increases, the charge approaches a maximum value of -1.

Electrophoresis

Electrophoresis is arguably the most important analytical tool in a biochemist's repertoire. Electrophoresis uses an electric field to separate amino acids (or proteins or nucleic acids) on the basis of their electrical charge and molecular weight. When electrophoresis is applied to nucleic acid samples, it is commonly called *DNA fingerprinting*. The sample is placed on a solid support. The solid support looks and feels like Jello. Agarose is often used for DNA, and polyacrylamide is often used for proteins. The gel is immersed in a buffer solution. The role of the buffer is not only to control the acid–base form of the amino acids (or whatever) present, but also to serve as an electrolyte (electrical conductor). A positive electrode is placed at one end of the gel, and a negative electrode is placed at the other end. If the species has a charge, it will migrate toward the oppositely charged electrode. Species with a greater net charge move farther on the gel. So, the closer the buffer pH is to the pI of the sample, the less it moves. Smaller molecules with the same charge migrate faster than larger molecules, in accordance with Newton's laws.

The Peptide Bond

A protein is a polypeptide chain consisting of amino acids joined by peptide bonds. Actually, though, a peptide bond is an amide bond formed between the amino group of one amino acid and the carboxyl group of another amino acid. For example, the reaction in Figure 12.50 shows formation of a dipeptide from two amino acids.

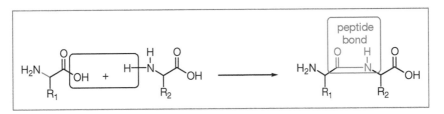

Figure 12.50 Formation of a Peptide Bond

Notice that there is a direction in a polypeptide chain. One end of the chain has a free amino group, which is called the N-terminus. The other end of the chain has a free carboxyl group, and this is called the C-terminus. Figure 12.51 shows the structure of a tripeptide chain consisting of alanine, valine, and phenylalanine. Alanine is the N-terminal amino acid, and phenylalanine is the C-terminal amino acid.

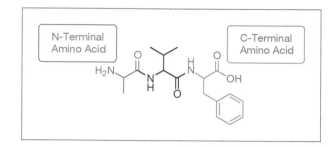

Figure 12.51 N-Terminal and C-Terminal Amino Acids

Drawing the complete structure of the polypeptide chain is very time consuming, so biochemists typically give the abbreviations of the amino acids. So, the above tripeptide can be expressed using either the three-letter or one-letter abbreviations: Ala-Val-Phe or A-V-F. When they are written this way, the N-terminal amino acid is always on the left.

Primary Structure of Proteins

Video 12.1

The primary structure of a protein is the sequence of amino acids in the peptide chain. The primary structure is immensely important, because it is the sequence of amino acids that determines the higher levels of protein structure and, consequently, the function of the protein. Small changes in the primary structure can cause a protein to be completely nonfunctional. For example, sickle cell anemia is caused by the substitution of a single amino acid in the hemoglobin chain.

Secondary Structure of Proteins

The secondary structure of a protein is how the polypeptide chain is "twisted." There are two common types of secondary structure: the alpha helix and the beta pleated sheet.

In an alpha helix, the polypeptide chain is twisted into a coil, as illustrated in Figure 12.52. The "alpha" means the coil twists in a clockwise direction. The black spheres represent carbon atoms, the red spheres represent oxygen atoms, and blue spheres represent nitrogen atoms. Hydrogen atoms are yellow. For the sake of clarity, only hydrogen atoms bonded to the amide nitrogen atoms are shown. The dashed lines represent hydrogen bonding between the hydrogen atoms on the nitrogen and a carbonyl oxygen.

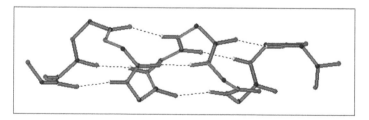

Figure 12.52 Alpha Helix Structure

The beta pleated sheet, or simply beta sheet, structure is illustrated in Figure 12.53. The carbon backbone in the beta sheet is fully extended, and adjacent chains are held together by a large number of hydrogen bonds.

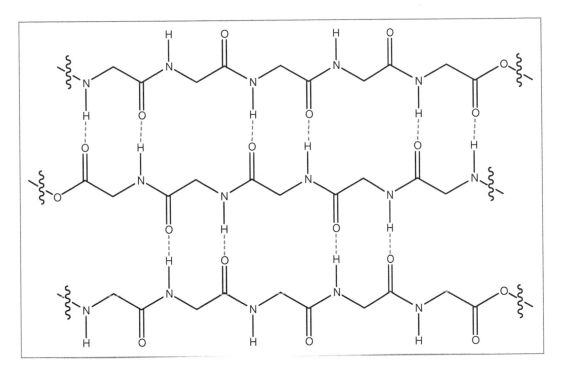

Figure 12.53 Beta Pleated Sheet Structure

Tertiary Structure of Proteins

The tertiary structure of a protein refers to how the alpha helices and beta sheet portions of a polypeptide chain are folded into a compact or globular structure. Two-dimensional representations of the three-dimensional tertiary structures of proteins are, well, pretty two-dimensional. If you want to better understand tertiary structure, you can view inter-active, three-dimensional models of a large variety of proteins at the RCSB Protein Data Base at http://www.rcsb.org.

In addition to hydrogen bonding, tertiary structure is maintained by London forces and, sometimes, disulfide bridges. London forces act between nonpolar side chains that fold into the protein's interior, away from the polar water environment.

Disulfide bridges occur between different cysteine residues in the chain, as shown in Figure 12.54. The functional group in the cysteine chain is a sulfhydril or thiol group. It is simply the sulfur analog of an alcohol. However, thiol groups are easily oxidized into disulfide groups that bridge two portions of the peptide chains, or between two different peptide chains.

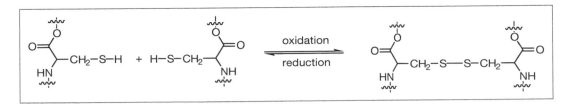

Figure 12.54 Formation of a Disulfide Bridge

By the way, this is the chemistry of a permanent hair wave. Hair is a protein rich in sulfur and has many disulfide bridges. First, hair is wrapped around a template, such as a curler. Then a mild reducing agent is applied to reduce some of the disulfide bridges present in hair. The physical stress of stretching the hair around a curler causes the polypeptide chains in hair to shift slightly with respect to each other. Finally, a mild oxidizing agent is applied, which creates new disulfide bridges that hold the hair in its deformed conformation.

The Folding Problem

The primary structure of the polypeptide chain determines the higher levels of structure. All proteins have a primary structure. Most proteins have a secondary structure, and compact globular proteins have a tertiary structure. Some proteins, such as hemoglobin, even have a quaternary structure, which is the association of globular proteins through noncovalent bonding interactions (e.g., hydrogen bonding, etc.)

We can make some generalizations about how proteins fold. For example, it is a stabilizing feature to get hydrogen-bonding portions of the chain in close proximity. Proteins typically fold with nonpolar side chains on the interior of the protein, away from water, and with polar side chains on the outside of the protein, where they can interact with water molecules. In spite of these (gross) generalizations, the problem of how and why polypeptide chains fold into functional proteins remains one of the fundamental unsolved problems in physical biochemistry.

Denaturing Proteins

Video 12.2

In order to be functional, globular proteins must retain their higher levels of structure. If something disrupts the higher levels of a protein's structure, the protein is *denatured* and usually comes crashing out of a solution. An egg white is largely a solution of a protein called albumin. When you cook an egg, the heat denatures the albumin, and the denatured protein comes out of a solution to form the cooked white of the egg. In addition to high temperatures, heavy metal ions, such as Ag^+ and Hg^{2+}, and changes in pH can cause a protein to denature.

Enzymes

All living organisms are chemical factories, and virtually every chemical reaction that occurs in a living system is catalyzed by special proteins called *enzymes*. All enzymes are globular proteins. Folding the peptide chains into a compact structure creates a chiral pocket. This is called the active site of the enzyme. The extraordinary specificity that enzymes show for their given substrate molecules is because the active site exactly matches the dimension and shape of the molecules upon which the enzyme acts. One reason enzymes speed reaction rates is that enzymes capture reacting molecules and hold them in place next to each other. Furthermore, key amino acid side chains are located in the active site of each enzyme. For example, if a reaction is catalyzed by acid, then an acidic side chain will be located in the active site, exactly where it is needed to catalyze the reaction.

NUCLEIC ACIDS

As we said earlier, nucleic acids are the architects and construction contractors for synthesizing proteins. There are two kinds of nucleic acids. DNA, or deoxyribonucleic acid, is the blueprint for synthesis of proteins. RNA, or ribonucleic acid, is the construction contractor. Messenger RNA reads the instructions for synthesis of a protein encoded on a strand of DNA and carries those instructions to the worksite, where transfer RNA brings the amino acids in for incorporation into the polypeptide chain. Now, let's take a closer look at the structures of DNA and RNA.

DNA

Deoxyribonucleic acid is a biopolymer of phosphate sugars. Additionally, each phosphate sugar carries a nitrogenous base. The bases pair through hydrogen bonding, establishing a chemical basis for the genetic code. Now, let's break down these components, so we can understand them.

2-Deoxyribose

2-Deoxyribose is an aldopentose that is the structural sugar in DNA. This sugar is called "deoxy" because it does not have a hydroxyl group on carbon 2. Deoxyribose cyclizes into a furanose (five-membered) ring system. The structures of D-2-deoxyribose and β-D-2-deoxyribofuranose are given in Figure 12.55.

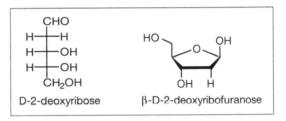

Figure 12.55 Deoxyribose

Nucleosides: Adding the Nitrogenous Bases

When we add a nitrogenous base to the sugar, we get a nucleoside, which is sometimes called a *nucleoside base*. DNA uses four bases, which fall into two categories: purines and pyrimidines. The four DNA bases are adenine, thymine, guanine, and cytosine.

Adenine and guanine are purines because they are derivatives of the heteroaromatic compound purine. Thymine and cytosine are pyrimidines, because they are derivatives of the heteroaromatic compound pyrimidine. The structures of these compounds are shown in Figure 12.56.

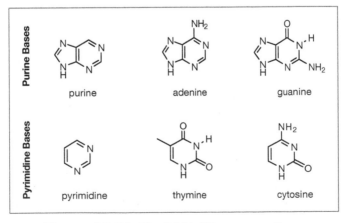

Figure 12.56 The DNA Nitrogenous Bases

Each base has a nitrogen atom capable of forming an acetal-like bond to the anomeric (or hemiacetal) carbon of deoxyribose. Recall that formation of an acetal from an alcohol and a hemiacetal involves elimination of water. We can envision that the water molecule is formed from the hydroxyl group of the hemiacetal and a hydrogen atom from the alcohol (or, in this case, the amine). Figure 12.57 illustrates this reaction for a purine base and for a pyrimidine base.

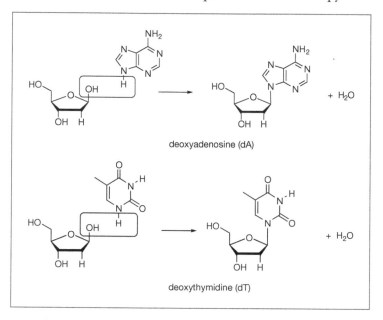

Figure 12.57 Formation of DNA Nucleoside Bases

So, there are four DNA nucleoside bases: deoxyadenosine, deoxythymidine, deoxycytidine, and deoxyguanosine. Usually, these bases are represented as dA, dT, dC, and dG, respectively. Often, the little d (which stands for deoxy) is omitted, and we simply use A, T, C, and G when describing the primary sequence of bases in DNA. Figure 12.58 gives the names and structures of these compounds.

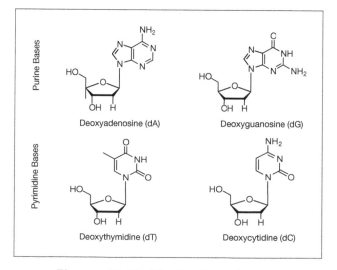

Figure 12.58 The DNA Nucleosides

Nucleotides: Adding the Phosphate Group

When we add a phosphate ester group to a nucleo<u>side</u>, we get a nucleo<u>tide</u>. The phosphate ester group is (conceptually) added by the reaction of the hydroxyl group on carbon 5 (the one not in the ring) to phosphoric acid. In the example in Figure 12.59, deoxyadenosine forms a phosphate ester to give deoxyadenosine monophosphate.

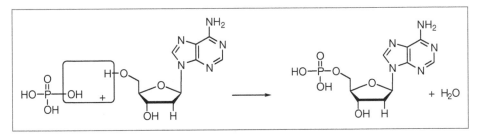

Figure 12.59 Formation of a Nucleotide

However, this reaction does not occur as written by simply mixing a nucleoside and phosphoric acid, and requires special reaction conditions. However, we are omitting those details for clarity. If you are interested, take a look in any modern biochemistry book. Phenomenonologically, however, just as in other esterification reactions, a hydroxyl group from the acid combines with a hydrogen atom from the alcohol to give a water molecule. The remaining fragments join to afford the ester.

When two acid molecules condense by elimination of a molecule of water, the product is called an *acid anhydride*, as can be seen in Figure 12.60. Acid anhydrides are always very reactive, or high-energy, compounds. When deoxyadenosine monophosphate forms an anhydride with phosphoric acid, we have deoxyadenosine diphosphate (dADP). Of course, if we add another phosphate group, we have deoxyadenosine triphosphate (dATP).

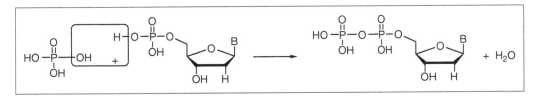

Figure 12.60 Formation of a Phosphate Anhydride

Forming the DNA Strand

The nucleosides are connected to the phosphate group by a phosphate ester functional group. Notice the phosphate group has additional OH groups, and those OH groups can be used to form additional phosphate ester bonds. Forming a new phosphate ester bond to the alcohol group on carbon 3 of another nucleotide strings the nucleotides into a strand of DNA.

Notice there is a directionality to this chain illustrated in Figure 12.61. At the left end, the leading phosphate (shown in red) is bonded to carbon 3 of a deoxyribose unit. On the right end, the terminal phosphate (shown in green) is bonded to carbon 5 of the deoxyribose. The structure in Figure 12.61 shows the polymer in the 3´→ 5´ direction. Therefore, this strand of DNA could be represented as:

$$(3´)C–T–A(5´)$$

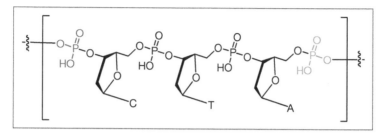

Figure 12.61 A Short Strand of DNA

This directionality is essential in transcription of DNA because it enables the transcription enzymes to read the genetic code in the proper direction. That is, C–T–A is not the same as A–T–C.

Base Pairing

The rigid rings in the bases hold hydrogen-bonding pieces in the exact location so that A always matches with T, and G always matches with C. The hydrogen bonds formed are illustrated in Figure 12.62.

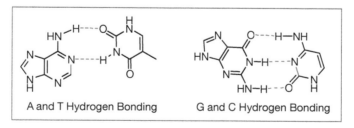

A and T Hydrogen Bonding   G and C Hydrogen Bonding

Figure 12.62 Base Pairing Through Hydrogen Bonding

Double-Stranded Helix

In most cases, DNA found in the nucleus of a cell is a helical double-stranded structure. Figure 12.63 represents the double-stranded structure of a short section of DNA. The double strand is rather like a ladder. The sides of the ladder are formed by the phosphate sugar backbone. The rungs of the ladder are formed by the nitrogeneous bases.

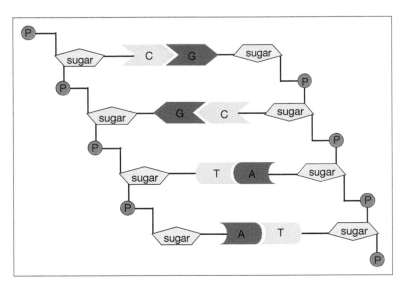

Figure 12.63 Model of Double-Stranded DNA

The shape of the deoxyribose units forces the double strand into a helical shape (Figure 12.64). Carbon atoms are black, oxygen atoms are red, and nitrogen atoms are blue. The ladder appears to have been twisted around a flagpole. The deoxyribose units are chiral, and their chirality is evident in the overall structure of the helix, which is also chiral. You can run your right hand along the helix (and stay in the groove), but not your left hand. Since the helix twists to the right, it is called an alpha helix.

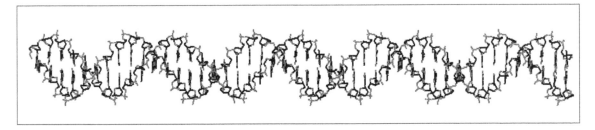

Figure 12.64 Helical Structure of DNA

In forming a double helix, two complementary strands of DNA come together. That is, wherever there is an A in the first strand, there is a T in the second strand, and so forth. The double strand is held together by hydrogen bonding between the base pairs. Figure 12.65 illustrates this base pairing. Each strand of DNA has two nucleotides, both with a base sequence of T-A. The hydrogen bonds between the base pairs are represented by dotted lines.

Why the double strand? The integrity of DNA is of utmost importance, and the double strand provides a mechanism for ensuring that the base sequence in each DNA strand remains intact. If an incorrect base is accidentally incorporated into one of the strands,

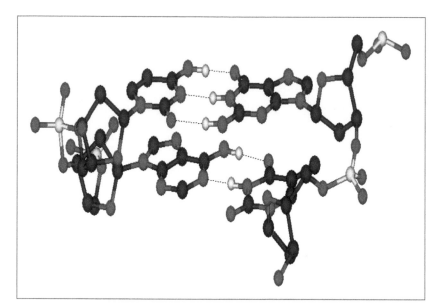

Figure 12.65 A Short Section of Double-Stranded DNA

it won't match its complement in the other strand, and repairs will be initiated to replace that base.

RNA

Ribonucleic acid has an identical structure to DNA, except for two features. The sugar backbone is comprised of ribose, not deoxyribose, and RNA does not contain the base thymine. Instead it uses the base uracil.

Ribose

Ribose is an aldopentose that is the structural sugar in RNA. Ribose cyclizes into a furanose (five-membered) ring system. The structures of D-ribose and β-D-ribofuranose are given in Figure 12.66.

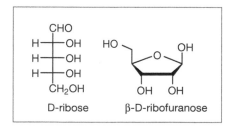

Figure 12.66 Ribose

RNA Nucleosides

The four RNA bases are adenine, uracil, guanine, and cytosine. RNA does not use thymine. The structures of these compounds are shown in Figure 12.67.

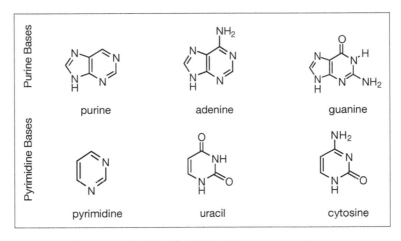

Figure 12.67 The RNA Nitrogenous Bases

Base Pairing

Just as with DNA, the bases have a geometry which ensures that A always matches with U, and G always matches with C. The hydrogen bonds that ensure this pairing are illustrated in Figure 12.68.

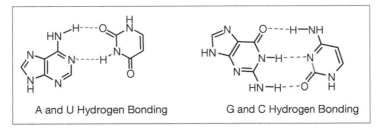

Figure 12.68 Base Pairing Through Hydrogen Bonding

If we attach the nitrogenous bases to a ribose, we have the RNA nucleosides. The four RNA nucleoside bases are adenosine, uridine, cytidine, and guanosine. Usually, these nitrogenous bases are represented as A, U, C, and G, respectively. Figure 12.69 gives the names and structures of these compounds. Addition of a phosphate group to carbon 5 of the ribose sugar affords the RNA nucleotide bases.

Forming the RNA Strand

Just as with DNA, phosphate esters link the RNA nucleotides into a strand of RNA.

One role of the bases in the RNA strand is to match up to the bases in a sequence of DNA. For example, in Figure 12.70, the RNA strand reads:

C–U–A

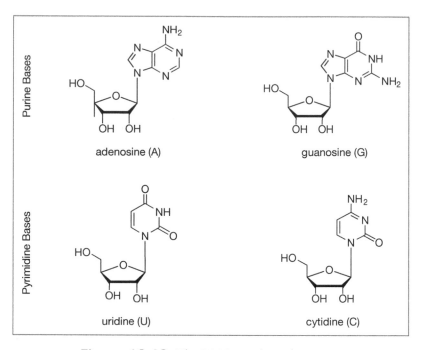

Figure 12.69 The RNA Nucleosides

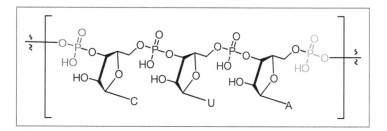

Figure 12.70 A Short Strand of RNA

This RNA strand would match DNA with the sequence:

G–A–T

Messenger RNA and Transfer RNA

Messenger RNA (m-RNA) is synthesized in the nucleus directly from DNA. Its job is to carry the instructions coded on the DNA out into the cytoplasm, where protein synthesis will occur. While m-RNA can adopt more complicated structures, we can think about m-RNA as a simple, straight chain of nucleotides.

Transfer RNA (t-RNA) is present in the cytoplasm. Each transfer RNA carries a single amino acid. Transfer RNA is a much more complicated structure than m-RNA. Figure 12.71 shows a model of t-RNA.

The blue line represents a chain of RNA nucleotides, and a sequence of three bases, known as the *anticodon*, is represented by the green shapes at the bottom of the structure. Each transfer

Figure 12.71 Transfer RNA

RNA has a specific amino acid attached to the 3´ terminal end. The anticodon determines which amino acid the t-RNA will pick up.

Protein Synthesis

Protein synthesis involves many steps, but the broad overview of the process begins with DNA. A section of DNA that contains enough information to make one protein is called a *gene*. The code is first transcribed from a strand of DNA into a strand of m-RNA. Then, the code now contained on the m-RNA strand is translated into a protein using t-RNA. The sequence of bases in DNA ultimately determines the sequence of amino acids in the peptide chain. Remember the structure and function of a protein is determined by its primary structure, and the primary structure of a protein is determined by the primary structure (base sequence) in DNA.

Each amino acid code consists of a three-base sequence. For example, the DNA sequence CGA codes for the amino acid alanine.[6] This sequence in DNA codes for the m-RNA sequence GCU. A three-base pair in m-RNA that codes for an amino acid is called a *codon*. An anticodon is the three-base sequence in t-RNA that matches a codon in m-RNA.

Transcription

Transcription is the process of forming a complementary m-RNA strand from the DNA. This process occurs in the cell nucleus. To begin, an enzyme helps the DNA double helix to break apart and uncoil slightly. This allows RNA nucleotide bases to form complementary hydrogen bonds to the DNA bases. In essence, the m-RNA is reading the DNA code. For example, let's imagine a short section of DNA with the following sequence:

A–A–T–G–C–C–G–A–A

---

[6] For the sake of clarity, we are being somewhat cavalier about directions on the DNA and RNA strands. Please consult a biochemistry text if you wish to explore this topic in greater depth.

First, an RNA base with a U matches up to the first base in the DNA strand (A). Remember that A always pairs with U, C always pairs with G, and T always pairs with A. Also, don't forget that DNA has the T base, and RNA has the U base. The RNA nucleotide is held in place by hydrogen bonding. Then a second RNA base (U) hydrogen-bonds to the second base on the DNA strand, and it is also held in place by hydrogen bonding. Then, an enzyme establishes a phosphate ester link between the two RNA nucleotides, forming an m-RNA strand containing two bases. This process is repeated until all nine RNA nucleotides are stitched together into an m-RNA strand. This is shown in Figure 12.72.

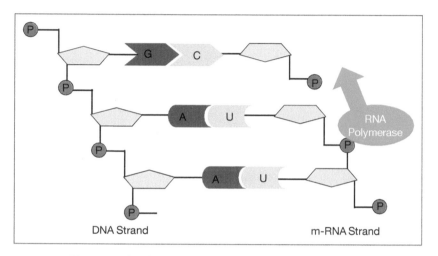

Figure 12.72 Transcription of DNA into m-RNA

So, the sequence in our m-RNA strand (and the complementary DNA strand) is illustrated in Figure 12.73.

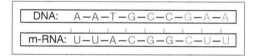

Figure 12.73 Transcribing DNA into m-RNA

This nine-base strand of m-RNA contains three codons: UUC, GGC, and AUU, and calls for the amino acids phenylalanine, glycine, and, isoleucine in that order.[7] Thus the order of the bases in DNA dictates a particular sequence in the m-RNA strand. When the m-RNA strand is complete, it leaves the nucleus and heads for the ribosomes in the cytoplasm.

Translation

Translation is the process of pairing an amino-acid-bearing t-RNA to the m-RNA (see Figure 12.74). The three-base codon sequence on the m-RNA must match the three-base

---

[7] We know this is confusing. m-RNA is read $5' \rightarrow 3'$, from right to left as we have written the strand.

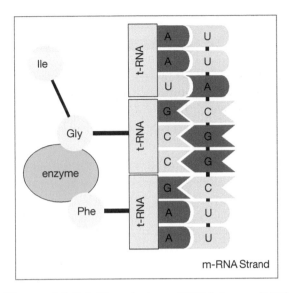

Figure 12.74 Translation of DNA into m-RNA

anticodon sequence on the t-RNA. If the anticodon matches the codon, the t-RNA, along with its amino acid, will be held in place on the m-RNA chain by hydrogen bonding. From our previous example, the first codon in the m-RNA is UUA. Only a t-RNA with the anticodon AAU will pair up to this codon. The t-RNA with this antidcodon carries the amino acid isoleucine. Then the next t-RNA settles into place, and an enzyme establishes a peptide bond between the amino acids bound to the t RNA. Forming the peptide bond requires the expenditure of an ATP molecule, so protein synthesis is energetically expensive. This is why we use our fat reserves, rather than muscle, when our energy expenditures exceed our food intake.

Once the amino acid has been cleaved from the t-RNA, the t-RNA is released from the m-RNA chain and proceeds back out into the cytoplasm to pick up another amino acid. This process continues until the synthesis of the polypeptide chain is complete.

Summary

- Biomolecules are the molecules that comprise living systems. Biomolecules include carbohydrates, lipids, proteins and amino acids, and nucleic acids.
- Carbohydrates are polyalcohol aldehydes or ketones.
- Monosaccharides are not easily broken down into simpler sugars by the action of dilute aqueous acid. Polysaccharides are broken down into monosaccharides by dilute aqueous acid.
- Most carbohydrates are chiral, which means they are not superimposable on their own mirror image. Naturally occurring monosaccharides belong to the D family, which means the next-to-last OH group in the Fischer projection is pointed to the right.
- Monosaccharides form cyclic hemiacetals or hemiketals by the addition of one of the alcohol functional groups across the carbonyl double bond. The resulting cyclic hemiacetals/ketals contain five- or six-membered rings.
- Monosaccharides can combine by the condensation of an alcohol functional group of one monosaccharide with the hemiacetal functional group of a second. The resulting acetal functional group is known as a glycoside bond.
- Lipids are characterized by a physical property. Lipids are more soluble in organic solvents, such as ether, than in water.
- Saponifiable lipids are broken down into fatty acids by the action of hot aqueous base. Triglycerides are a common saponifiable lipid.
  - Triglycerides are tri-esters of glycerin and three fatty acids.
  - A fatty acid is a carboxylic acid having a carbon backbone containing 12 to 18 carbon atoms.
  - Fatty acids are saturated if there are no carbon–carbon double bonds.
  - Fatty acids having one or more carbon–carbon double bonds are termed unsaturated.
  - Naturally occurring unsaturated fatty acids contain *cis* double bonds.
- Non-saponifiable lipids include steroids and prostaglandins.
- Carbohydrates and fatty acids are the principle energy sources. Polysaccharides are broken down into glucose. Glycolysis converts this six-carbon sugar into a pair of 3-carbon sugar-acids (lactic acid or pyruvic acid) plus some energy. Pyruvic acid is converted into acetylcoenzyme A, which feeds into the Krebs Cycle. Fatty acids are broken down acetylcoenzyme, which is also fed into the Krebs Cycle. The Krebs Cycle is the most efficient energy production pathway, giving some ATP, but more importantly, producing reduced coenzymes, such as $NADH_2$. The electron-transport chain uses $NADH_2$ to make ATP.
- Amino acids contain two organic functional groups: a carboxylic acid and an amine functional group. Amino acids also contain a side chain group, which largely determines the physical and chemical properties of the amino acid.
- Side chains may be non-polar hydrocarbon functional groups, polar alcohol or amide functional groups, or ionic acidic/basic functional groups. Non-polar side chains interact by London forces, and are often located in the interior of a protein, away from the polar water solvent. Polar side chains interact by dipole–dipole interactions and/or hydrogen bonding. Ionic side chains interact by ion–ion and ion–dipole interactions.

- Amines are bases, so the amine functional group is converted into its positively charged conjugate acid at low pH. The carboxylic acid functional group is converted into its negatively charged conjugate base at high pH. At intermediate pH, an amino acid is electrically neutral. This pH is called the isoelectric point, or pI.

- Proteins are formed by condensing the amine functional group of one amino acid with the carboxylic acid functional group of another. The resulting amide functional group is called a peptide bond. The protein is sometimes called a polypeptide.

- The primary structure of a polypeptide chain is the sequence of the amino acids in the chain. The primary structure is ultimately responsible for all higher levels and structure, and therefore, the function of the protein.

- There are two common secondary structures for a polypeptide chain. The alpha helix has the chain coiled into a helix. The beta pleated sheet has the polypeptide chain fully extended. Both the alpha helix and the beta pleated sheet are maintained by hydrogen bonding that occurs between the hydrogen atom on an amide nitrogen with an oxygen in a carbonyl group.

- The tertiary structure of a polypeptide chains describes how the chain is folded into a compact, globular structure. The tertiary structure is maintained by hydrogen bonding and London forces. Addition of an organic solvent, changing the pH, or adding a heavy metal ion, disrupts these stabilizing forces and causes the protein to lose its tertiary (and secondary) structure. This is a called denaturing. Denatured proteins are inactive and often precipitate out of solution.

- Nucleic acids are polymers consisting of five-carbon sugar molecules (ribose in RNA or deoxyribose in DNA) lined together by phosphate groups. Each sugar molecule also carries a nitrogenous base.

- A nucleoside is a sugar (ribose or deoxyribose) bonded to a nitrogen base.

- There are four nucleoside bases in DNA: adenosine (A), guanosine (G), thymidine (T), and cytidine (C). In RNA, thymidine is replaced with uridine (U).

- Condensation of a phosphate group with the alcohol functional group of carbon five of the sugar in a nucleosides gives a phosphate ester called a nucleotide.

- The phosphate group can form more than one ester functional group. So, condensation of the phosphate in one nucleotide with the alcohol functional group on carbon three of a second nucleotide joins the two nucleotides into a dinucleotide. RNA and DNA are polymers that contain thousands of nucleotide bases, joined together by phosphate ester bridges.

- DNA consists of two strands of polynucleotide chains. The double strands are held together by base-pairing.

- Because of the molecular geometry, A fits with T to form two hydrogen bonds. Likewise, C and G fit together to form three hydrogen bonds. Base-pairing means the A in one strand of DNA will always pair with a T in the other strand. Likewise, C always pairs with G.

- The primary structure of DNA is the sequence of bases, and the sequence of bases forms the genetic code, and contains the information needed to synthesize proteins.

- There are two kinds of RNA: messenger RNA and transfer RNA.

- Transfer RNA is a polynucleotide chain that transcribes the genetic code from DNA. The structure of transfer RNA is the same as DNA, except the sugar is ribose, not deoxyribose. Transcription of the genetic code is achieved through base-pairing. If the DNA has a C base, the corresponding transfer RNA chain will have a G base. If the DNA has a G base, the corresponding RNA will have a C base. If the DNA has a T base, the RNA will have an A base. In RNA, T has been replaced with U, so if the DNA has an A base, the RNA will have a U base. So, transcription of the DNA sequence GATACA leads to synthesis of the messenger RNA sequence CUAUGU. A codon is a three-base sequence on the messenger RNA chain that codes for a specific amino acid.

- Translation of the genetic code is achieved by matching transfer RNA molecules onto the messenger RNA chain. Transfer RNA carries an amino acid and a three-base sequence called an anticodon. Base pairing ensures that the three-base sequence in a codon on the messenger RNA chain will specifically match to a three-base sequence in the anticodon in a transfer RNA molecule that carries a specific amino acid. As each sequential transfer RNA fits its anticodon to the codons on the messenger RNA, the amino acids carried by the transfer RNA are fit together in specific sequence. Thus, the sequence of bases in DNA is translated and then transcribed into a sequence of amino acids in a polypeptide chain.

Review Questions for Biochemistry

1. Which of these objects are chiral? An ear, a glove, a bolt, a femur, a soda can (ignore the writing), a human body (ignore minor differences on the right and left sides and the internal anatomy).

2. Alpha linolenic acid is an omega-3 unsaturated fatty acid. What does omega-3 mean? If linolenic acid has 18 carbon atoms and the positions of the double bonds are $\Delta^{9,12,15}$, what is the structure of alpha linolenic acid?

3. Draw the structure of α-D-ribofuranose.

4. As a polypetide chain is folding into a compact structure, where do you expect the side chains on amino acids such as valine and phenylalanine will wind up? Where will the side chains on amino acids such as serine wind up?

5. Wool and silk are both proteins. One is predominately β-pleated sheet and the other is predominately α-helix. Which is the α-helix and which is the β-sheet?

6. You doubtless learned in a nutrition course that fats have twice the energy content of carbohydrates. The molecular formula of glucose is $C_6H_{12}O_6$. The molecular formula of stearic acid is $C_{18}H_{36}O_2$. How many ATPs are produced per gram of glucose, and how many ATPs are produced per gram of stearic acid?

7. If the base sequence in a section of DNA is A–T–C, what is the base sequence in the complementary strand of DNA? What is the sequence of the m-RNA? What is the sequence of the t RNA?

8. The order of amino acids in a peptide chain is the:
   a. primary structure
   b. molecular formula
   c. codon code

9. Which are held together by peptide bonds?
   a. Nucleic acids
   b. Polysaccharides
   c. Proteins

10. The molecule that carries the genetic message from the nucleus to the ribosomes is:
    a. DNA
    b. m-RNA
    c. t-RNA

11. A polymer of β-glucose is:
    a. starch
    b. cellulose
    c. amylose

12. Which of these biomolecules performs the function of catalyzing biochemical reaction?
    a. proteins
    b. enzymes
    c. DNA

13. All naturally occurring amino acids belong to which optical family?
    a. L
    b. D
    c. some are D and some are L

14. Carbohydrates that are hydrolyzed into simpler carbohydrates by aqueous acid are called:
    a. complex carbohydrates
    b. polymers
    c. polysaccharides

15. Sucrose is a disaccharide comprised of
    a. two glucoses
    b. maltose and lactose
    c. glucose and fructose

16. The group of biomolecules characterized by their solubility in organic solvents are the:
    a. lipids
    b. carbohydrates
    c. nucleic acids

17. A triglyceride is composed of glycerin and three:
    a. terpenes
    b. sugars
    c. fatty acids

18. The pH at which the net charge on an amino acid is zero is called the:
    a. isoelectric pH
    b. zero charge pH
    c. neutrality point

19. A protein is spotted onto a piece of filter paper wetted with a pH buffer lower than the pI of the protein. This paper is placed in a DC electric field. The protein will:
    a. migrate toward the positive electrode
    b. move toward the negative electrode
    c. not move

20. Virtually all naturally occurring sugars are:
    a. L-sugars
    b. D-sugars
    c. sweeter than glucose

21. Poly alpha glucose is more commonly known as:
    a. starch
    b. sucrose
    c. cellulose

22. An unsaturated triglyceride has:
    a. a lower melting point
    b. few double bonds
    c. a higher smoke point

23. Heating most globular proteins will cause them to lose their three-dimensional shape. This is called:
    a. denaturing
    b. coagulation
    c. frying

24. Which nucleotide base is present in DNA but absent in RNA?
    a. Adenine
    b. Thymine
    c. Uracil

25. Select the true statement:
    a. Transfer RNA transfers the genetic code from the nucleus to the site of protein synthesis.
    b. The transfer RNA chain base sequence is directly coded from DNA.
    c. The primary structure of RNA is determined by the DNA.

26. The DNA sequence GAT requires which anticodon sequence in transfer RNA?
    a. CUA
    b. GAU
    c. GTU

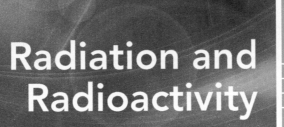

Radiation and Radioactivity

In 1896 Antoine Henri Becquerel was experimenting with uranium-containing crystals (potassium uranyl sulfate) when he discovered that they emitted some type of rays that were able to expose photographic film and even penetrate solid matter. Ernest Rutherford studied these rays and determined that at least three types were emitted from the uranium-containing material. He named them alpha rays, beta rays, and gamma rays.

Since these classic experiments were performed, much has been learned about radiation and radioactivity, including how we can use it to benefit humankind. To many, however, radiation and radioactivity remain mysteries and even something beyond comprehension, perhaps only something to be feared.

Radiation, *radioactivity*, and *nuclear* are some of the most misunderstood, misused, and sometimes mispronounced words in the English language. Note the last word is pronounced "new-klee-ar" not "new-kue-ler." These words tend to have negative connotations. Justifiably, nuclear weapons have a negative reputation; however, nuclear power gets a bum rap. Currently, ~20% of the electricity generated in the United States is produced by 104 commercial nuclear power plants. Other countries such as France generate a much larger percentage of their electricity using nuclear reactors. Nuclear power plants have been safely generating large amounts of electricity around the globe. Radioactive materials are frequently used to diagnose and treat diseases. In addition, radioactive materials have many different industrial applications including radioluminescent (i.e., glow in the dark) aircraft dials and watch faces; smoke detectors; radioisotope thermoelectric generators for deep space missions where solar panels are not effective; neutron activation analysis for extremely sensitive trace element analyses; oil exploration aids; and many others. Radiation is also used to kill bacteria on food and extend its shelf life. For instance, in 2008 the Food and Drug Administration (FDA) approved the irradiation of fresh iceberg lettuce and fresh spinach. Finally, the detection of radiation from space enables us to study the beauty and workings of our universe.

RADIATION

Radiation is frequently associated with that "bad nuclear stuff." However, the general scientific meaning of this word is much broader. In Chapter 4 we discussed the concept of heat, or the process by which energy is transferred from a hotter body to a colder one. What wasn't discussed was how energy gets exchanged between objects. Heat exchange can occur via *conduction*, *convection*, and *radiation*.

Conduction

Conduction is accomplished when atoms or molecules collide. There is an old saying in glass blowing that states "hot glass looks the same as cold glass." Imagine heating a glass rod in a flame and setting it down on a table. Out of curiosity, your unsuspecting colleague sees the rod and picks it up, resulting in a wicked skin burn. This type of energy transfer occurs by physical contact between the hot molecules in the glass rod and the cooler molecules in your colleague's skin. The glass molecules are vibrating more vigorously (that is, they have a greater amplitude of vibration) because of their temperature. The hot glass molecules physically interact with the molecules in the skin causing them to vibrate with greater amplitude, thus raising their temperature and causing the burn.

When a patient coming out of anesthesia gets the chills, what do you do? Put a warm blanket on them, of course. The energy in the warm blanket transfers energy to the patient, at least partially through conduction. Countercurrent heat exchange between veins and arteries is an important temperature regulation mechanism in the human body, which also relies on conduction.

Convection

Heat transfer due to the movement of matter from one location to another is called *convection*. You have probably placed your hand over an open flame of one type or another. Even though there is no physical contact between your hand and the flame, you can still feel the warmth of the flame. Uneven heating is generally the cause of convection in a fluid. For example, the air near an open flame will be heated and become less dense than the air higher above the flame. The heated air will rise while the cooler denser air will fall. The rising warm air will subsequently come into contact with your hand and heat it. Convection also causes the motion of the tectonic plates and air circulation patterns of our planet.

Radiation

Heat transfer involving electromagnetic waves is called *radiation*. Electromagnetic waves, which include radio waves, microwaves, visible light, and others, will be discussed in the next section, so don't worry about them right now. This type of energy transfer requires no physical medium or physical contact between the objects. For example, we can see the Sun, which is 96 million miles from us, because visible light reaches our eyes. This visible light must travel through the vacuum of space in order to reach us.

All objects emit energy via radiation. Hopefully, you now realize that radiation is a very common process that is all around us. Without radiation, life on Earth would not be possible because the Sun's energy would have no way of reaching us! However, there are some types of radiation that *can* be harmful. We'll get into the specifics of this later.

ELECTROMAGNETIC RADIATION

Production of Electromagnetic Waves

Electromagnetic waves are produced when charged particles are accelerated. These waves, which are very important in our modern electronic-filled world, are composed of mutually perpendicular electric and magnetic fields which are constantly varying but remain in phase. This means that when the electric field is at a maximum the magnetic field is at a maximum, when the electric field is at a minimum the magnetic field is at a minimum, and so forth. Thus, electromagnetic waves are self-propagating and require no medium through which to travel. An electromagnetic wave is illustrated in Figure 13.1, with the electric field shown in medium blue and the magnetic field shown in light blue. Notice that the direction of propagation, represented by the v above the arrow along the x-axis in the figure, is also perpendicular to both the electric field and the magnetic field.

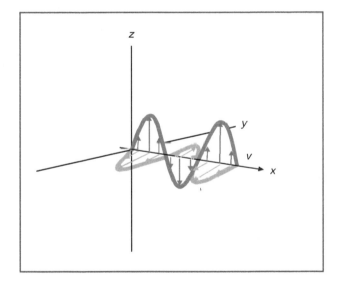

Figure 13.1 An Electromagnetic Wave

Wave Properties of Electromagnetic Waves

As the name implies, electromagnetic waves exhibit all of the classical properties of waves. Figure 13.2 illustrates the various features of a simple wave. The wavelength, λ (lowercase Greek letter lambda), is the *distance* required for a wave to repeat itself. For instance, it is the distance between adjacent peaks (or crests) and also the distance between adjacent troughs. Wavelength is usually measured in meters. The period, T, is the *time* required for a wave to repeat itself.

The period is measured in seconds. The frequency, f, describes the number of wave cycles or oscillations that occur in one second. Frequency is sometimes represented with the lowercase Greek letter nu (ν). However, this letter looks very similar to an italicized lowercase v (v),

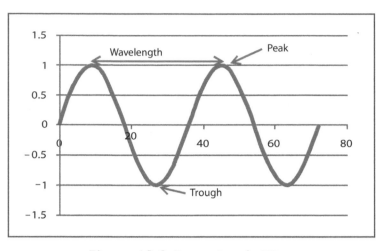

Figure 13.2 Properties of a Wave

which is usually used to represent velocity. Thus, in this text we will always use f to represent frequency. The frequency is also the inverse of the period, so we can write

$$f = \frac{1}{T}$$

Therefore, the unit for frequency is reciprocal seconds (s⁻¹), which is given the name hertz (Hz) in the SI.

For *all* waves, the velocity is related to period, frequency, and wavelength by the following relationship:

$$v = \frac{\lambda}{T} = \lambda f$$

If official SI units are used for the wavelength (m) and period (s) or the wavelength (m) and frequency (s⁻¹), the velocity will have units of meters/second (m/s).

Particle Properties of Electromagnetic Waves

It turns out that electromagnetic waves exhibit properties of *both* waves and particles, or equally valid, electromagnetic waves are *neither* waves nor particles. This fundamental paradox is at the heart of quantum theory. You can perform experiments that unequivocally demonstrate light is definitely a wave. You can also perform experiments that unequivocally demonstrate light is definitely a particle. Nonetheless, there is one important relationship that allows the energy of electromagnetic radiation to be calculated if the frequency or wavelength is known:

$$E = \frac{hc}{\lambda} = hf$$

In this equation, E has units of joules (J), h is Planck's constant (6.626×10^{-34} J·s), λ is the wavelength in meters, c is a constant called the speed of light in a vacuum (3.00×10^8 m/s), and f is the frequency in Hz. Inspection of the energy equation above reveals to us that the energy of electromagnetic radiation is *inversely* proportional to the wavelength and *directly* proportional to the frequency. Since electromagnetic radiation has wave *and* particle properties, the energy equation can be viewed as describing the energy of a discrete unit or *photon* of electromagnetic radiation.

All electromagnetic waves move at the same speed through a vacuum. We represent this speed with the symbol c and call it the speed of light or the speed of light in a vacuum. The speed of light is approximately equal to 3.00×10^8 m/s. For comparison, sound moves at about 343 m/s in air, depending on the air temperature. The large difference in these two values is exactly why you see lightning before you hear the thunder clap. Light waves travel at the fastest possible velocity. Nothing goes faster than light in a vacuum, which is the foundation of Einstein's Theory of Relativity.

RADIOACTIVE MATERIALS—AN INTRODUCTION

Law of Energy Conservation Revisited

In Chapter 4 we discussed the fact that the amount of energy in the universe is constant and that energy is continually converted from one form to another, but not created or destroyed. Well, there is more to this story that we must now discuss. It turns out that not only can energy be converted from one form to another, matter can be converted to energy and vice versa. Einstein's famous equation,

$$E = mc^2$$

gives us the equivalence between a rest mass m, rest energy E, and the speed of light c. Since $c = 3.00 \times 10^8$ m/s, a very small amount of mass has a huge energy equivalence. For example, 1.00 g of water has a rest energy of:

$$E = mc^2 = (0.00100 \text{ kg})\left(3.00 \times 10^8 \tfrac{\text{m}}{\text{s}}\right)^2 = 9.00 \times 10^{13} \text{ J}$$

It is presently thought that the early universe was mostly energy with very little matter. Today, our universe is dominated by matter. It is the hope of chemists and physicists to be able to account for and track all the different forms of energy present in a given system along with the mass in order to better understand the chemical and physical processes that occur in our universe.

The Law of Energy Conservation needs to be modified when we move outside of the realm of classical physics, which covers large, slow-moving objects only, in order to reflect the equivalence between matter and energy. Since matter and energy can interconvert, we should really say that the *total energy*, consisting of the kinetic energy and the mass equivalent energy ($E = mc^2$), remains constant for an isolated system.

Some Important Terminology

The periodic table provides the chemist with a very brief and simple representation of the elements. Recall that most of the mass of an individual atom is contained in the nucleus. In addition, a nucleus contains one or more protons and, with the exception of ordinary hydrogen (^1H), one or more neutrons. As discussed in Chapter 2, the mass number (A) is an integer and is equal to the number of protons (Z) plus the number of neutrons (N) in the nucleus. It is the number of protons that determines what element a particular atom is. The general term *nucleon* is used to describe both neutrons and protons, or particles found in the atomic nucleus.

The word *nuclide* is used to describe an atom in the most general sense. A nuclide is an atomic species defined by its atomic number, number of neutrons, and the energy state of the nucleus. The word "nuclide" is used to describe a nucleus in the most general sense. A nuclide is a nucleus with a particular or unique number of protons and neutrons. For example, we could say that ^{99}Tc and ^{235}U are both nuclides. ^{99}Tc has 43 protons and 56 neutrons, while ^{235}U has 92 protons and 143 neutrons. *Isotopes* are nuclides that have the same Z but different N (and therefore different A). Thus, if two atoms are isotopes, they are both the same element. As an example, ^{233}U and ^{238}U are isotopes. Isotopes have the same Z and hence identical electron configurations; thus, we expect their chemical behavior to be very similar. For example, both ^{239}Pu and ^{241}Pu form insoluble fluoride compounds. *Isotones* have the same number of neutrons but different numbers of protons. Since isotones are not the same element, we would not expect them to behave the same chemically. ^{65}Ni ($A = 65$, $Z = 28$, $N = 37$) and ^{66}Cu ($A = 66$, $Z = 29$, $N = 37$) are two examples of isotones. Finally, isobars have the same mass number but different proton number. ^{90}Sr ($A = 90$, $Z = 38$, $N = 52$) and ^{90}Y ($A = 90$, $Z = 39$, $N = 51$) are isobars. These terms are summarized in Table 13.1.

Table 13.1 NUCLEAR TERMS

TERM	A	Z	N
Nuclide	Unique	Unique	Unique
Isotope	Different	Same	Different
Isotone	Different	Different	Same
Isobar	Same	Different	Different

RADIOACTIVITY

The number of nuclides currently known to exist exceeds 3000. Most nuclides are not stable and will decay by splitting apart into two or more pieces, emitting small subatomic particles and/or emitting electromagnetic radiation. We'll say more about these processes later. Only a relatively small number of nuclides, 266 to be exact, are stable and do not decay. In actuality, some of the "stable" nuclides have exceedingly long half-lives, which for our purpose, we can consider infinitely long. In any nucleus, there is a balancing act between attractive and repulsive forces. The positively charged protons repel each other. On the other hand, an inherently attractive force known as the strong nuclear force binds nucleons together like the world's most perfect glue. However, the strong nuclear force operates only over extremely short distances; so two nucleons have to be essentially in contact in order to participate in this attractive force. When the repulsive forces exceed the attractive forces, the nucleus is unstable and undergoes radioactive decay.

RADIOACTIVE DECAY

Nuclides that decay are aptly named *radionuclides*. Radioactive decay is the spontaneous process by which an unstable nucleus goes from a level of higher energy to one of lower energy. Radioactive decay can be achieved by several different means including fragmentation (fission), emission of subatomic particles, and/or emission of electromagnetic radiation. There are many different forms of radioactive decay, and a complete discussion would be outside the scope of this text. Hence, we'll limit our discussion to only five types of radioactive decay: alpha decay, beta-minus decay, positron decay, electron capture decay, and gamma ray emission.

In radioactive decay, the decaying nucleus is called the *parent*, while the nucleus produced by the decay is called the *daughter*:

$$\text{parent} \xrightarrow{\text{radioactive decay}} \text{daughter}$$

Alpha Decay

An alpha particle consists of two protons and two neutrons. From that standpoint, an alpha particle is a helium nucleus. So, is that helium balloon full of alpha radiation? Of course not. A helium atom also includes two electrons, so it is electrically neutral and chemically inert. An alpha particle has a net charge of $+2$, and it has vastly more kinetic energy.

Alpha decay is characterized by the emission of an alpha particle from the parent nucleus. In this process, energy is released in the form of kinetic energy of the escaping alpha particle and the recoiling daughter nucleus. For example:

$$^{235}_{92}\text{U} \rightarrow {}^{231}_{90}\text{Th} + {}^{4}_{2}\text{He} + \text{energy}$$

Note that the daughter nucleus has two fewer protons and two fewer neutrons than the parent, resulting in a different element. In this particular case, uranium has decayed to thorium. In radioactive decay equations, the total mass number A on the left side must equal the total mass number on the right. In the example above, $A_{\text{left,total}} = 235$, while $A_{\text{right,total}} = 231 + 4 = 235$. In addition, the total proton number on the left side of the equation must be equal to total proton number on the right side of the equation ($92 = 90 + 2$). So, in any nuclear reaction, there is conservation of mass number and charge.

So why does alpha decay occur? It occurs because it is energetically favorable and results in the alpha decay daughter being in a lower energy state. Alpha decay is more common among very heavy nuclei ($A \geq 210$) where the repulsion between the positively charged protons in the nucleus makes elimination of an alpha particle energetically advantageous. The recoiling products from each decay, especially the alpha particle, carry huge amounts of energy and do much damage to the surrounding material by depositing this energy in very small volumes (e.g., cellular structures).

Beta Decay

Beta decay is a general term applied to radioactive decay processes that result in the mass number A remaining constant while the atomic number Z changes. There are three types of beta decay: beta-minus (β^-) decay, positron (β^+) decay, and electron capture decay. It should be mentioned that β^- decay is often referred to as just beta decay, which is not strictly correct, because it is only one type of beta decay.

Beta-Minus Decay

Beta-minus decay is the radioactive decay process in which a nucleus emits an electron (also known as a beta particle, β^-, or e^-) and an anti-neutrino ($\bar{\nu}$), which is a very weakly interacting particle with an extremely small mass. By weakly interacting, we mean neutrinos are so aloof from ordinary matter that they can fly unimpeded through several trillion miles of lead. Energy is also released in the β^- decay process. The net effect of β^- decay is shown in the equation below:

$$n \rightarrow p^+ + e^- + \bar{\nu} + \text{energy}$$

where n is the neutron and p^+ is a proton.

Technetium-99m ($^{99}Tc^m$) is a radionuclide that finds many applications in nuclear medicine. Virtually all technetium used in nuclear medicine labs is prepared synthetically from other radioactive materials. $^{99}Tc^m$ is produced by the β^- decay of ^{99}Mo as illustrated in the reaction below. ^{99}Mo is produced through fission of ^{235}U or via the capture of a neutron by ^{98}Mo.

$$^{99}_{42}Mo \rightarrow \,^{99}_{43}Tc^m + \,^{0}_{-1}\beta^- + \bar{\nu} + \text{energy}$$

Note that the mass numbers of the parent and daughter are identical but the atomic numbers are different. Thus a new element, with a Z one *greater* than the parent nucleus, results from β^- decay. It is interesting to note that free neutrons are not stable and decay via β^- emission.

The daughter, $^{99}Tc^m$, is a "metastable" product which has excess energy that it releases by gamma decay. More on gamma decay in a moment. In the nuclear medicine lab, a $^{99}Tc^m$ generator consists of a "cow" or column that contains a solid support such as aluminum oxide (alumina gel). The column is charged with sodium molybdate ($NaMoO_4$) containing radioactive ^{99}Mo. Notice that the chemical form of a radioactive element has essentially no impact on the radioactive properties. ^{99}Mo decays to $^{99}Tc^m$ whether it is present as molybdenum metal or sodium molybdate. The molybdate has a strong affinity for the column and is effectively immobilized. The ^{99}Mo in the molybdate ion decays to ^{99}Tc^m, resulting in the pertechnetate ion (^{99}Tc^{m}O$_4$$^-$), which is not strongly adsorbed onto the alumina. Therefore, eluting the column with a solvent such as normal saline selectively removes pure pertechnetate. The apparatus is commonly called a cow, while elution is known as milking. ^{99}Tc^m is a useful radionuclide because it forms the pertechnetate ion (^{99}Tc^{m}O$_4$$^-$) in normal saline. This ion behaves in a similar manner to I$^-$ in the human body, concentrating mainly in the thyroid, salivary glands, gastric mucosa, and choroid plexus. In addition, ^{99}Tc^m has a reasonably short half-life (6 hours), emits a gamma ray with an energy of 140 keV, and emits no beta particles. These properties all result in a relatively low absorbed radiation dose to the patient and decent tissue penetration.

Why does β^- decay occur? Well, stable nuclei have stable ratios between the proton number and neutron number. These ratios are nontrivial to predict and are arcane to the extent that they include the concept of "magic numbers."[1] Beta-minus emission occurs because some nuclei have too many neutrons; therefore β^- decay is energetically favorable, resulting in a reduction in the neutron-to-proton ratio.

Positron Decay

Some nuclei have too many protons and thus positron decay becomes energetically favorable. For smaller radionuclides, the repulsion between the positively charged protons makes the nucleus unstable, but the strong nuclear force is too great to allow the loss of four nucleons. So, positron emission is another route to reduce the proton number. The net result of positron decay is the transformation of a proton into a neutron, a positron (β^+ or e$^+$), and a neutrino (ν). A positron is the antiparticle of an electron. Antiparticles are an example of antimatter. Antiparticles have identical masses but the opposite charge of the corresponding particle. For example, an antiproton has the same mass as a proton, but has a negative charge. Positrons have the same properties as electrons with the exception of charge. A positron has a charge of $+e$ ($+1.60 \times 10^{-19}$ C), whereas an electron has a charge of $-e$ (-1.60×10^{-19} C). In symbols, positron decay can be represented as follows:

$$p^+ \rightarrow n + e^+ + \nu + \text{energy}$$

Carbon-11 is a radioactive source used in positron emission tomography (PET) scans. The equation for this process is:

$$^{11}_{6}\text{C} \rightarrow\ ^{11}_{5}\text{B} +\ ^{0}_{1}\beta^+ + \nu + \text{energy}$$

Just like β^- decay, the mass numbers of the parent and daughter in positron decay are identical but the atomic numbers are different. However, in positron decay, the daughter has a Z that is one *less* than the parent nucleus. Positron decay becomes energetically possible only when the decay energy exceeds 1.02 MeV, a limit established by the Einstein equation $E = mc^2$. Since

---

[1] Nucleons have a quantum ordering, analogous to the quantum energy levels that electrons occupy. However, the quantum rules for protons and neutrons are much more complicated and beyond our purposes here.

1 eV = 1.602×10^{-19} J, 1 million electron volts (MeV) is equal to 1.602×10^{-13} J. Thus, 1.02 MeV corresponds to a decay energy of 1.634×10^{-13} J.

Electron Capture Decay

Electron capture decay is a competing process to positron decay and thus results in an increase in the neutron-to-proton ratio in the nucleus. In this process, a bound, inner orbital electron is captured by the nucleus, resulting in the conversion of a proton into a neutron, the emission of a neutrino, and, if the daughter nucleus is left in an excited state, the emission of one or more gamma rays. The net reaction is shown below:

$$p^+ + e^- \rightarrow n + \nu + \text{energy}$$

Since there will be a vacancy or hole where the captured electron was previously located, an electron from another energy level will fill the hole resulting in the emission of an X-ray and the creation of another hole. This second hole will be filled by another electron, resulting in an X-ray of different energy, and so on. Thus, several characteristic X-rays can be emitted. It should be said that instead of X-ray emission, an internal photoelectric process can result in the emission of an orbital electron known as an Auger electron. The photoelectric process was discussed in Chapter 10. As an example of electron capture decay, ^{57}Co is used in liver imaging and decays according to the following equation:

$$^{57}_{27}\text{Co} + e^- \rightarrow \, ^{57}_{26}\text{Fe} + \nu + \text{energy}$$

Note that just like β^+ decay, the daughter nucleus has a Z that is one less than the parent nucleus, and the mass number A of the parent is the same as the mass number of the daughter.

Gamma Ray Emission

A nucleus in an excited state can de-excite and emit a photon of electromagnetic radiation called a *gamma ray*. When *electrons* fall from a higher energy state to a lower one, the extra energy is often emitted as visible light. The same physics also applies to nucleons. In a gamma decay event, a *nucleon* in the excited nucleus makes a transition from a higher energy state to a lower energy state. The emitted high-energy photon carries off an amount of energy equal to the energy difference between the two levels. Gamma ray emission frequently accompanies other types of radioactive decay because those decay modes can leave a daughter nucleus in an excited state. For example, when ^{99}Mo undergoes β^- decay, the daughter is metastable (or excited) ^{99}Tc^m. The extra energy contained in the ^{99}Tc^m is emitted as a gamma ray photon (γ).

$$^{99}_{43}\text{Tc}^m \rightarrow \, ^{99}_{43}\text{Tc} + \gamma$$

Internal Conversion

Frequently after beta decay processes, the daughter nucleus will be left in an excited state. As discussed in the previous section, the nucleus can de-excite via gamma ray emission.

Alternatively, internal conversion can occur in which the excitation energy is transmitted to an orbital electron, resulting in the ejection of the electron. The ejected electron has an energy equal to the difference between the excitation energy of the nucleus and the binding energy of the electron.

DECAY RATE

Decay rate is a way to quantify radioactive decay and is equal to the number of radioactive decays or disintegrations occurring per unit time. The official SI unit of decay rate is the *becquerel* (Bq), defined to be:

$$1 \text{ Bq} = 1 \, \frac{\text{disintegration}}{\text{s}}$$

The curie (Ci) is a much older and historical unit of decay rate that is still widely in use today. It is defined to be

$$1 \text{ Ci} = 3.7 \times 10^{10} \text{ Bq} = 3.7 \times 10^{10} \, \frac{\text{disintegrations}}{\text{s}} = 2.22 \times 10^{12} \, \frac{\text{disintegrations}}{\text{min}}$$

Note that the curie is a much larger unit than the becquerel and SI prefixes are commonly used with both units to signify ordinary working levels of radioactivity. For example, environmental levels of certain radionuclides may be in the mBq or pCi range.

The decay rate (or the number of disintegrations a given radionuclide undergoes per unit time) depends on two factors. The first is how much material is in the sample; let's call that quantity A. The second is a characteristic constant (λ). The value of λ is unique to each radionuclide, and the values of λ show enormous variation. For example, the decay constant for ^{239}Pu is 3.28×10^{-9} hr^{-1}, while the decay constant for ^{99}Tc^m is 0.115 hr^{-1}. A larger value of λ indicates that a given quantity of material is "more radioactive"; that is, it undergoes a greater number of disintegrations per unit time. The decay rate is thus given by:

$$Rate = \lambda A$$

Imagine having a large collection of ^{239}Pu atoms in a test tube. We can "watch" these radionuclides decay using an alpha particle detector. We observe that the number of ^{239}Pu atoms decreases in an exponential manner over time. The decay constant (λ) relates the initial amount of radionuclide (A_0) and the amount that will remain (A) at some time (t) in the future. The units of A and A_0 can be any quantity related to amount (e.g., grams, moles, or atoms). For instance, if we want to follow the number of atoms N present at time t, given an initial number N_0, we can use methods of calculus to get:

$$N = N_0 e^{-\lambda t}$$

The details of the math used to obtain this equation are outside the scope of this text, but the interested student is encouraged to consult a nuclear chemistry textbook. Similarly, we can follow the activity A (in Ci or Bq for instance) at time t, given an initial activity A_0 using

$$A = A_0 e^{-\lambda t}$$

For example, suppose the eluant milked from the $^{99}Tc^m$ generator is calibrated at 8:00 to contain 5 mCi/mL of eluant. If a physician orders a procedure to be performed at 12:00, what is the activity of the eluant? Well, $A_0 = 5$ mCi/mL, and the time is 4 hr. From our previous discussion, the decay constant is 0.115 hr^{-1}. So we find the activity of the eluant has fallen by about 40%.

$$A = \left(5\ \frac{mCi}{mL}\right)e^{-(0.115\ hr-1)(4\ hr)} = 3.16\ \frac{mCi}{mL}$$

Back to our ^{239}Pu example: The decay behavior indicates that the radioactive decay process is random. In fact, no matter what radionuclide we choose, we would observe the same exponential decrease. If we physically had the technology and capability to view each individual radioactive atom, we could not predict which nucleus would decay first, second, third, and so on. This is not unlike microwave popcorn. We cannot predict which kernel pops first, second, third, and so on, but we *can* make a prediction that most of the kernels will pop within 2 minutes of being placed in a microwave oven at full power. Thus, the alpha decay of ^{239}Pu (as well as other types of radioactive decay of other radionuclides) is a random process where we cannot link the decay of one specific ^{239}Pu nucleus to the decay of another ^{239}Pu nucleus.

HALF-LIFE

Since the decay of a single radionuclide in a large collection of atoms is not readily predictable, we must come up with some type of measurement tool for radioactive decay. The concept of half-life allows us to do this. The half-life of a given radionuclide is the time required for an initially large number of atoms to decay such that only half of the initial number of atoms is left. For instance, if we start with N_0 atoms, after one half-life $1/2 N_0$ remain. After two half-lives, $1/4 N_0$ are left, $1/8 N_0$ after three half-lives remain, and so forth. Note that after seven half-lives, only $(1/2)^7 N_0$ remain, which corresponds to 0.78% of the original amount.

There is a simple relationship between the half-life $(t_{1/2})$ and the decay constant (λ). The product of these two quantities equals the natural log of 2 (0.693). Notice we're using natural logs (base e) and not common logs (base 10).

$$(\lambda)(t_{1/2}) = \ln 2 = 0.693$$

So, if the decay constant of $^{99}Tc^m = 0.115$ hr^{-1}, what is the half-life?

$$t_{\frac{1}{2}} = \ln 2/\lambda = 0.639/0.115\ hr^{-1} = 6.03\ hr$$

One important aspect of half-life for a radionuclide is that it remains constant over time. For instance, the half-life of ^{239}Pu is approximately 2.41×10^4 years. If we have some radioactive waste containing ^{239}Pu, we can thus predict with relative certainty what the ^{239}Pu content will be at some point in the future. This type of information is vital for the proper long-term isolation and storage of radioactive waste. Some nuclides have extremely long half-lives such as naturally occurring ^{232}Th ($t_{1/2} = 1.40 \times 10^{10}$ years), while other nuclides have extremely short half-lives like ^{269}Ds ($t_{1/2} = 179$ μs).

In a clinical setting, the *effective* half-life of a radionuclide depends on two processes. First is the physical half-life. But we can naturally excrete foreign substances, so there is also a biological half-life. For example, the physical half-life of ^{99}Tc^m is about 6.0 hours, and the biological half-life is 24 hours. So, how long does it take us to lower the amount of ^{99}Tc^m by half? Well, the total rate of decay is the sum of the two rates:

$$Rate_{effective} = Rate_{radioactive\ decay} + Rate_{biological}$$

Each rate is equal to the product of the amount of material times its decay constant:

$$[^{99}Tc^m](\lambda)_{effective} = [^{99}Tc^m](\lambda)_{radioactive\ decay} + [^{99}Tc^m](\lambda)_{biological}$$

Well, the amount of ^{99}Tc^m is a common term to both sides, so we can divide it out. Second, λ is equal to ln 2 divided by $t_{1/2}$, so we can write:

$$\frac{\ln 2}{\left(t_{\frac{1}{2}}\right)_{effective}} = \frac{\ln 2}{\left(t_{\frac{1}{2}}\right)_{radioactive}} + \frac{\ln 2}{\left(t_{\frac{1}{2}}\right)_{biological}}$$

Again, we have a common term (ln 2), and we can divide it out, so the effective half-life is given by:

$$\frac{1}{\left(t_{\frac{1}{2}}\right)_{effective}} = \frac{1}{\left(t_{\frac{1}{2}}\right)_{radioactive}} + \frac{1}{\left(t_{\frac{1}{2}}\right)_{biological}}$$

So, what is the effective half-life of ^{99}Tc^m? If we substitute the known values into the above equation,

$$\frac{1}{\left(t_{\frac{1}{2}}\right)_{effective}} = \frac{1}{6.0\ hr} + \frac{1}{24\ hr} = 0.208\ hr^{-1}$$

Taking the inverse of 0.208 hr^{-1}, we find the effective half life is 4.80 hours.

IONIZING RADIATION VERSUS NONIONIZING RADIATION

The effect of electromagnetic radiation on matter will depend on the photon's energy, which is, as we previously discussed, given by:

$$E = hf$$

Lower energy photons, including those from the microwave, infrared, visible, and ultraviolet regions, can be absorbed by matter, resulting in an increase in internal energy. This type of radiation is called *nonionizing* because it results in rotation, vibration, or excitation of electrons within atoms or molecules. However, note that this lower frequency radiation lacks sufficient energy to remove any electrons, so no ions are formed. Hence, this frequency range of electromagnetic radiation is termed nonionizing. The specific changes caused by each type of photon are given in Table 13.2.

Table 13.2 INTERACTION OF ELECTROMAGNETIC RADIATION WITH MATTER

	ATOMS	MOLECULES
Microwaves	N/A	Rotation
Infrared	N/A	Rotation and vibration
Visible	Rearrangement of electrons	Rearrangement of electrons
Ultraviolet	Rearrangement of electrons	Rearrangement of electrons

Alpha particles, beta particles (β^- and β^+), gamma rays, and X-rays are known as ionizing radiation since they have sufficient energy per photon or particle to remove electrons from atoms and molecules. In addition, since covalent bonds are shared pairs of electrons, ionizing radiation can break bonds in molecules. As previously discussed, X-rays are high-energy photons released from electron transitions involving the innermost or core electrons, whereas gamma rays, which are also high-energy photons, are produced from the transitions of nucleons from a higher quantum energy level to a lower one.

Ionizing radiation is inherently much more destructive than nonionizing radiation. Don't be fooled, though. Nonionizing radiation can also have harmful effects. A classic example is exposure to ultraviolet radiation, which can lead to skin cancer and cataracts.

SOURCES OF RADIOACTIVE MATERIALS

Since radioactive materials are ubiquitous on Earth, it is very helpful to categorize the sources from which these materials come. Traditionally, the sources of radioactive materials have been divided into three categories: (1) *primordial*, (2) *cosmogenic*, and (3) *anthropogenic*. Primordial radionuclides have been around since the earliest times of the solar system (and Earth), and their long half-lives enable us to detect them even today. Examples of primordial radionuclides include ^{238}U ($t_{1/2} = 4.47 \times 10^9$ y) and ^{40}K ($t_{1/2} = 1.25 \times 10^9$ y). Primordial radionuclides and their progeny make excellent time-measurement tools because of their known half-lives and decay properties. Thus, they are used for determining the age of very old geologic formations. This process is known as *radiometric dating*. For instance, ^{87}Rb ($t_{1/2} = 4.89 \times 10^{10}$ y), and its daughter ^{87}Sr ($t_{1/2}$ = stable) can be used to date geologic formations by simply looking at the ^{87}Rb/^{87}Sr ratio. The ratio becomes smaller as the sample gets older.

Cosmogenic radionuclides are formed when radiation from space (or secondary radiation produced by the radiation from space), called *cosmic rays*, interacts with matter. Cosmogenic radionuclides are formed continuously and include things such as ^{3}H (also known as tritium) and ^{14}C. Carbon dating uses ^{14}C to date anthropological artifacts up to approximately 100,000 years old. Since living organisms continuously take in ^{14}C while alive, the ratio of ^{14}C to ^{12}C will be relatively constant while the organism is living and is equal to the ratio observed in the atmospheric carbon dioxide. After the organism dies, the ^{14}C/^{12}C ratio will decline as a result of the decay of ^{14}C with a 5730-year half-life. Thus, by measuring the ^{14}C/^{12}C ratio, the age of an artifact from an organism that was previously living can be readily established.

And finally, as the name implies, anthropogenic radionuclides occur in the environment as a result of human activities including nuclear medicine, electricity generation by nuclear power stations, and nuclear weapons testing. Anthropogenic radionuclides include things like ^{90}Sr ($t_{1/2} = 29$ y), ^{137}Cs ($t_{1/2} = 30.2$ y), and ^{241}Pu ($t_{1/2} = 14.4$ y).

RADIATION EXPOSURE

We are constantly being exposed to radioactive materials, whether it is primordial, cosmogenic, or anthropogenic in origin. What makes matters worse is that the human senses are not very adept at detecting ionizing radiation. In fact, it's only under extreme conditions (i.e., very high radiation levels) that we would have any physiological indication of the presence of ionizing radiation. Because ionizing radiation from radioactive materials can be so destructive to matter, and the fact that our senses cannot detect the radiation, it is vital that we have a means to monitor and control our exposure.

We must therefore rely on detectors of one sort or another to determine the amount of ionizing radiation present. It is outside the scope of this book to discuss the wide variety of radiation detectors. Suffice it to say, there are many types of devices for detecting and quantifying the various types of ionizing radiation. The interested student should consult a modern nuclear chemistry textbook for more details regarding radiation detection and instrumentation.

Just because we are exposed to ionizing radiation does not mean that the energy from the radiation will be *absorbed* by our bodies. Some ionizing radiation, like gamma rays, has the ability to penetrate thick layers of matter, such as tissue, without significantly interacting with the material. In cases like this, very little energy gets deposited by the radiation. Other types of radiation, such as alpha particles and β^- particles, will significantly interact with the matter when they happen to be passing through, resulting in a deposition of a large amount of energy in a very small volume.

The Roentgen—A Unit of Radiation Exposure

Because the amount of energy deposited in a given material will depend on the type of ionizing radiation, we must separately quantify the *radiation exposure* and the energy deposited (or the *energy absorbed*) by the radiation. The roentgen (R) is the unit used to quantify the amount of *radiation exposure* and technically only applies to X-rays and gamma rays. If we expose a sample of air to 1 R of photons, an extremely large number of ion pairs would be produced (1.61×10^{15} ion pairs per kilogram of air). The resulting charge density would be 2.58×10^{-4} C/kg, where C represents the unit of charge called the coulomb, which was discussed in Chapter 10.

Units of Absorbed Dose

The quantity of energy from radiation absorbed per unit mass is called the *absorbed dose*. We would expect the absorbed dose to have units of joules per kilogram (J/kg). This combination of units is called the gray (Gy). Thus,

$$1 \text{ Gy} = 1 \frac{J}{kg}$$

The *rad* (radiation-absorbed *dose*) is an older unit of absorbed dose. One rad of radiation impinging on a sample of matter deposits 100 ergs/g of material. Since $1 \text{ erg} = 1 \times 10^{-7}$ J, one rad is equal to:

$$1 \text{ rad} = 100 \frac{erg}{g} \times \frac{1 \times 10^{-7} J}{1 \text{ erg}} \times \frac{1000 \text{ g}}{1 \text{ kg}} = 1 \times 10^{-2} \text{ J/kg} = 1 \times 10^{-2} \text{ Gy}$$

Units of Dose Equivalent

When considering the interaction and subsequent effects of radiation on biological systems, we must be concerned not only with the fact that the radiation is ionizing, but also with the total energy deposited by the radiation and the resulting *energy density*. For example, an alpha particle, which has a relatively large mass and charge, can deposit a large amount of energy in a very small volume, resulting in a rather large energy density.

Scientists studying the interaction of high-energy photons (gamma rays and X-rays) and charged particles (alpha particles, β^- particles, positrons, etc.) with matter use a quantity called the *linear energy transfer* (LET) to describe the amount of energy lost by the ionizing radiation and transferred to the matter. Photons and fast electrons have much smaller LETs than alpha particles. An *equivalent dose* can be calculated by the use of an appropriate weighting factor to account for the varying LET values of the different types of radiation. Using official SI units to calculate the equivalent dose in sieverts (Sv) for a given tissue, one simply multiplies the absorbed dose in grays by the appropriate weighting factor, known as the *radiation weighting factor* (W_R):

equivalent dose (Sv) = absorbed dose (Gy) × radiation weighting factor (W_R)

Table 13.3 lists the radiation weighting factors for three types of radiation.

Table 13.3 WEIGHTING FACTORS FOR RADIATION

RADIATION TYPE	W_R
Gamma rays	1
Beta (β^-) particles	1
Alpha particles	20

Note that the radiation weighting factor for alpha particles is 20 times that of gamma rays and β^- particles. With the above information, we can calculate the tissue equivalent dose H_T using:

$$H_T = \sum_R W_R D_{T,R}$$

where $D_{T,R}$ is the averaged absorbed dose of the tissue or organ T due to radiation R.

The *roentgen equivalent man* or *rem* has traditionally been used as the unit of dose equivalent, especially (as the name implies) in humans. To calculate the dose equivalent in rems, the absorbed dose in rads is multiplied by an appropriate term, formerly called the *quality factor* or *QF*, which accounted for the level of damage inflicted by the specific type of radiation. The concept of quality factor is similar to that of the more modern radiation weighting factor. *QF* values of 1 were assigned to X-rays, gamma rays, and β^- particles, while values of 1–20 were assigned to alpha particles. Since dose equivalent is calculated by multiplying the absorbed dose by the appropriate radiation weighting factor, and we know that 1 Gy = 100 rad, it follows that 1 Sv = 100 rem. Since the quality factor for X-rays is 1, radiology folks tend to use rads and rems interchangeably, although this is not technically correct.

EFFECTS OF IONIZING RADIATION ON BIOLOGICAL SYSTEMS

At the most basic level, ionizing radiation can remove electrons from atoms and molecules of a given material. The resulting products of the ionizations can cause larger scale disturbances within the material. Isolated ionizations may not lead to any significant damage, but absorption of larger amounts of ionizing radiation can lead to significant levels of ionization and subsequently greater disruptions to the material matrix.

The interaction of ionizing radiation with biological material can lead to many different outcomes depending on the type of radiation, the exposure level, and the type of biological material. Since cells are composed mostly of water, a large fraction of the interactions will be between the incoming radiation and water. Products of these radiation–water interactions include ions, reactive species with an unpaired electron known as free radicals, and atoms and molecules in excited states. These products can lead to damage in the DNA. It should be noted that these types of DNA changes are accomplished by an indirect route. Alternatively, the radiation can interact directly with the DNA and also cause damage.

Of course, when changes occur in the DNA, serious problems can result. Ionizing radiation can damage organelles or other essential structures in a cell. Biological effects due to radiation exposure are termed somatic when they occur in the exposed organism. In some cases, the damage can be repaired, and in other cases permanent damage to the cell can occur. For instance, an extreme somatic effect is death. DNA damage to a cell involved with reproduction can result in an effect or a mutation passed on to future generations. Such changes are called *genetic* effects.

Different types of cells display varying sensitivities to radiation. For instance, those cells that divide frequently such as bone marrow, embryos, and the male gonads, are highly sensitive to radiation. Since tumors are made of cells that divide rapidly, radiation can be used to kill tumor cells in a relatively selective manner. In addition, organs and tissues having cells that are not replaced quickly also exhibit an enhanced sensitivity to radiation. Thus, ovaries and parts of the central nervous system are highly sensitive to radiation. Owing to the variability in radiation sensitivity, tissue weighting factors (W_T) have been developed to help trained professionals such as health physicists quantify the effects of therapeutic and/or occupational doses on specific tissues and organs. Example tissue weighting factors are given in Table 13.4.

Table 13.4 TISSUE WEIGHTING FACTORS

TARGET TISSUE OR ORGAN	W_T
Gonads	0.20
Bone marrow	0.12
Lung	0.12
Bladder	0.05
Breast	0.05
Liver	0.05
Thyroid	0.05
Skin	0.01

Tissue weighting factors are used with the tissue dose equivalent H_T to calculate the *effective dose equivalent* H_E:

$$H_E = \sum_T W_T H_T$$

As an example, let's say that a person received a dose equivalent (of some radionuclide like ^{131}I) to the thyroid of 400 mSv, 20 mSv to the breast, and a whole body dose of 2.00 mSv. To calculate the effective dose equivalent, we multiply each dose equivalent by the appropriate weighting factor and sum these quantities. Looking at Table 13.4, we see that both the thyroid and the breast have a weighting factor of 0.05 each; therefore the weighting factor for the remaining whole body will be $1 - (0.05 + 0.05) = 0.90$:

$$H_E = \sum_T W_T H_T = (0.05)(400 \text{ mSv}) + (0.05)(20 \text{ mSv}) + (0.90)(2.00 \text{ mSv}) = 22.8 \text{ mSv}$$

MEDICAL USES OF RADIONUCLIDES (NUCLEAR MEDICINE)

Diagnostic Uses

The field of nuclear medicine actively uses several different techniques involving radioactive materials and high-energy electromagnetic radiation to effectively diagnose and treat disease. Our treatment of nuclear medicine presented here is very brief and far from comprehensive.

The interested student should refer to a modern nuclear chemistry or nuclear medicine text-book for more details of the various techniques discussed in this section.

In addition to the ^{99}Mo-^{99}Tc^m generator discussed previously, there are several other commercially available generators for the production of medical radionuclides, including ^{62}Zn-^{62}Cu, ^{68}Ge-^{68}Ga, ^{81}Rb-^{81}Kr^m, ^{82}Sr-^{82}Rb, and ^{113}Sn-^{113}In^m. In each case, the radioactive daughter is used for a specific imaging or therapeutic application via a variety of delivery techniques. For instance, a radionuclide may be administered in its elemental form, as in the case of gaseous ^{81}Kr^m, or the radionuclide may be incorporated into a more complex chemical species. For example, ^{62}Cu is incorporated into pyruvaldehyde bis (N^4-methylthiosemicarbazone) for use in PET scanning.

For many years now, X-rays have been used by both physicians and dentists to effectively see underlying anatomical structures through the overlying layers without the need to remove any overlying materials. The resulting X-ray image is caused by differences in attenuation of the X-rays by the various materials with which the X-rays interact. Years of continuous use have shown that the benefits of using X-rays for medical or dental purposes far outweigh any risks of radiation exposure.

Gamma cameras, also called Anger scintillation cameras, are used to detect gamma rays emitted from injected radionuclides. The Anger gamma scintillation camera is named in honor of its inventor, Hal Anger, who developed the device in the late 1950s. The camera unit or head typically consists of a large-area sodium iodide crystal activated with a small quantity (< 1%) of thallium. The crystal is optically coupled to a group of photomultiplier tubes (PMTs), and the detector/PMT assembly is connected to the appropriate electronics and a computer. In some instances, a gamma camera can have two or more heads arranged in either a fixed or variable geometry. An extensive lead collimator is positioned between each detector crystal and the patient. Several types of collimators are available, including parallel-hole, converging, diverging, and pinhole. The type of collimator actually used in a specific exam will depend upon the clinical requirements. For example, collimators designed for higher energy gamma rays (≤ 364 keV) are thicker and have larger holes compared to collimators designed for lower energy gamma rays (≤ 140 keV). The interaction of gamma rays with the NaI(Tl) crystal produces photons of light that are subsequently detected by the PMTs. The lead collimator ensures that only those photons traveling in the proper direction actually reach the detector. Photons with improper trajectories simply hit the collimator and therefore do not interact with the scintillating crystal. Computerized logic algorithms are consequently used to locate the position from which the gamma ray originated in the patient, allowing a usable image to be created and displayed.

Transmission computerized tomography (TCT) uses external radiation sources, such as X-ray tubes, along with radiation detectors to create slices or image planes through the human body by simply moving the sources and detectors relative to the patient. The radiation is allowed to pass through the body to a series of detectors located on the opposite side of the body. A computer uses the attenuated radiation signals from the detectors to construct slices and produce a useable image. The resolution of such techniques is ≤ 1 mm. Since a real knife is not involved, computerized tomography can be called *virtual slicing*.

If radionuclides are actually injected into the body, the technique becomes emission tomography. Emission tomography can result in either two-dimensional (known as PLANAR) or three-dimensional images. The three-dimensional images are obtained using a technique known as *single-photon emission computerized tomography* (SPECT) which uses injected gamma ray emitting radionuclides for imaging. If a positron emitter is injected, the technique is called *positron emission tomography* (PET). A process known as *annihilation* results when a positron, which is the antiparticle of an electron, interacts with an electron. Annihilation results in the production of two characteristic gamma rays emitted 180° apart. These two gamma rays are subsequently detected by separate radiation detectors located in a ring around the patient. The location of the annihilation event can easily be determined from the simultaneous gamma ray information. The results of multiple events lead to a complete image.

Since emission tomography requires some type of internal radioactive sources, a large number of radionuclides have been developed and produced specifically for these applications. Gamma emitters are used for SPECT, while positron emitters are used for PET. The chemical form of a radionuclide is tailored specifically for a given target (tissue, bone, or organ). Compounds labeled with radionuclides for administration to patients are known as *radiopharmaceuticals*.

Therapeutic Uses

The effects of ionizing radiation on biological tissue have been used to our advantage when it comes to treating disease. Both internal and external radioactive sources can be used for treatments. One of the biggest challenges for the therapeutic use of radiation is how to target the diseased tissue without adversely affecting healthy tissue. Unfortunately, no "silver bullet" exists to magically deliver a radioactive source or electromagnetic radiation to a target without any collateral damage. An example of the use of an internal radioactive source is ^{131}I ($t_{1/2} = 8.04$ d). This radionuclide has been used to successfully treat the thyroid. A particularly clever external technique is used with gamma ray sources, known as the *gamma knife*. This device uses many tens of collimated gamma ray beams directed at a predetermined location within the patient. Taken individually, the gamma ray beams are relatively innocuous. However, a large dose is produced where the beams intersect. The overlapping beams can be manipulated to correspond to the volume and shape of the target tissue.

WORKING WITH RADIOACTIVE MATERIALS

Generally Accepted Practices

When we put in a hard day's work, we want to be assured that we aren't being exposed to any unnecessary risks. Let's face it, at the end of our shift we all want to return safely to our homes in good health. Those working with radioactive materials therefore need to follow a few simple guidelines to help protect themselves from potential dangers. First and foremost, we must have a healthy respect for radioactive materials and radiation. Note that respect is different from fear. This respect will ensure that we do not get careless or complacent in our day-to-day activities. Second, we must maintain our exposure to radioactive materials in accordance with the ALARA principle. That is, we must maintain our exposure *as low as reasonably achievable*. Third, three words will help enforce the ALARA principle: *time*, *distance*, and *shielding*. We should minimize our time working with radioactive materials, maximize the distance between us and the source of radiation, and use proper shielding to minimize our exposure.

Finally, we must make sure that we understand and follow all the facility procedures and regulations regarding the use of radioactive materials so that our safety and that of our patients are assured.

Monitoring Occupational Exposure

If we have a reasonable chance of being exposed to radioactive materials and radiation while on the job, the employer must monitor our occupational exposure. Limits on dose have been established by various government entities for your protection. Dose is monitored using a variety of devices, including pocket dosimeters, thermoluminescent dosimeters, electronic dosimeters, film badges, finger rings, and others. The particular dosimeter used will depend upon the type of radiation and the specific situation at hand. Proper placement of the dosimeter is essential to accurate readings. Thus it is vital that you carefully follow the placement and usage instructions for each dosimeter.

Summary

- Heat exchange can occur via conduction, convection, and radiation.
- Conduction is heat transfer that occurs by physical contact between objects.
- Convection is heat transfer due to the movement of matter from one location to another.
- Radiation is heat transfer involving electromagnetic waves.
- Electromagnetic waves are produced when charged particles accelerate. They are composed of continuously varying, mutually perpendicular electric and magnetic fields.
- Electromagnetic waves exhibit wave-like properties such as frequency, wavelength, and interference. Electromagnetic waves also exhibit particle-like properties. The energy of electromagnetic radiation is given by:

$$E = \frac{hc}{\lambda} = hf$$

- The equivalence between the rest mass, rest energy, and speed of light is given by:

$$E = mc^2$$

- Radioactive decay is a spontaneous process by which an unstable nucleus goes from a level of higher energy to a level of lower energy.

$$\text{parent} \xrightarrow{\text{radioactive decay}} \text{daughter}$$

- Alpha decay is the emission of an alpha particle (helium nucleus) from the parent nucleus. In this process, energy is released in the form of kinetic energy of the escaping alpha particle and the recoiling daughter nucleus.
- Beta decay is the general term used to describe radioactive decay processes that result in the mass number remaining constant while the atomic number changes.
- Beta-minus decay:

$$n \rightarrow p^+ + e^- + \bar{\nu} + \text{energy}$$

- Positron decay:

$$p^+ \rightarrow n + e^+ + \nu + \text{energy}$$

- Electron capture decay:

$$p^+ + e^- \rightarrow n + \nu + \text{energy}$$

- Gamma ray emission occurs when a nucleus in an excited state de-excites and emits a photon of electromagnetic radiation.

- The decay rate is equal to the number of disintegrations occurring per unit time:

$$1 \text{ Bq} = 1 \frac{\text{disintegration}}{\text{s}}$$

$$1 \text{ Ci} = 3.7 \times 10^{10} \text{ Bq} = 3.7 \times 10^{10} \frac{\text{disintegrations}}{\text{s}} = 2.22 \times 10^{12} \frac{\text{disintegrations}}{\text{min}}$$

- Decay rate:

$$Rate = \lambda A$$

- The number of atoms present at time t, given an initial number N_0 of atoms:

$$N = N_0 e^{-\lambda t}$$

- The activity present at time t, given an initial activity A_0:

$$A = A_0 e^{-\lambda t}$$

- Half-life of a radionuclide:

$$t_{\frac{1}{2}} = \ln 2/\lambda$$

- Electromagnetic radiation can be divided into two categories: (1) ionizing and (2) nonionizing. X-rays and gamma rays are considered ionizing radiation. Although they are not electromagnetic radiation, alpha particles and beta particles (β^- and β^+) are also considered ionizing radiation.
- The sources of radioactive materials are: (1) primordial, (2) cosmogenic, and (3) anthropogenic.
- The roentgen (R) is the unit used to quantify radiation exposure from X-rays and gamma rays.
- Units of absorbed radiation dose:

$$1 \text{ Gy} = 1 \frac{\text{J}}{\text{kg}}$$

$$1 \text{ rad} = 100 \frac{\text{erg}}{\text{g}} = 1 \times 10^{-2} \text{ Gy}$$

- Equivalent dose:

$$\text{equivalent dose (Sv)} = \text{absorbed dose (Gy)} \times \text{radiation weighting factor } (W_R)$$

- Tissue equivalent dose:

$$H_T = \sum_R W_R D_{T,R}$$

■ Effective dose equivalent:

$$H_E = \sum_T W_T H_T$$

■ Radionuclides are commonly used in medicine for both diagnostic and therapeutic purposes.

■ The ALARA principle: We must maintain our exposure to radioactive materials as low as reasonably achievable. This is accomplished through minimizing time, maximizing distance, and using shielding.

Review Questions for Radiation and Radioactivity

1. Calculate the frequency of an electromagnetic wave that has a period of 5.0 ms.

2. Calculate the wavelength in meters and the energy in joules of the electromagnetic wave in Problem 1.

3. A radiowave propagates through space at a speed of 3.00×10^8 m/s. If its frequency is 2.0×10^8 Hz, calculate its wavelength in meters.

4. Calculate how long it would take a sound wave to travel around the Earth one time at the equator (total distance of travel $= 6.37 \times 10^6$ m), assuming it travels with a speed of 343 m/s. Compare this with the time taken by an electromagnetic wave to travel the same distance.

5. Calculate the rest energy equivalence (rest energy) of 5.0 g of uranium.

6. Define nuclide, isotope, isotone, and isobar.

7. Write balanced equations for the following:
   a. Beta-minus decay of ^{241}Pu
   b. Alpha decay of ^{248}Cm
   c. Positron decay of ^{22}Na
   d. Electron capture decay of ^{85}Sr

8. How many ^{3}H ($t_{1/2} = 12.3$ y) atoms remain after 61.5 years, assuming 1.0×10^{19} atoms were present initially?

9. Perform the following conversions:
   a. 15 mCi = _____ Bq
   b. 25 kBq = _____ Ci
   c. 60 rad = _____ Gy

10. Calculate the ^{99}Tc^m activity that remains after 7.2 hr if the initial activity is $\overline{2}0$ GBq.

11. Why is radiation effective at killing or reducing tumors?

12. Use the Einstein equation ($E = mc^2$) to calculate the energy (in MeV) to create a positron and an electron. Each has a mass of 9.11×10^{-31} kg. 1 MeV $= 1.602 \times 10^{-13}$ J, and c $= 3.00 \times 10^8$ m/s.

13. Explain why ^{99}Tc^m is a widely used radionuclide in the field of nuclear medicine.

14. Why are collimators used in conjunction with Anger scintillation cameras?

Answers to End-of-Chapter Review Questions

CHAPTER 1: MEASUREMENT

1. a. 10^7
    b. 10^{-3}
    c. 10^{703}
    d. cm^3

2. a. 2.34×10^1
    b. 2.64×10^{-3}

3. a. 2.60×10^{11}
    b. 1.33×10^{-25}

4. A significant figure is a digit in a measurement that has a physical meaning and can be reproducibly determined.

5. An exact number is a small number that can be reproducibly determined by counting or one that is defined to a particular value. Exact numbers have infinite precision and significant figures. Exact numbers are not obtained using measuring devices.

6. Accuracy is agreement between a measured quantity and the accepted value. Precision is the agreement between replicate measurements. Precision is assessed (but not necessarily improved) by calculating the standard deviation of the measurements. Precision is improved through careful laboratory technique and/or by using more precise measuring devices. Accuracy is improved by replicate measurements and assessed by the percent error of the measurement.

7. $2\overline{0}0$ N-m

8. $0.061 \, in^3 = 1 \, cm^3$

9. Density is unit dependent and equal to the mass of the substance divided by the volume of the substance. The value of the density, therefore, depends on the units of the mass and the volume. Specific gravity is equal to the density of the substance divided by the density of water (in the same units) and is, therefore, a constant numerical value.

10. $1\overline{2}00 \, lb/ft^3$

11. 56.9 kg

12. $R = PV/nT$

13. $T_1 = \dfrac{1}{\left[\dfrac{1}{T_2} - \ln\left(\dfrac{P_1}{P_2}\right)\dfrac{R}{\Delta H}\right]}$

14. $k = 1.04 \times 10^{-19}$

15.

MULTIPLIER	ABBREVIATION	MATHEMATICAL MEANING
micro	μ	10^{-6}
milli	m	10^{-3}
centi	c	10^{-2}
kilo	k	10^{3}
nano	n	10^{-9}
mega	M	10^{6}

16. 22 L

17. c.

18. b.

19. b.

20. b.

21. a.

22. a.

23. b.

24. c.

25. b.

26. *Volume* = 0.0822 (*Temperature*) + 22.4

27. a. 0.45 m
    b. 0.036 km
    c. 450 mL
    d. 9.7×10^{-4} s
    e. 1.7×10^{-5} g

28. a. 2.6×10^{10} mg
    b. 2.6×10^{-7} mm
    c. 380 cL
    d. 7.2×10^4 nm
    e. 0.055 gigabytes

29. a. 3.0×10^1 cm
    b. 140 L
    c. 25 kg
    d. 53 mi
    e. 0.91 gr

30. a. 197 cm^3
    b. 26 mi^2
    c. 1.2×10^5 ft/min^2
    d. 26 W/km^2
    e. 1.4 g/cm^3

31. a. 100 mEq
    b. 6700 mi
    c. 0.51 gal
    d. 230 min
    e. 2.1×10^9 s (68 y)

32. –40 degrees

33. Mathematically, this requires a temperature of –654 degrees. However, this is impossible since there are no negative Kelvin temperatures.

34. 4.4 g/cm^3

35. 2.8 lb/ft^3

36. 0.0036 kg/m^3

CHAPTER 2: CHEMISTRY

1. Analytical chemistry develops methods for determining the identity and quantity of the components in a sample; physical chemistry develops unifying theories and laws for all of chemistry; inorganic chemistry studies the properties of all elements except carbon; organic chemistry studies compounds based on carbon; biochemistry studies chemical processes in living organisms.

2. Matter can be either a pure substance or a mixture. Pure substances cannot be further broken down into simpler components through physical processes and can be either elements (one type of atom) or compounds (more than one type of atom). Mixtures can be homogeneous (aka, solutions) or heterogeneous. Heterogeneous mixtures exhibit phase boundaries or sharp demarcations where the chemical and/or physical properties of the sample change. Mixtures are separable into pure substances through physical processes.

3. A chemical change results in the formation of a new chemical substance. A physical change changes the state of a substance, but it is still the same substance.

4. Physical: density is 1.0 g/mL; water is clear and colorless, etc. Chemical: water can be decomposed into 1 part oxygen and 2 parts hydrogen; water is the product of metabolizing food, etc.

5. Intensive properties do not depend on sample size (e.g., density, color, etc.). Extensive physical properties depend on sample size (e.g., mass, volume, etc.).

6. Intensive: density, colorless, etc.; extensive: mass, volume, etc.

7. Density is intensive, even though it relates two extensive properties.

8. An atom is the fundamental building block of all matter. Molecules are aggregates of atoms chemically bonded into a discrete unit. Compounds are comprised of two or more kinds of atoms. Mixtures are aggregates of two or more pure substances that can be separated through physical means.

9. Isotopes have the same atomic number but different neutron numbers.

10. Yes. For example, oxygen is an element because it consists of only oxygen atoms. Oxygen gas is a molecule that consists of two atoms of oxygen chemically bound into an O_2 molecule.

11. H: mass number = 1; $Z = 1$; $P = 1$; $N = 0$; $E = 1$; He: mass number = 4; $Z = 2$; $P = 2$; $N = 2$; $E = 2$; Li: mass number = 6; $Z = 3$; $P = 3$; $N = 3$; $E = 3$; Be: mass number = 9; $Z = 4$; $P = 4$; $N = 5$; $E = 4$.

12. Mass number is equal to the proton number plus the neutron number. The average atomic mass appears on the periodic table and is the mass of each naturally occurring isotope of an element weighted by the fractional abundance of the isotope.

13. 80 amu.

14. In the ionization sector, an electron is removed from the sample molecule, resulting in a positive charge on the molecule, called a molecular ion. In the acceleration sector, charged plates accelerate the molecular ion into the magnetic sector. The magnetic field of an external magnet exerts a force on the moving charged particle. The direction of this magnetic force is perpendicular to both the magnetic field and the direction of travel of the molecular ion. This results in a curved path. The curvature of the path depends on the mass of the molecular ion.

15. Periods are horizontal rows on the table. Each successive element has one additional proton and one additional electron. Each period represents filling a quantum energy level on that series of atoms. Elements at the end of a period have filled electron energy shells and are especially stable. Groups are vertical columns. Members of the same group have similar chemical and physical properties. In the older A/B numbering system, representative elements have an A designation, and the group number represents the number of valence electrons available for chemical reaction/bonding.

16. Metals are located on the left side of the periodic table. Metals tend to form cations, are generally ductile and malleable, and are good electrical and thermal conductors.

Nonmetals are located on the right side of the periodic table. Nonmetals tend to form anions and have a wide variety of physical properties. Metalloids look like metals but have electrical conductivity intermediate between metals and nonmetals. For this reason, metalloids are called semiconductors.

17. An ion is an atom or group of atoms with a net electrical charge. Anions are negatively charged ions that are formed by the addition of electrons. Cations are positively charged ions that are formed by the loss of electrons.

18. Metals (left side of the table) tend to form cations; nonmetals (right side of table) tend to form anions.

19. For the representative metals, the cationic charge equals the group number. For representative nonmetals, the anionic charge equals the group number minus 8. Transition metals frequently form more than one cation.

20. Hydrates are compounds that absorb water into their crystalline structure. Formation of a hydrate is generally reversible.

21. Ionic compounds generally consist of a metal and one or more nonmetals, whereas molecular compounds contain only nonmetals.

22. Na^+; P^{3-}; Ba^{2+}

23. The representative elements comprise the two "high rises" on either side of the periodic table. In the older numbering system, the representative elements belong to groups 1A through 8A. The transition elements comprise the center connection between the representative elements. In the A/B numbering system, representative elements belong to the B groups.

24. Since the Na^+ ion has a +1 charge, the lactate ion must have a −1 charge, $C_3H_5O_3^{-1}$.

25. Since glucose contains only nonmetals, it is a molecular compound.

26. $2(14) + 16 = 44$ g/mol

27. 20 g (1 mol/44 g) = 0.46 mol; 0.46 mol (6.02×10^{23} molecules/mol) = 2.7×10^{23} molecules

28. A period is a horizontal row that represents filling electron quantum energy levels. A group is a vertical column that contains elements with similar chemical and physical properties.

29. a. $NaHCO_3$
    b. KCl
    c. NH_3
    d. Magnesium sulfate
    e. CCl_4
    f. Ammonium nitrate
    g. $FeSO_4$
    h. Magnesium phosphate

    i. Copper(II) carbonate
    j. Phosphate ion
    k. Iron(III) ion or ferric ion
    l. Sodium monohydrogenphosphate
    m. Ammonium ion
    n. CO
    o. Phosphorous trichloride
    p. Potassium chloride
    q. N_2O
    r. LiH_2PO_4
    s. NO_2
    t. Aluminum chloride
    u. $CaCO_3$
    v. NaF
    w. O_3
    x. Barium sulfate

30. b.

31. d.

32. a.

33. e.

34. e.

35. c.

36. a. Al^{3+}
    b. Rb^+
    c. Mg^{2+}
    d. Li^+
    e. Ga^{3+}

37. a. S^2
    b. I
    c. N^3
    d. P^3
    e. Se^2

38. a. Mg_3N_2: magnesium nitride
    b. Al_2S_3: aluminum sulfide
    c. LiI: lithium iodide
    d. K_2O: potassium oxide
    e. CaC_2: calcium carbide

39. a. carbonate
    b. sulfate
    c. nitrite

d. hydroxide

e. ammonium

f. iron (III) or ferric ion

g. phosphate

h. hydrogen sulfite or bisulfite

i. dihydrogen phosphate

40. a. Fe^{2+}

b. OH^-

c. PO_4^{3-}

d. SO_4^{2-}

e. HCO_3^-

f. CO_3^{2-}

g. SO_3^{2-}

h. HSO_4^-

i. $H_2PO_4^-$

41. a. $NaHCO_3$

b. K_3PO_4

c. Li_2CO_3

d. NH_4Cl

e. BaO

f. $Cu(C_2H_3O_2)_2$

g. $FeSO_4$

h. NiO

i. NH_4NO_3

j. $AgCl$

42. a. CS_2

b. ICl_3

c. NI_3

d. P_2O_5

e. NH_3

f. OF_2

g. NO_2

h. $NO;$

i. N_2O

j. SCl_4

43. negative one

44. $C_{18}H_{36}O_2^-$; $Fe(C_{18}H_{36}O_2)_3$

45. a. hydrogen

b. argon

c. bromine

d. lithium

e. arsenic

f. carbon

g. arsenic

46. TcO_4^-

47. WO_4^{2-}

48. a. yes
    b. no
    c. no
    d. yes
    e. yes

49. a. true
    b. false: this process gives a new chemical substance
    c. false: lidocaine is not an ionic compound
    d. true

50. a. false: Dalton did not know about isotopes
    b. true
    c. true
    d. false: atoms are comprised of protons, neutrons, and electrons
    e. true

CHAPTER 3: PHYSICS PART 1

1. Velocity is the change in position with respect to time. Acceleration is the change in velocity with respect to time.

2. A force is a push or a pull.

3. $\vec{F} = m\vec{a}$, or force is equal to mass times acceleration. A bigger force causes a greater acceleration when applied to a mass. $\vec{F}$ and $\vec{a}$ are vectors, but mass is not.

4. Mass describes the amount of material or the inherent reluctance to undergo a change in motion (Newton's Second Law). Weight is a force caused by gravity acting on a mass.

5. Weight is measured in newtons, and mass is measured in kilograms.

6. 670 N

7. 68 N/cm²

8. $10\overline{0}$ mmHg = 0.132 atm = $10\overline{0}$ torr = 13.3 kPa

9. Mass = $(11.2 \text{ g/cm}^3)(27 \text{ cm}^3)$ = 302.4 g; Weight = $(0.3024 \text{ kg})(9.8 \text{ m/s}^2)$ = 2.964 N; Area = $(3.00 \text{ cm})(3.00 \text{ cm})$ = 9.00 cm² = 9.00×10^{-4} m². $P = \dfrac{F}{A} = \dfrac{2.964 \text{ N}}{9.00 \times 10^{-4} \text{m}^2} = 3.29 \times 10^3$ Pa.

10. Atmospheric pressure supports a column of liquid in a tube closed at the top. The height of the column of liquid exerts a pressure at its base equal to the atmospheric pressure.

11. A Bourdon gauge contains a curved metal tube connected to a pointer. Increasing pressure in the tube causes it to unwind slightly, and this causes the pointer to move around a numerical scale.

12. The absolute pressure of the gas in a cylinder represents the force per unit area exerted by the gas as the gas molecules collide with the inner surface of the cylinder. Gauge pressure is registered by a Bourdon gauge on a gas cylinder and represents the difference in pressure between the pressure exerted by the gas in the cylinder and the atmospheric pressure.

13. Circle 1: area = 0.00785 m²; circle 2: area = 0.00196 m²; radius 1/radius 2 = 2; area 1/area 2 = 4; dividing the diameter by two reduces the area by a factor of 4

14. $P_1/P_2 = \frac{892\ \text{Pa}}{3565\ \text{Pa}} = 0.25$; if the diameter of the circle is reduced by a factor of 2, the pressure increases by a factor of 4

15. 24.7 psi

16. 1.97 atm or 1.50×10^3 mmHg

17. 2.2×10^5 Pa

18. 3 m/s²

19. $m = F/a$; units: kg = (kg m/s²)/(m/s²)

20. 29 in = 737 mm, so 29 in Hg is less than 1 atm

21. a. foot · pounds is work
    b. inches/day² is acceleration
    c. kg · m²/s³ is power
    d. liter · atmosphere is work

22. d.

23. c.

24. d.

25. c.

26. c.

27. b.

28. a.

29. c.

30. c.

31. a.

32. c.

33. c.

34. c.

35. c.

CHAPTER 4: PHYSICS PART 2

1. Work is the expenditure of energy; energy is the ability to do work.

2. $W = (force)(distance)$; unless the object moves some finite distance, no work has been done.

3. J or N · m

4. 44 kJ

5. $W = P\Delta V = 1.950 \text{ m}^3 \times 101 \text{ kPa} = 197 \text{ kJ}$

6. Kinetic energy is the energy a mass possesses by virtue of being in motion. A car going down the highway and nitrous oxide molecules racing through a tube toward a patient both have kinetic energy.

7. Potential energy is stored energy. Compressed gases and chemical bonds represent potential energy.

8. Heat is a process of energy transfer caused by a difference in temperature, not an amount of energy itself. In practice, however, heat and "heat energy" are used interchangeably.

9. The internal energy is the total of the kinetic and potential energies of the molecules in an object. The temperature is related to the average kinetic energy of the molecules in the sample.

10. Heat is a process of energy transfer caused by a difference in temperature. Internal energy is the sum of the kinetic and potential energies of the particles in a sample. The First Law of Thermodynamics states that the change in internal energy is equal to the sum of heat and work processes done by/on the system.

11. The change in internal energy ΔU is the sum of the heat processes (Q) and work processes (W) done on or accomplished by the system.

12. In an endothermic process, heat flows from the surroundings into the system ($Q > 0$). In an exothermic process, heat flows from the system out into the surroundings ($Q < 0$).

13. As heat flows into a system from the surroundings, this energy is absorbed into the total internal energy of the system. The key here is that the energy absorbed through heat is not stored as heat but as internal energy. Some of the kinetic energy of the molecules in the system and/or surroundings is absorbed into the internal energy of the system. This means the average kinetic energy of the remaining molecules decreases, so the temperature decreases.

14. A state function is a function that is completely described by a set of parameters. Changing any of the parameters in any order results in the same change in the state function as long as the initial states and final states are the same.

15. The specific heat is the energy necessary to change the temperature of 1 g of a material by 1°C. The heat capacity is the energy necessary to change the temperature of the entire sample by 1 degree. Specific heat is an intensive property, while heat capacity is extensive.

16. 20.9 kJ

17. Power is the rate of using energy or the rate of doing work. The unit of power is the watt (W), and 1 W equals using energy at a rate of 1 J/s.

18. 42 MW

19. b.

20. b.

21. d.

22. b.

23. a.

24. c.

25. c.

26. b.

27. 121 W; a.

28. c.

29. d.

30. a.

31. a.

32. 1b; 2k; 3j; 4d; 5l; 6f; 7i; 8c; 9o; 10g; 11m; 12a; 13n; 14p; 15e; 16h

33. 125 W

CHAPTER 5: FLUIDS

1. A fluid has the ability to flow and may be either a liquid or a gas. Hydrostatics describes fluids that are not moving, while hydrodynamics describes fluids that are in motion.

2. 550,000 Pa

3. Pascal's Principle states that a change in pressure on an enclosed, incompressible fluid is transferred equally throughout the body of the fluid.

4. An object placed into a fluid experiences buoyancy, or a buoyant force, because of the fluid it displaces. Points of the object deeper in the fluid experience a greater force from the fluid, and this unbalanced force pushes the object upward.

5. An object totally immersed into a fluid will displace a volume of fluid equal to the volume of the object. An object floating on a fluid will displace a weight of fluid equal to the weight of the object.

6. 1.1×10^6 N

7. In laminar flow, the molecules are all moving in the same direction with no eddies or cross currents. You can envision the fluid as consisting of smooth microscopic layers, and the layers don't mix or interfere with each other. Turbulent flow has eddies and is chaotic.

8. The flow rate is equal to the volume of fluid moving past a point per unit time.

9. 1.5 L/s or 0.0015 m^3/s

10. The flow rate is proportional to the cross-sectional area.

11. 1.27 m/s

12. 127 m/s

13. The pressure exerted by a moving fluid decreases as the velocity of the fluid increases.

14. The velocity increases by a factor 1.0×10^4. The pressure drop is 9.0×10^7 Pa.

15. A Venturi tube measures the pressure difference between two tubes of different diameter through which a fluid is flowing. The pressure difference can be used to calculate the flow rate.

16. Viscosity is the resistance of a fluid against flow. Viscosity is measured by calculating the time it takes for a fluid to flow through a tube of known length and diameter.

17. The flow rate of a fluid is inversely proportional to the visocity of the fluid and the length of the tube through which it flows. The flow rate is directly proportional to the fourth power of the radius of the tube.

18. 4.4×10^{-8} m^3/s

19. Raising the IV bag higher increases the pressure at the needle according to $P_2 = P_1 + \rho g h$, where P_2 is the pressure at the exit of the needle. This in turn increases the speed of the fluid exiting the needle. The flow rate is directly proportional to the fluid's speed.

20. 2450 Pa: 2.9×10^{-6} m^3/s or 1.7×10^2 mL/min
    4900 Pa: 5.8×10^{-6} m^3/s or 3.5×10^2 mL/min

CHAPTER 6: GASES

1. Boyle's Law states the volume of a fixed quantity of gas is inversely proportional to the pressure (assuming the temperature remains constant). Charles's Law states the volume of a fixed quantity of gas is directly proportional to the absolute temperature (assuming the pressure remains constant). Avogadro's Law states the volume of a sample of gas is directly proportional to the number of gas molecules (assuming pressure and temperature are constant). J. L. Gay-Lussac actually first reported Charles's Law, but Gay-Lussac is often credited with the law that states pressure and absolute temperature are directly proportional.

2. The ideal gas law states the product of volume (V) and pressure (P) is equal to the product of a constant (R, the universal gas constant), the amount of gas (n), and the Kelvin temperature. The ideal gas law applies exactly only to ideal gases (which don't exist). However, real gases approach ideal gas behavior as pressure decreases and temperature increases.

3. The ideal gas law is most useful when the problem describes only a single state (temperature, pressure, and/or volume). The empirical gas laws are useful for changing the state of a gas (P, n, V, and/or T change).

4. Gases are comprised of infinitely small particles in constant random motion. The gas molecules collide with each other and with the sides of the container with no attractive or repulsive forces. The average kinetic energy is related to the temperature of the system.

5. He, because a helium atom is much smaller than a sulfur hexafluoride molecule, which better approximates the first tenet of the kinetic molecular theory of gases.

6. R is the universal gas constant, which has units of energy/mol/K. R is truly universal and appears in many physics equations that relate energy/work to amount of material and temperature. Some common units include 0.0821 L atm/mol/K and 8.31 J/mol/K.

7. Standard temperature and pressure. $T = 0°C$ and $P = 1$ bar $= 100$ kPa. Older texts set standard pressure as 1 atm, but this was revised by the IUPAC.

8. The SMV of an ideal gas is the volume occupied by one mole of gas at STP. Under the current definition of STP, the SMV $= 22.7$ L. The older definition of STP (273 K and 1 atm) gives an SMV of 22.4 L.

9. 83.14 mL bar/mol/K

10. 25.5 L

11. 1.6 atm

12. 174°C

13. 8.31 Pa·m^3/mol/K $=$ 8.31 J/mol/K

14. Increasing the pressure will decrease the volume, while increasing the temperature will increase the volume. Therefore, the change in volume will depend on the relative sizes of the changes in temperature and pressure.

15. Path 1: 2.0 L expands upon heating at constant pressure to 2.8 L, which is then compressed by an increase in pressure to a final volume of 0.19 L. Path 2: 2.0 L is compressed by increasing pressure to 0.13 L, which then expands upon heating to a final volume of 0.19 L.

16. 1.5×10^6 mol

17. Helium has a density of 0.18 g/L. The density of air is 1.3 g/L. Since He is less dense, it floats on air.

18. The total pressure in a mixture of gases is equal to the pressure that each component gas would exert in the absence of all of the others. This individual pressure is called the partial pressure of the gas. $P_{total} = P_1 + P_2 + \ldots + P_n$.

19. 0.8

20. 0.90

21. 0.029

22. Real gases most closely approximate ideal behavior at high temperatures and low pressures.

23. The Van der Waals equation takes into account the deviations of real gases from the Kinetic Molecular Theory of Gases (nonzero molecular volume and nonelastic collisions).

24. 141 torr

25. As the volume of a fixed sample of gas is decreased at a constant temperature, the pressure increases. Since the temperature is constant, the average velocity of the gas particles remains constant. Constrained to a smaller volume, the collision frequency of the molecules with the walls of the container increases. Therefore, the pressure increases.

26. 6.0 L

27. b.

28. b.

29. a.

30. d.

31. c.

32. a.

33. c.

34. d.

35. 1a; 2c; 3f; 4e; 5b; 6g

36. 1. 169 L
    2. 0.425 atm
    3. 0.00729 mol
    4. 145K
    5. 24.5 L
    6. 4.14×10^4 bar
    7. 0.0428 g
    8. 654K
    9. 22.4 L
    10. 22.4 L

37. 1. 85.7 mL
    2. 133 mL
    3. 100 mL
    4. 180 mL
    5. 75.0 mL
    6. 50.0 mL

38. a. O_2= 12.3 L; N_2=36.9 L
    b. 49.3 L
    c. 0.25
    d. 0.50 atm
    e. 25%

39. a. 425 torr
    b. 21.5 L
    c. 0.50
    d. 50%

40. 24.5 L

41. 62,400 mL·torr/mol/K

42. 0.0821 L·atm/mol/K

43. 0.50 each

44. 0.840

45. 42.1%

46. 20%

47. 754 torr

48. 53%

49. a. 158 torr
    b. 25.6 torr
    c. 724 torr
    d. 152 torr

50. a. 0.45 torr
    b. 0.0026 mol
    c. 0.048 g

51. 412 m/s

52. 483 m/s

53. 0.85 m

CHAPTER 7: STATES OF MATTER AND CHANGES OF STATE

1. Dipolar attractions result from the attraction between permanent molecular dipoles. Hydrogen bonding is a special type of dipolar interaction, but only occurs when a hydrogen atom is directly bonded to O, F, or N. London forces are temporary dipolar attractions that result from nonsymmetrical distribution of electrons in an atom or molecule.

2. Solids are characterized by definite volumes and shapes due to strong intermolecular forces that lock the molecules into a rigid crystalline lattice. Liquids are characterized by a definite volume, but have no definite shape. Intermolecular attractions in liquids are sufficient to keep the molecules in contact (a condensed state of matter) but not strong enough to prevent the molecules from sliding around past each other. Thus, liquids have the ability to flow. Gases have neither a definite volume nor shape because there are negligible intermolecular forces between gas molecules.

3. Helium gas, because water can form hydrogen bonds.

4. Solids are characterized by definite volumes and shapes due to strong intermolecular forces that lock the molecules into a rigid crystalline lattice. Liquids are characterized by a definite volume, but have no definite shape. Intermolecular attractions in liquids are sufficient to keep the molecules in contact (a condensed state of matter) but not strong enough to prevent the molecules from sliding around past each other. Thus, liquids have the ability to flow. Gases have neither a definite volume nor shape because there are negligible intermolecular forces between gas molecules.

5. Covalent bonds result from the Coulombic attraction for one or more shared pairs of electrons between two nuclei. Ionic bonds result from the Coulombic attraction between oppositely charged ions.

6. The Valence Shell Electron Pair Repulsion Theory states that all electrons in a molecule mutually repel each other and achieve a geometry so that the bonding pairs and lone pairs of electrons are as far apart in space as possible. This theory allows one to predict the shape and geometry of a molecule.

7. a. $\ddot{\text{O}} = \text{C} = \ddot{\text{O}}$

    b. $\text{H} - \ddot{\text{O}} - \text{H}$

    c. $\ddot{\text{C}}\text{l} - \overset{\overset{\textstyle \text{H}}{|}}{\underset{\underset{\textstyle :\ddot{\text{C}}\text{l}:}{|}}{\text{C}}} - \ddot{\text{C}}\text{l}:$

    d. $:\ddot{\text{C}}\text{l} - \overset{\overset{\textstyle :\ddot{\text{C}}\text{l}:}{|}}{\underset{\underset{\textstyle :\ddot{\text{C}}\text{l}:}{|}}{\text{C}}} - :\ddot{\text{C}}\text{l}$

    e. $:\ddot{\text{O}} = \ddot{\text{S}} = \ddot{\text{O}}:$

8. Water can form hydrogen bonds that can hold the water molecules into a liquid state even when the average kinetic energy of the molecules is relatively high.

9. When NaCl dissolves in water, it separates into Na^+ ions and Cl^- ions. These ions are surrounded by a poorly defined sphere of polar water molecules. Organic solvents are not sufficiently polar to solvate the ions and keep them away from each other, allowing them to settle back into the solid state.

10. CHI_3 (iodoform), because it contains the larger iodine atoms and, therefore, experiences more London forces.

11. As temperature increases, the vapor pressure increases. As intermolecular forces decrease, the vapor pressure increases.

12. Bent

13. Linear

14. 360 torr

15. Freezing is an exothermic process. So, when a thin layer of water on the fruit freezes, it delivers up its heat of fusion, and that helps prevent the fruit from freezing.

16. Surface active agents, commonly known as soaps and detergents, have an ionic (or highly polar) hydrophilic end that allows the surfactant to interact with water molecules. The surfactant also has a hydrophobic tail that prevents the water molecules from interacting with each other. This reduces the surface tension and cohesion of the water. Cleaning and preventing the collapse of a premature baby's lungs are two applications of surfactants.

17. a. 1.0×10^5 torr
    b. 0.17
    c. 86,000 torr
    d. 0.56

18. Since the oxygen is less polar than the nitrous oxide, it is less likely to dissolve in water.

19. Nitrous oxide has a small molecular dipole, but it is not very polar compared to water. Therefore, nitrous oxide is expected to be preferentially soluble in adipose tissue.

20. a.

21. b.

22. The gas state. Since the molecules are not in contact with each other, the volume of the gas will be greater, and as the volume of a sample increases, the density decreases.

23. a. H—Ö—Cl̈:

    b. H—Ö—Cl̈—Ö:

    c. H—Ö—Cl̈—Ö: (with :Ö: above Cl)

    d. H—Ö—Cl̈—Ö: (with :Ö: above Cl and :Ö: below Cl)

e. 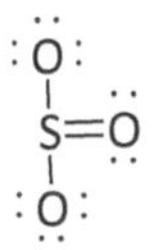
$$\ddot{O}$$
|
S=Ö
|
:Ö:

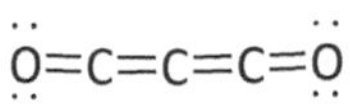

f. :F̈—Ö—F̈:

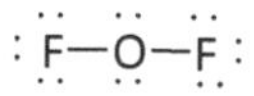

g. 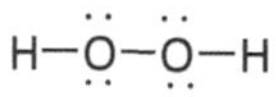 H—Ö—Ö—H

h. Ö=C=C=C=Ö

i. 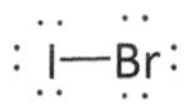 :Ï—B̈r:

j.  H—C≡N:

k. 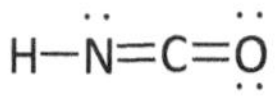 H—N̈=C=Ö

l. 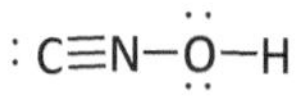 :C≡N—Ö—H

m. 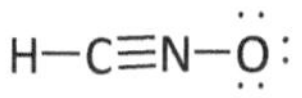 H—C≡N—Ö:

24. a. O=bent
    b. O=bent; Cl=bent
    c. O=bent; Cl=pyramidal
    d. O=bent; Cl=tetrahedral
    e. S=trigonal planar
    f. O=bent
    g. both O=bent
    h. All C=linear
    i. n/a
    j. C=linear
    k. N=bent; C=linear
    l. N=linear; O=bent
    m. C=linear; N=linear

25. a. dipolar
    b. London
    c. London
    d. hydrogen bonding
    e. dipolar
    f. London
    g. hydrogen bonding
    h. dipolar
    i. hydrogen bonding
    j. hydrogen bonding
    k. dipolar
    l. London

26. a. CH_3CH_2OH; hydrogen bonding
    b. CH_3OCH_3; dipolar

    c. HF; hydrogen bonding
    d. IBr; dipolar (and London)
    e. SO_2; dipolar
    f. NaCl; ionic
    g. Na_2O; ionic
    h. Ar; London

27. (16,000 Pa) (0.010m) = 160 N/m

28. 80 N/m

29. 80 N/m

30. a. 156 torr
    b. 195 torr
    c. 301 torr
    d. set P = 760 torr and solve for T; 59°C
    e. 57°C
    f. 60°C

31. a. sevoflurane
    b. sevoflurane
    c. sevoflurane

32. slope = −1671; intercept = 8.075. A is the intercept and B is the slope

33. 258 torr

34. 1 = e
    2 = c
    3 = b
    4 = a
    5 = d

CHAPTER 8: SOLUTIONS

1. A solution is a homogeneous mixture that has one or more solutes dispersed at a molecular or ionic level throughout a medium called the solvent. The dispersed phase in a colloid is much larger than a typical molecule. For this reason, colloids exhibit the Tyndall effect, or the ability to trace out a ray of light shown through the colloid.

2. a. rubbing alcohol is 80% isopropanol and 20% water
    b. normal saline is 0.8% NaCl in water
    c. club soda has carbon dioxide dissolved in water
    d. an alloy is stainless steel

3. As the concentration of solute particles increases, the vapor pressure decreases, the boiling point increases, the freezing point decreases, and the osmotic pressure increases.

4. Osmotic pressure

5. 55.6 M

6. Two liquids are miscible if they are soluble in each other in all proportions.

7. This depends on the relative magnitudes of the solvation energy and the lattice energy of the species. As the magnitude of the solvation energy increases, the heat of solution becomes more exothermic. As the magnitude of the lattice energy increases, the heat of solution becomes more endothermic.

8. 0.26 m; –0.50°C

9. The solubility of solid and liquid solutes typically increases with higher temperatures, while the solubility of gaseous solutes always decreases with increasing temperatures.

10. 0.141 g/L

11. As the prime "escape sites" near the surface of the solution are occupied by solute molecules, fewer solvent molecules can jump to the gas phase. Hence there is a decrease in the vapor pressure. This means a higher temperature is needed to drive the vapor pressure up to the ambient pressure. The change in boiling point is proportional to the molality of all solute particles.

12. You want water to osmose OUT of the potatoes to minimize hot oil spattering, to season the potatoes, and to concentrate the potato flavor.

13. Since the sharks are isotonic to sea water, their cells would burst from the hypotonic environment of a fresh water lake.

14. These terms are relative between two specified solutions and have no meaning outside of a definite context. The hypertonic of two solutions has a higher solute particle concentration. Two solutions are isotonic if they have equal solute particle concentrations.

15. a.

16. b.

17. d; normal boiling point vp = 1 atm

18. d.

19. a.

20. a.

21. b.

22. d.

23. c.

24. a.

25. c.

26.

SOLUTION NUMBER	DENSITY [g/cm³]	% (w/w)	% (w/v)	MOLARITY	MOLALITY
1	1.0021	1.000%	1.002	0.02927	0.02951
2	1.0060	2.000%	2.012	0.05877	0.05961
3	1.0279	7.500%	7.709	0.2252	0.2369
4	1.0381	10.00%	10.38	0.3032	0.3246
5	1.0810	20.00%	21.62	0.6315	0.7303

27.

SOLUTION NUMBER	PERCENT BY WEIGHT	DENSITY [g/cm³]	MASS OF NACL IN 100 GRAMS SOLUTION	MASS OF NACL IN 100 ML SOLUTION	MASS OF SOLUTION THAT CONTAINS 100.0 GRAMS OF NaCl	ML OF SOLUTION THAT CONTAINS 100.0 GRAMS OF NaCl
1	1.000%	1.0020	1.000	1.002	100.0	99.80
2	2.500%	1.0078	2.500	2.520	40.00	39.69
3	5.000%	1.0175	5.000	5.088	20.00	19.66
4	10.00%	1.0375	10.000	10.38	10.00	9.639

28. a. 0.995% (*w/w*)
    b. 1.00% (*w/v*)

29. a. 1.00% (*w/w*)
    b. 1.01% (*w/v*)

30. 0.15 ppm; 0.53 mg

31. 4.5 mL

32. a. 1 Eq/mol

    b. 2 Eq/mol

    c. 3 Eq/mol

33. 539 g

34. 250 mL

35. The absolute value of the heat of solvation is less than the absolute value of the lattice energy.

36. a. exothermic
    b. |heat of solvation| > |lattice energy|
    c. endothermic

37. a. solubility will increase
    b. little change in solubility
    c. solubility will increase

38. a. solubility will decrease
    b. solubility will increase
    c. solubility will increase

39. a. 0.0083
    b. 0.0042

40. a. 0.0019
    b. 0.0085

41. a. $-3.7°C$
    b. $-7.4°C$
    c. $-5.6°C$

42. $-0.19°C$

43. 5.4 m

44. a. $102°C$
    b. $104°C$
    c. $106°C$

45. a. 24 atm
    b. 49 atm
    c. 73 atm

46. 22.9 torr

47. a. 0.50
    b. 380 torr
    c. 178 torr
    d. 558 torr
    e. above

CHAPTER 9: ACIDS AND BASES

1. In a dynamic equilibrium, the population of two or more states is free to move between states. However, the ratio of the populations of each state is a constant value.

2. When a system at equilibrium is disturbed, it will react so as to minimize the change.

3. K represents the balance between products and starting materials in a system at equilibrium. K is the ratio of the molar concentrations of products divided by the molar

concentrations of the starting materials, each raised to the power of their stoichiometric coefficients. As K increases, the reaction becomes increasingly product-favored.

4. Since ionization of water is endothermic, we can think of heat as a reactant. At higher temperatures, therefore, the position of the equilibrium shifts toward the products H^+ and OH^-. Therefore, the pH decreases.

5. An acid is a species that donates a hydrogen ion (proton) to a base. When the base accepts the proton, it is converted into its conjugate acid.

6. O^{2-}. It is a strong base.

7. NaCl (neutral), NaI (neutral), NH_4Cl (acidic), cocaine hydrochloride (acidic), sodium acetate (basic), sodium palmitate (basic).

8. a. 3.37
   b. 8.37
   c. 10.63
   d. 5.63

9. This equilibrium fixes the ratio of the H^+ and the OH^- ions. This establishes the maximum and minimum acid strengths that can exist in water and establishes 7.0 as the neutrality point.

10. The pH scale describes the level of acidity or alkalinity in a solution

11. The p-function means "take the log of a number, and then change the sign."

12. pH = 1

13. An amphoprotic species can behave as either an acid or a base

14. A buffer is a solution that contains a weak acid and its conjugate base or a weak base and its conjugate acid. This system resists changes in pH.

15. 0.82

16. $pOH = 6.70$; $[H^+] = 5.0 \times 10^{-8}$ M; $[OH^-] = 2.0 \times 10^{-7}$ M

17. Salt and water

18. 2.44

19. 3.89

20. 8.44

21. 3.89

22. a. 95%
    b. 85%
    c. 43%

23. The weak acid (HA) reacts with the added hydroxide ion to produce A^-, thereby trading a strong base for a weak one.

24. a.

25. c.

26. a.

27. d.

28. a.

29. d.

30. b.

31. c.

32. c.

33. b.

34. a.

35. a.

36. b.

37. a. $\dfrac{[H^+][CN^-]}{[HCN]}$ d. $\dfrac{[H^+][HPO_4^{2-}]}{[H_2PO_4^-]}$

　　b. $\dfrac{[HC_5H_5N^+][OH^-]}{[C_5H_5N]}$ e. $\dfrac{[OH^-][H_2PO_4^-]}{[HPO_4^{2-}]}$

　　c. $\dfrac{[H^+][HCO_3^-]}{[CO_2]}$

38. a. products
    b. starting material
    c. starting material
    d. starting material
    e. products

39. a. starting material
    b. products
    c. starting material
    d. starting material
    e. starting material

40.

a. NO_2^-	b. F^-	c. HO_2^-	d. CNO^-
e. CO_2H^-	f. BrO_3^-	g. $H_2PO_4^-$	h. HPO_4^{2-}
i. SO_4^{2-}	j. H_2O	k. N_2H_4	l. C_5H_5N

41.

a. NH_4^+	b. H_2O	c. $HCH_3NH_2^+$	d. $HN_2H_4^+$
e. $HC_{10}H_{10}N_2^+$	f. HCO_3^-	g. H_2CO_3	h. HSO_4^-
i. H_2SO_3	j. HCH_3OH^+	k. NH_3	l. $H_2NO_3^+$

42. a. acid: $H_2O \rightleftharpoons H^+ + OH^-$; base: $H_2O + H_2O \rightleftharpoons H_3O^+ + OH^-$

b. acid: $H_2PO_4^- \rightleftharpoons H^+ + HPO_4^{2-}$; base: $H_2PO_4^- + H_2O \rightleftharpoons H_3PO_4 + OH^-$

c. HCO_3^- acid: $HCO_3^- \rightleftharpoons H^+ + CO_3^{2-}$; base: $HCO_3^- + H_2O \rightleftharpoons H_2CO_3 + OH^-$

d. acid: $HC_2H_4NO_2 \rightleftharpoons H^+ + C_2H_4NO_2^-$; base: $HC_2H_4NO_2 + H_2O \rightleftharpoons H_2C_2H_4NO_2^+ + OH^-$

43. a. $HNO_2 = H^+ + NO_2^-$; $\dfrac{[H^+][NO_2^-]}{[HNO_2]}$

b. $HF = H^+ F^-$; $\dfrac{[H^+][F^-]}{[HF_2]}$

c. $HCO_2H = H^+ + CO_2H^-$; $\dfrac{[H^+][CO_2H^-]}{[HCO_2H]}$

d. $HCNO = H^+ CNO^-$; $\dfrac{[H^+][CNO^-]}{[HCNO]}$

44. a. $+ H_2O = HCH_3NH_2^+ + OH^-$; $\dfrac{[NO_2^-][OH^-]}{[HNO_2]}$

b. $N_2H_4 + H_2O = HN_2H_4^+ + OH^-$; $\dfrac{[HN_2H_4^+][OH^-]}{[N_2H_4]}$

c. $CO_3^{2-} + H_2O = HCO_3^- + OH^-$; $\dfrac{[HCO_3^-][OH^-]}{[H_2CO_2]}$

d. $HSO_3^- + H_2O = H_2SO_3 + OH^-$; $\dfrac{[H_2SO_3][OH^-]}{[HSO_3^-]}$

45.

SOLUTION	pH	POH	$[H^+]$	$[OH^-]$
1	6.52	7.48	3.0×10^{-7}	3.3×10^{-8}
2	9.36	4.64	4.4×10^{-10}	2.3×10^{-5}
3	1.28	12.72	5.2×10^{-2}	1.9×10^{-13}
4	9.75	4.25	1.8×10^{-10}	5.6×10^{-5}
5	3.11	10.89	7.8×10^{-4}	1.3×10^{-11}
6	8.19	5.81	6.5×10^{-9}	1.5×10^{-6}
7	3.70	10.30	2.0×10^{-4}	5.0×10^{-11}
8	1.28	12.72	5.2×10^{-2}	1.9×10^{-13}

46.

WEAK ACID	CONJUGATE BASE	K_a FOR ACID	pK_a FOR ACID	K_b FOR BASE	pK_b FOR CONJUGATE BASE
$HC_2H_3O_2$ (acetic acid)	$C_2H_3O_2^-$ (acetate)	1.74×10^{-5}	4.759	5.75×10^{-10}	9.241
NH_4^+ (ammonium)	NH_3 (ammonia)	5.62×10^{-10}	9.250	1.78×10^{-5}	4.750
$HC_6H_5O_2$ (benzoic acid)	$C_6H_5O_2^-$ (benzoate)	6.25×10^{-5}	4.204	1.60×10^{-10}	9.796
HF (hydrofluoric acid)	F^- (fluoride)	6.76×10^{-4}	3.170	1.48×10^{-11}	10.830
HCN (hydrocyanic acid)	(cyanide)	6.17×10^{-10}	9.210	1.62×10^{-5}	4.790
lactic acid	$C_3H_5O_2^-$ (lactate)	1.38×10^{-4}	3.86	7.25×10^{-11}	10.140

47. a. 1.82
    b. 2.49
    c. 0.79
    d. 2.32

48. a. 12.18
    b. 11.51
    c. 13.21

49.

SOLUTION	pH
1	2.68
2	3.18
3	3.68
4	5.04
5	5.54
6	6.04
7	2.60
8	3.10
9	3.60
10	2.65

11	2.58
12	3.06

50.

SOLUTION	WEAK BASE
1	9.08
2	8.58
3	8.08
4	11.21
5	10.71
6	10.21
7	8.60
8	8.10
9	7.60
10	8.21
11	8.28
12	7.80

51. a. neutral
    b. neutral
    c. acidic
    d. basic
    e. basic
    f. basic
    g. basic
    h. acidic

52. a. 5.46
    b. 4.67
    c. 4.76
    d. 4.76

53. a. 7.20
    b. 7.50
    c. 6.90
    d. 7.20

54. a. 1:1
    b. 1:2
    c. 1:0.63

CHAPTER 10: ELECTRICITY

1. Charge is a quantized property of nonneutral matter. Charge can be positive or negative. Like charges repel and unlike charges attract. Electrical potential is $V = \frac{U}{q_0} = k\frac{q}{r} =$ electric potential energy per unit charge. Current is $I = \frac{\Delta Q}{\Delta t}$. Resistance is $R = \frac{V}{I}$. Power $= P = \frac{\Delta U}{\Delta t} = IV$.

2. $F = k\frac{|q_1||q_2|}{r^2}$. The force, F is directly proportional to the magnitude of each charge and inversely proportional to the square of the distance between them.

3. An electrical insulator has a high resistance to electricity because its molecular structure does not allow the movement of electrons along its surface, whereas conductors allow charges to easily move.

4. $V = IR$, or the voltage driving electrons through a circuit is directly proportional to the volume of electron flow and the resistance of the medium to that flow.

5. 3.0×10^5 V

6. 2.4 A

7. 15 Ω

8. 0.15 W

9. 8.3 A

10. $15\bar{0}$ W

11. 17.5 Ω

12. 1.43 Ω

13. Semiconductors are materials with electrical conducting properties somewhere between those of insulators and conductors. p-type semiconductors are semiconductors doped with an element having fewer valence electrons than silicon or germanium. n-type semiconductors are semiconductors doped with an element having more valence electrons than silicon or germanium.

14. A diode is a circuit element having adjacent n-type and p-type regions. Diodes have large conductances in one direction and small conductances in the reverse direction. Diodes are used as light sources in many electronic circuits and are known as light emitting diodes or LEDs. A triode is made from two closely situated p–n junctions. Triodes can be used to amplify signals in electrical circuits.

15. 3.1×10^4 Ω

16. 2.0 mA

17. The return electrode is used to help establish a safe return path for the current. The large area will ensure that the current disperses, maintaining a low current density, thus protecting tissue as the current travels through the patient's body.

CHAPTER 12: BIOCHEMISTRY

1. Chiral: ear, glove, bolt, and femur. Nonchiral: soda can and human body (because each has a mirror plane of symmetry).

2. CH_3CH_2-CH=CH-CH_2-CH=CH-CH_2-CH=CH-$(CH_2)_7$COOH. Alpha linolenic is an omega-3 acid because there is a double bond on the third carbon from the end (away from the carboxyl group).

3.

4. The nonpolar side chains (valine and phenylalanine, for instance) tend to wind up inside the globular protein, away from the polar water molecules.

5. Wool stretches and is primarily comprised of α-helices. Silk, which does not stretch, is primarily comprised of β-sheets.

6. 1 g $C_6H_{12}O_6$ = 5.6 × 10^{-3} mol glucose (32 ATP/glucose) = 0.18 mol ATP. 1 g stearic = 3.5 × 10^{-3} mol glucose (120 ATP/glucose) = 0.42 mol ATP. Stearic acid (a fat) has 0.42/0.18 > 2 times the energy production as glucose.

7. TAG, UAG, AUC

8. a.

9. c.

10. b.

11. b.

12. b.

13. a.

14. c.

15. c.

16. a.

17. c.

18. a.

19. b.

20. b.

21. a.

22. a.

23. a.

24. b.

25. c.

26. b.

CHAPTER 13: RADIATION AND RADIOACTIVITY

1. $2\bar{0}0$ Hz

2. 1.5×10^6 m; 1.3×10^{-31} J

3. 1.5 m

4. 18,600 s versus 0.0212 s

5. 4.5×10^{14} J

6.

TERM	A	Z	N
Nuclide	Unique	Unique	Unique
Isotope	Different	Same	Different
Isotone	Different	Different	Same
Isobar	Same	Different	Different

7. a. $^{241}_{94}\text{Pu} \rightarrow {}^{241}_{95}\text{Am} + \beta^- + \bar{\nu} + \text{energy}$

   b. $^{248}_{96}\text{Cm} \rightarrow {}^{244}_{94}\text{Pu} + \alpha + \text{energy}$

   c. $^{22}_{11}\text{Na} \rightarrow {}^{22}_{10}\text{Ne} + \beta^+ + \nu + \text{energy}$

   d. $^{85}_{38}\text{Sr} + e^- \rightarrow {}^{85}_{37}\text{Rb} + \nu + \text{energy}$

8. 3.1×10^{17} atoms

9. a. 15 mCi = 5.6×10^8 Bq
   b. 25 kBq = 6.8×10^{-7} Ci
   c. 60 rad = 0.60 Gy

10. 8.7 GBq

11. Cells that rapidly divide are very sensitive to radiation. Since tumors are made of cells that divide rapidly, radiation can be used to kill tumor cells in a relatively selective manner.

12. 1.02 MeV

13. $^{99}Tc^m$ in normal saline forms the pertechnetate ion, which behaves in a similar manner to I^- in the human body, concentrating mainly in the thyroid, salivary glands, gastric mucosa, and choroid plexus. This nuclide's short half-life, low-energy gamma ray, and lack of beta emission result in a relatively low absorbed radiation dose to the patient and decent tissue penetration.

14. Lead collimators ensure that only those photons traveling in the proper direction actually reach the NaI(Tl) detector. Collimators will absorb photons with improper trajectories.

Formulas and Constants

NAME	SYMBOL	NUMERIC VALUE	UNITS
Acceleration due to gravity	g	9.8	$\dfrac{m}{s^2}$
	g	32.2	$\dfrac{ft}{s^2}$
Atomic mass unit	amu	1.66×10^{-27}	kg
Avogadro's number	N_A	6.02×10^{23}	$\dfrac{particles}{mol}$
Boltzmann constant	$k = \dfrac{R}{N_A}$	1.38×10^{-23}	$\dfrac{J}{K}$
Coulomb's Law constant	k	8.99×10^9	$\dfrac{N \cdot m^2}{C^2}$
Ideal gas constant	R	8.314	$\dfrac{J}{mol \cdot k}$
	R	0.08205	$\dfrac{L \cdot atm}{mol \cdot k}$
Magnitude of electron charge	e	1.602×10^{-19}	C
Planck's constant	h	6.626×10^{-34}	J·s
Speed of light in a vacuum	c	3.00×10^8	$\dfrac{m}{s}$
Universal gravitational constant	G	6.67×10^{-11}	$\dfrac{N \cdot m^2}{kg^2}$

Glossary

Absolute zero is the coldest possible temperature. Absolute zero corresponds to 0 K or –273 °C.

Acceleration is the change in velocity with respect to time. Acceleration is a vector, so an object undergoes acceleration when it speeds up, slows down, or changes direction.

Accuracy is the agreement between an experimental value and the accepted value.

Acetal/Ketal functional groups are characterized by two ether functional groups on the same carbon atom. Acetals are derived from aldehydes, and ketals are derived from ketones.

ALARA principle (As Low As Reasonably Achievable) states that workers should limit exposure to radioactivity as far below the dose limits as is practical.

Alcohols are organic functional groups characterized by a hydroxyl (-OH) group. A generic alcohol may be represented as ROH.

Aldehydes are organic functional groups characterized by a carbonyl group (C=O) that has a hydrogen bonded to the carbonyl carbon. A generic aldehyde may be represented as RCHO.

Alkanes are organic functional groups characterized by all carbon–carbon single bonds.

Alkenes are organic functional groups characterized by carbon–carbon double bonds.

Alkyl groups are "stick on" groups derived from alkanes. An alkyl group is an alkane minus one hydrogen atom. A generic alkyl group is abbreviated by R.

Alkynes are organic functional groups characterized by carbon–carbon triple bonds.

Alpha decay is a radioactive decay mode in which an alpha particle is ejected from a radioactive nucleus. An alpha particle consists of two protons and two neutrons. After alpha decay, the atomic number decreases by two and the mass number decreases by four.

Amides are organic functional groups that are characterized by a nitrogen bonded to a carbonyl group. A generic amide may be represented as $RCONR_2$.

Amines are organic functional groups characterized by a nitrogen bonded to one, two, or three alkyl groups. A generic amine may be represented as R_3N.

Amino acids are the building blocks of proteins. Amino acids are relatively simple organic molecules that contain an amine functional group and a carboxylic acid functional group.

Ampere (A) is the measure of electric current, or the "volume" of electrons flowing through a circuit. An ampere is equal to one coulomb of electrons (6.24×10^{18} electrons) passing a point each second.

Amphiprotic species. An amphiprotic species has the capacity to either donate a hydrogen ion (react as an acid) or accept a hydrogen ion (react as a base).

Anticodons are three base sequences carried by transfer RNA. Anticodons match to codons in messenger RNA.

Archimedes' Principle states an object floating in a fluid will displace a volume of fluid equal to the weight of the object. If the object sinks, it displaces a volume of fluid equal to the volume of the object.

Aromatics (arenes) are functional groups that are characterized by a benzene ring.

Atmospheric pressure is the result of gravity pulling down on the gases in the atmosphere. At sea level, the weight of the air over each square inch is about 14.7 pounds, so the air pressure is 14.7 psi.

Atomic mass number (A) is equal to the sum of the number of protons plus the number of neutrons in the nucleus of an atom.

Atomic number (z) is equal to the number of protons in the nucleus of an atom.

Average atomic weight, or more precisely, average atomic mass, is the average mass of the atoms of an element that is comprised of the natural abundance of the isotopes of that element.

Avogadro's Law states that the volume of a gas is directly proportional to the number of moles of the gas, as long as the pressure and temperature remain constant. Symbolically, Avogadro's Law is $\dfrac{V_1}{n_1} = \dfrac{V_2}{n_2}$.

Barometers are devices that measure atmospheric pressure. A mercury barometer consists of a tube of mercury inverted into a container of mercury. The height of the mercury column in the barometer is proportional to the atmospheric pressure. An aneroid barometer uses an evacuated container. A change in atmospheric pressure causes a proportional distortion of the container.

Base pairing among nucleic acids results from the molecular geometry of the genetic bases. The bases A and T (adenine and thymine) form two hydrogen bonds, while the bases C and G (cytosine and guanine) match to form three hydrogen bonds. In RNA, T is replaced with U (uracil).

Becquerel (Bq) is a unit of radioactivity that is equal to one nuclear decay per second.

Beer's Law states the amount of light captured by a sample (the absorbance, A) is proportional to the concentration of the absorbing species (c) and the length of the sample through which the light travels (b). The proportionality constant is called the absorptivity or the extinction coefficient (a). Symbolically, Beer's Law is given as $A = abc$.

Bent geometry is observed around an atom that has two ligands and one or two lone pairs of electrons, AX_2E or AX_2E_2.

Bernoulli's equation describes how the pressure a fluid exerts changes with the velocity of the fluid. Symbolically, Bernoulli's equation can be stated as: $P_1 + \frac{1}{2}\rho v_1^2 = P_2 + \frac{1}{2}\rho v_2^2$ (assuming there is no change in gravitational force or density).

Beta (minus) decay is a type of radioactive decay in which a neutron in a radioactive nucleus decays into a proton and an electron. The electron is ejected from the nucleus, and the ejected electron is called a beta particle (more precisely, a beta-minus) particle. After beta-minus decay, the atomic number increases by one and the mass number remains unchanged.

Biomolecules are the molecules used to form living systems, and include carbohydrates, lipids, amino acids, and nucleic acids.

Boiling point is the temperature at which the vapor pressure of a liquid equals the applied pressure. The normal boiling point is the temperature at which the vapor pressure of a liquid is equal to 1 atm. Alternatively, the boiling point is the temperature at which a liquid converts reversibly into a gas.

Boltzmann distribution gives the number of atoms or molecules at a given kinetic energy in a sample at a given temperature. The average kinetic energy in the Boltzmann distribution is proportional to the absolute temperature of the sample.

Bond dissociation energy is the energy necessary to homolytically break a chemical bond.

Bond-line structures represent organic compounds as molecular graphs. The lines represent bonds between carbon atoms. There is a carbon atom located at each point that the molecular graph changes. It is assumed that each carbon has sufficient hydrogens to give a total of four bonds, but heteroatoms (atoms other than C or H) must be drawn.

Bourdon gauges measure gas pressure relative to atmospheric pressure. The heart of a Bourdon gauge is a hollow metal coil that expands or contracts in response to changes in gas pressure.

Boyle's Law states the volume of a gas (V) is inversely proportional to the pressure of the gas (P), presuming that the temperature and amount of gas remain constant. Symbolically, Boyle's Law is $P_1 V_1 = P_2 V_2$

Brønsted acids are hydrogen ion donors. When a Brønsted acid donates its proton to some base, the acid is transformed into its conjugate base.

Brønsted bases are hydrogen ion acceptors. When a Brønsted base accepts a proton from some acid, the base is transformed into its conjugate acid.

Brownian motion describes the small, erratic movements of small particles suspended in a fluid. Albert Einstein proposed Brownian motion as the result of collisions of the fluid molecules with the particle. Einstein's theory led to the common acceptance of atoms as a reality, rather than a theoretical model.

Buffer capacity describes the amount of strong acid or base that a buffer solution is able to absorb. The capacity of a buffer is at a maximum when the concentration of the weak acid in the buffer is equal to the concentration of the weak base. Buffers have a reasonable capacity at a pH within one unit of the pK_a of the acidic component of the buffer.

Buffers, specifically pH buffers, are solutions that resist changes in pH. A buffer contains a weak acid and its conjugate base or a weak base and its conjugate acid. The pH of a buffer depends most strongly on the pK_a of the weak acid or the pK_b of the weak base.

Buoyant force is the force exerted on an object floating or immersed in a fluid. The buoyant force is equal to the volume of the object immersed in the fluid (V), times the acceleration due to gravity (g), times the density of the fluid (ρ). Symbolically, the buoyant force is $F_b = \rho\, g\, V$.

Calories and calories (Cal, cal) are units of energy. The calorie (with a lowercase "c") is the amount of energy needed to raise the temperature of one gram of water by one degree Celsius. The Calorie (with an uppercase "C") is a kilocalorie or 1000 calories. The energy content of food is given in Calories.

Carbohydrates are biomolecules characterized as polyalcohol aldehydes or ketones. Glucose (formerly known as dextrose) is the most common carbohydrate.

Carboxylic acids are organic functional groups characterized by a hydroxyl group bonded to a carbonyl group. A generic carboxylic acid may be represented as $RCOOH$ or RCO_2H.

Celsius temperature scale was originally defined by the freezing point and normal boiling point of water (0°C and 100°C), with the temperature difference between these two points equally divided into 100 degrees. Now the Celsius temperature scale is defined by the triple point of water (273.16°C) and absolute zero (–273.15°C).

CGS (centimeter/gram/second) system uses centimeters to measure length, grams to measure mass, and seconds to measure time. The CGS system also has derived units to measure force and energy (dynes and ergs). The CGS system is rarely used today, in favor of the MKS (meter-kilogram-second) system.

Charge (electrical charge, q) is a physical property of matter. Electrical charges are described as either "positive" or "negative." An object carrying a charge will experience an attractive force when placed near an object with the opposite charge and a repulsive force when placed near an object with the same charge. The SI unit of charge is the Coulomb. The magnitude of the charge on an electron is equal to the magnitude of the charge on a proton, which is numerically equal to 1.602×10^{-19} C. A Faraday is the charge on a mole of electrons and is equal to 96,500 C.

Charles's Law states the volume of a gas (V) is directly proportional to the absolute temperature of the gas (T, measured in Kelvins), as long as the pressure and amount of gas remain constant. Symbolically, Charles's Law is $\dfrac{V_1}{T_1} = \dfrac{V_2}{T_2}$.

Chemical changes convert a substance into a different substance.

Chemical equilibrium. A system is in a state of chemical equilibrium when it contains both reactants and products that are interconverting, and the concentrations of the reactants and products remain constant.

Chemical properties. Observation of a chemical property results in changing the chemical identity of the substance.

Chemistry is the study of matter and the changes it undergoes. Matter is anything that has mass and occupies space.

Chirality is a topological property of an object wherein the object cannot be superimposed on its mirror image. Chirality can be thought of as handedness, so chiral objects can exist as either a right-handed form or a left-handed form. Most biomolecules are chiral.

Cis/trans isomerism is exhibited by vicinally di-substituted alkenes (i.e., each carbon in the alkene double bond has one hydrogen ligand and one nonhydrogen ligand). If both hydrogen ligands are on the same sides of the double bond, it is the cis isomer. If the hydrogen ligands are on opposite sides of the double bond, it is the trans isomer.

Clausius–Clapeyron equation describes how the vapor pressure (P) of a liquid depends on the heat of vaporization of the liquid (ΔH_{vap}), the ideal gas constant (R), and the Kelvin temperature (T). Symbolically, the Clausius–Clapeyron equation is $\ln(P) = -\dfrac{\Delta H_{vap}}{R}\left(\dfrac{1}{T}\right) + c$, where C is a constant unique to each liquid. Collecting the constants gives a simplified equation: $\log(P) = A + \left(\dfrac{B}{T}\right)$, where A and B are constants unique to each liquid.

Codon is a three-nucleic base sequence in a strand of messenger RNA that codes for an amino acid.

Colligative properties of solutions depend only on the concentration of solute particles, not on the identity of the solute. As solute concentrations increase, the vapor pressure decreases, the boiling point increases, the freezing point goes down, and there is an increase in osmotic pressure.

Colloids consist of one substance microscopically dispersed throughout a second substance. Colloids are not true solutions because the dispersed phase is not distributed at the ionic or molecular level or because the dispersed particles are much larger than a typical atom or molecule. Colloids exhibit the Tyndall effect and disperse light traveling through the colloid.

Combined gas law expresses the interdependence of the four state variables of a gas: pressure (P), volume (V), temperature (T), and moles of gas (n). Symbolically, the combined gas law is $\dfrac{P_1 V_1}{n_1 T_1} = \dfrac{P_2 V_2}{n_2 T_2}$.

Compounds are pure chemical substances that are comprised of two or more elements in a constant ratio by mass.

Condensation is the physical change wherein a gas changes to a liquid. Vaporization is the opposite of condensation.

Condensation reactions are chemical reactions that join two reactants into a larger molecule, with the concurrent expulsion of a small molecule, such as water. Formation of esters and amides are condensation reactions.

Condensed states of matter include the liquid state and the solid state.

Conductance (G) is a physical property that describes how well an object conducts electricity. Conductance is the reciprocal of resistance and is measured in siemens (S).

Conduction is a mode of transmitting energy ("heat"). Conduction occurs as higher energy molecules in an object at a higher temperature collide and transfer some of their kinetic energy to the lower energy molecules in a lower temperature object.

Conductors and insulators are relative terms that describe the electrical conductance properties of materials. Conductors have a low resistance (and high conductivity). Insulators have a high resistance (and a low conductivity). Conductors and insulators may also describe thermal conductivity. Insulators have a high specific heat and require a large amount of energy to undergo a given temperature change. Thermal conductors have a low specific heat and require a small amount of energy to change temperature.

Conjugate acid–base pairs differ by one hydrogen ion. The acid form has the proton, and the charge is one greater than the conjugate base form.

Continuity equation describes the flow of a fluid moving through a pipe. The flow rate is directly proportional to both the cross-sectional area (A) of the pipe and to the speed of the fluid (v). Symbolically, the continuity equation is Flow rate = $A \cdot v$. If the fluid moves from one size vessel to another, the continuity equation can be recast as $A_1 v_1 = A_2 v_2$.

Convection is a mode of transmitting energy ("heat"). Convection occurs as a higher temperature material (most commonly a gas or a liquid) moves into another area that is at a lower temperature.

Coulomb (C) is an amount of electrical charge. One coulomb is the amount of charge that passes a point each second when the current is one amp. One coulomb is also equal to the charge contained in 6.24×10^{18} electrons.

Coulomb's Law describes the force exerted between two charged particles. The magnitude of the force (F) is proportional to the absolute value of the charges on the particles (q_1 and q_2) and inversely proportional to the square of the distance (r) between the two particles. The proportionality constant (k) has a numerical value of 8.99×10^9 N·m²/C². Symbolically, Coulomb's Law is $F = k\dfrac{|q_1||q_2|}{r^2}$.

Covalent bonds result from sharing one or more pairs of electrons between two atoms.

Cryoscopic constant (freezing point constant, k_{fp}) is the proportionality constant that relates the freezing point depression of a liquid (ΔT_{fp}) to the molal concentration of all solutes (m_{total}). The complete equation is given by $\Delta T_{fp} = k_{fp} m_{total}$.

Curie (Ci) is an amount of radioactivity that equals 3.7×10^{10} nuclear disintegrations per second. One Ci is also equivalent to 3.7×10^{10} Bq.

Dalton's Atomic Theory consists of three tenets. First, all matter is comprised of tiny, particles called atoms, and all atoms of a given element are identical and unique to that element. Compounds are formed by bonding atoms together in a fixed ratio by mass. Finally, chemical reactions neither create nor destroy atoms.

Dalton's Law of Partial Pressures states the partial pressure of a gas (P_i) is equal to the total pressure (P_{total}) multiplied by the mole fraction of the gas (χ_i). Alternatively, the total pressure in a mixture of gases is equal to the sum of the partial pressures of each component gas. Symbolically, these equations are cast as: $P_i = \chi_i(P_{total})$ and $P_{total} = \sum P_i$.

Denaturing proteins occurs when changing conditions (e.g., changing pH, changing temperature, or adding heavy metal ions) interrupt the noncovalent forces (e.g., hydrogen bonding and London forces) that maintain the higher levels of protein structure (i.e., the secondary and tertiary structure). Denatured proteins are biologically inactive and usually precipitate out of solution.

Density (d or ρ) is an intensive physical property that equals mass divided by volume.

Deposition is a change of state wherein a gas is converted into a solid. Deposition is the opposite of sublimation.

Desiccants are chemical species that absorb water.

Detergents are synthetic surfactants that consist of a hydrophilic polar "head" and a hydrophobic non-polar hydrocarbon "tail." The polar head may be positively charged (cationic detergent), negatively charged (anionic detergent), or electrically neutral (non-ionic detergent).

Dextrorotatory isomers are chiral molecules that rotate plane-polarized light in a clockwise direction.

Dextrose is the old name for glucose. Solutions of glucose rotate plane-polarized light in a clockwise direction, and this is the origin of the term *dextrorotatory*.

Diodes (solid state) consist of a p-type semiconductor physically bonded to an n-type semiconductor. In diodes that are forward biased, current will readily flow, whereas under reverse bias, current does not flow.

Dipole–dipole attraction is an attractive force that results from the attraction between opposite partial charges in polar molecules.

Dynamic equilibrium describes a system in which the components of a system are allowed to change states, but the overall composition of the system is static. In a dynamic chemical equilibrium, reactants are continually becoming products, while products are continually reverting into reactants. However, the concentration of all reactants and products remains constant when the system has attained equilibrium.

Ebullioscopic constant (boiling point constant, k_{bp}) is the proportionality constant that relates the freezing point depression of a liquid (ΔT_{bp}) to the molal concentration of all solutes (m_{total}). The complete equation is given by $\Delta T_{bp} = k_{bp} m_{bp}$.

Electric current (I) is the flow of electrons through a conductive material.

Electric field is the region of space surrounding a charged particle. In an electric field, a test charge would feel a force that results from the Coulombic attraction (or repulsion) between the charged particle and the test charge.

Electric potential ("voltage," V) is equal to the electric potential energy (U) of a system of two point charges (q and q_0) divided by q_0. Symbolically, this relationship is $V = U/q_0$. Electrical potential is measured in volts.

Electric potential energy (U) is the potential energy that results from placing two charged particles at a distance r from each other. Symbolically, $U = k\dfrac{qq_0}{r}$, where k is Coulomb's constant, 8.99×10^9 N·m^2/C^2.

Electrical energy (E) is equal to the electrical power (P) multiplied by time (t), $E = P \cdot t$.

Electrical power (P) is equal to the square of the current (I) multiplied by the resistance (R). Symbolically, $P = I^2 R$. Electrical power is also equal to the product of current (I) and voltage (V); $P = IV$.

Electromagnetic force is one of the four fundamental forces in the standard model of physics. The electromagnetic force is the result of the interactions between electrically charged particles.

Electromagnetic radiation (aka "light") is produced when charged particles are accelerated. Electromagnetic radiation has properties of waves (e.g., wavelength and frequency) and travels at the speed of light (c; 3.0×10^8 m/s in a vacuum). These waves propagate through space as an oscillating electric field that is perpendicular to an oscillating magnetic field. Both fields are mutually perpendicular to the direction of propagation. Electromagnetic radiation can also be described as particles called photons. The energy of a photon (E) is proportional to its frequency (f), $E = hf$, where h is Planck's constant, 6.626×10^{-34} J·s.

Electron capture decay (EC) is a type of radioactive decay wherein an electron from the first quantum shell is captured by the atomic nucleus. The captured electron combines with a proton to form a neutron. After EC, the atomic number decreases by one, and the mass number remains unchanged.

Electronegativity is a property that describes an atom's affinity for electrons. Fluorine is the most electronegative atom, and atoms on the periodic table that are closer to fluorine are more electronegative.

Electrophoresis is a technique used to separate polypeptides and nucleic acids based on their molar mass and net charge. A buffered solution maintains a fixed net charge on the sample which is placed on an inert support, such as agarose or polyacrylamide. The support is placed into an electric field, and the fragments migrate in response to the net force they experience. Fragments with larger charges and/or smaller molar masses migrate faster.

Elements are pure substances that consist of a single type of atom. There are currently 118 reported elements. Elements are normally atomic or are found in nature as atoms; however, some elements are found in nature as molecules. The common molecular elements include H_2, N_2, O_2, F_2, Cl_2, Br_2, and I_2.

Emission tomography includes a collection of medical imaging techniques that are based on injection of radioactive materials into a patient. Techniques include *PLANAR* (production of a two-dimensional image), *SPECT* (single-photon emission computerized tomography), and *PET* (positron emission tomography).

Emulsions are colloids that consist of droplets of one liquid dispersed throughout a second liquid.

Enantiomers are mirror-image isomers of each other. Enantiomers are chiral and optically active. Enantiomers rotate plane-polarized light in equal and opposite directions. Generally, living organisms use only one enantiomer of a chiral material. The other enantiomer may be inert, but is sometimes deleterious to the organism.

Energy is the ability to do work. Energy cannot be created or destroyed, but it can be transformed into other types of energy. Common types of energy include kinetic energy and potential energy. Energy is measured in joules (J) or calories (cal).

Enzymes are proteins that catalyze chemical reactions.

Equilibrium constants are the ratios of the concentrations of products to starting materials in a chemical equilibrium. A large equilibrium constant means the system favors products, while a small equilibrium constant means the system favors reactants.

Esters are organic functional groups that are characterized by an alkoxy group (OR) bonded to a carbonyl group. A generic ester may be represented as RCO_2R' or RCOOR'. Esters are condensation products of carboxylic acids and alcohols. Esters can be hydrolyzed by aqueous acid into a carboxylic acid and an alcohol. Esters can be saponified by aqueous base into an alcohol and the conjugate base of a carboxylic acid.

Ethers are organic functional groups that are characterized by an oxygen bonded to two alkyl groups. Ethers may be represented as ROR'. Many inhaled anesthetic reagents contain an ether functional group.

Evaporation is a physical change wherein a liquid converts into a gas without reaching equilibrium.

Exact numbers are numbers that are either defined (e.g., 1 in is exactly defined as 2.54 cm) or are easily counted (e.g., there are five digits on a hand). Exact numbers have an infinite number of significant digits.

Fatty acids are carboxylic acids in which the R groups are long chains of carbon atoms. The R groups in common fatty acids contain 11 to 17 carbon atoms. If the R group contains one or more carbon–carbon double bonds, it is an unsaturated fatty acid. If there are only carbon–carbon single bonds, it is a saturated fatty acid.

First Law of Thermodynamics is a statement of the conservation of energy. In one formulation, the first law states that the change in the internal energy of a system (ΔU) is equal to the sum of the work (W) done on the system plus any heat (Q) gained by the system, or $\Delta U = Q + W$. In this definition, $W < 0$ when a gaseous system expands.

Flow rate is the volume of fluid that flows past a point per unit time.

Fluids are substances that have the ability to flow. Fluids may be either liquids or gases.

Forces are vector quantities that describe a push or a pull experienced by an object. A net force causes acceleration of the object. Force is measured in newtons (N).

Freezing is a physical change that describes the transition of a substance converting from a liquid into a solid. Freezing is the opposite of melting.

Freezing point (melting point) is the temperature at which a liquid converts reversibly into a solid.

Functional groups are specific sets of atoms bonded together in a specific way. For example, a carbon doubly bonded to another carbon is the functional group of an alkene.

Gamma ray emission is a mode of radioactive decay involving a quantum energy transition of one of the nucleons in a radioactive nucleus. The energy is given off in the form of a high-frequency photon called a gamma ray. After gamma decay, both the mass number and the atomic number remain the same.

Gaseous state of matter has neither a definite shape nor a definite volume. The energy of molecules or atoms in the gas phase is sufficiently great so as to allow the molecules or atoms to act essentially as independent particles.

Gauge pressure is the pressure in a gas cylinder that is registered by some pressure measuring device, such as a Bourdon gauge. The gauge pressure is equal to the total absolute pressure of the gas in the cylinder minus the atmospheric pressure.

Genes are sections of DNA that contain sufficient information to synthesize a protein.

Glycolysis is an energy production pathway that converts a six-carbon sugar (glucose) into a pair of three-carbon sugar acids (anaerobic metabolism = lactic acid; aerobic metabolism = pyruvic acid).

Glycoside bonds are the ketal/acetal functional groups that join disaccharides or polysaccharides. Glycoside bonds do not require a great deal of energy to hydrolyze.

Graham's Law of Effusion states that the rate of effusion (diffusion of a gas through a pinhole into a vacuum) is inversely proportional to the square root of the molar mass of the gas (M), or $\dfrac{\text{Rate}_1}{\text{Rate}_2} = \sqrt{\dfrac{M_2}{M_1}}$.

Gravity is one of the four fundamental forces in the standard model of physics. Gravity is the universal attraction of something with mass for everything else with mass. The acceleration of objects near the Earth's surface is commonly called gravitational acceleration, or simply gravity, and is symbolized by $g = 9.8$ m/s^2.

Half-life ($t_{1/2}$) is the amount of time required for 50% of a radioactive sample to decay.

Heat (Q) is an energy transfer process that occurs spontaneously from a warm object to a cold object. Heat is often (incorrectly) described as energy. In this faulty but useful construct, "heat energy" can be thought of as the sum of the kinetic energies of the particles in the sample.

Heat capacity (C) is an extensive physical property of a sample that relates the amount of heat (Q) applied to the sample and the temperature change it undergoes, $Q = C\Delta T$. Heat capacities are measured in J/degree or cal/degree. The specific heat of a substance is equal to the heat capacity divided by the mass of the substance.

Heat of fusion (molar heat of fusion) is the amount of energy required to melt one mole of a pure solid substance at its normal melting point.

Heat of vaporization (molar heat of vaporization) is the amount of energy required to vaporize one mole of a pure liquid substance at its normal boiling point.

Henderson–Hasselbalch equation is used to calculate the pH of a buffer and states the pH of a buffer is equal to the pK_a of the weak acid plus the log of the ratio of the concentration of the conjugate base, [A⁻], divided by the concentration of the acid, [HA], or

$$pH = pK_a + \log\left(\frac{[\text{A}^-]}{[\text{HA}]}\right).$$

Henry's Law states the solubility of a gas in a liquid (S) is equal to the Henry's Law constant (k_H) multiplied by the partial pressure of the gas (P_{gas}). The value of k_H depends on both the gas and the liquid solvent. Symbolically, Henry's Law is $S = k_H P_{gas}$. Alternatively, Henry's Law is used to calculate the solubility when the pressure is changed, $\dfrac{S_1}{S_2} = \dfrac{P_1}{P_2}$.

Homogeneous mixtures consist of two or more pure substances that do not exhibit phase boundaries between the constituents.

Hydrogen bonding results from the attraction between a hydrogen atom bonded to an oxygen, nitrogen, or fluorine and another electronegative atom. Hydrogen bonding is normally the strongest of the noncovalent intermolecular attractive forces.

Hydrometers are devices used to measure the specific gravity of a liquid. Hydrometers sink into a liquid until the buoyant force exerted by the liquid equals the weight of the hydrometer.

Ideal gas law relates the four state variables of a gas, pressure (P), volume (V), moles (n), and Kelvin temperature (T). Symbolically, the ideal gas law is $PV = nRT$, where R is the universal gas constant.

Ideal gases obey the ideal gas law under all conditions. Ideal gases do not exist. Real gases approximate ideal gas behavior at lower pressures and higher temperatures.

Internal energy (U) is equal to the sum of the kinetic and potential energies of the particles in the sample.

Ion-dipole attraction is the attractive force between an ion and molecules having permanent dipole moments.

Ionic bond is the attraction of positively charged cations for negatively charged anions.

Ionic compounds are pure substances containing cations and anions. Ionic compounds are almost always solids under normal conditions.

Ionizing radiation can remove an electron from matter. Ionizing radiation includes radioactive particles (alpha, beta, and gamma) as well as high-frequency photons (X-rays, gamma rays).

Ions are atoms or groups of atoms covalently bonded together that have a charge. Cations have a positive charge and anions have a negative charge.

Isobars have constant pressure.

Isoelectric point (pI) is the pH at which an amino acid (or protein, or nucleic acid) has no net charge. If the pH is less than pI, the amino acid has a net positive charge. If the pH is greater than pI, the amino acid has a net negative charge.

Isomers are different compounds that have the same molecular formula.

Isotones are nuclides that have the same number of neutrons.

Isotopes are nuclides that have the same atomic number (number of protons).

Joule (J) is the amount of energy expended by applying a force of one Newton over a distance of one meter.

Kelvin scale, or the absolute temperature scale, begins at absolute zero. A change in temperature of one kelvin is equal to a change in temperature of 1°C.

Ketones are organic functional groups characterized by two alkyl groups bonded to a carbonyl group. A generic ketone may be represented as $R_2C=O$.

Kinetic energy (*KE*) is the energy a mass has by virtue of being in motion. *KE* is equal to one-half times the mass (*m*) of the object times the square of its speed (*v*), $KE = \frac{1}{2}mv^2$.

Kinetic Molecular Theory of Gases describes gases as independent particles in constant, random motion that experience no forces for each other or the walls of the container. Although the molecules have a range of velocities defined by a Boltzmann distribution, the average kinetic energy depends only on the Kelvin temperature of the sample.

Laminar flow can be modeled as a set of layers all moving in the same direction, with no turbulence.

LaPlace's Law states that the surface tension (γ) on a fluid depends on the pressure of the fluid (ΔP) and the radius of the fluid's container or shape (*R*). If the fluid is in a cylindrical blood vessel, LaPlace's Law takes the form $\gamma = \Delta PR$. In a spherical container, $\gamma = \frac{1}{2}\Delta PR$.

Law of Conservation of Electric Charge states the total electrical charge in the universe is constant.

Law of Conservation of Energy states that energy cannot be created or destroyed, but only converted into other types of energy.

Law of Conservation of Mass states that mass–energy content of the universe is constant. On a macroscopic level, mass is neither created nor destroyed during the course of a chemical reaction. Although Einstein's Special Theory of Relativity shows this law to be untrue, the change in mass is so small as to be imperceptible.

Law of Definite Proportion states the elements in a compound are present in a fixed ratio by mass.

Le Châtelier's Principle states "equilibrium is good." If a system in a state of equilibrium is disturbed, the system will adjust so as to reestablish the equilibrium state.

Levorotatory isomers are enantiomers that rotate plane-polarized light in a counterclockwise direction.

Lewis structures represent covalently bonded species. Shared pairs of electrons are represented by lines, and lone pairs of electrons are represented by dots.

Linear geometry is observed around atoms that have two ligands and no lone pairs of electrons, AX_2.

Lipids are biomolecules that are more soluble in organic solvents than in water.

Liquid state of matter is characterized by a definite volume but no definite shape. In the liquid state, molecules are held in contact by intermolecular forces, but the kinetic energy of the molecules is sufficiently high so as to allow the molecules to slide around past each other.

London forces result from the temporary dipole moments that arise from an uneven distribution of electrons.

Manometers measure the difference in pressure between the test system and a reference system. Commonly, a manometer consists of a U-shaped tube filled with a liquid. One side is connected to the test system, and the other side is open to the atmosphere. The difference in heights of the liquid in the two tubes is proportional to the pressure difference.

Mass is a measure of an object's resistance to acceleration.

Matter is anything that has mass and occupies space.

Melting is the change of state when a solid converts into a liquid. Melting is the opposite of freezing.

Melting point is the temperature at which a solid reversibly converts into a liquid. The freezing point and melting point of a substance represent the same temperature.

Micelles are spherical aggregations of surfactant molecules.

MKS (meter/kilogram/second) system comprises a standard set of units that are used in physics. All units are derived from length measured in meters, mass measured in kilograms, and time measured in seconds.

Molality (m) is a concentration unit defined as moles of solute divided by kilograms of solvent,

$$m = \frac{\text{moles solute}}{\text{kg solvent}}.$$

Molar mass (molecular weight) is the mass in grams of one mole of material. The molar mass of an element is equal to the atomic mass in grams.

Molarity (M) is a concentration unit defined as moles of solute divided by liters of solution,

$$M = \frac{\text{moles solute}}{\text{liter solution}}.$$

Mole is an amount of material that contains as many particles as exactly 12 g of carbon-12. One mole of material contains 6.02×10^{23} particles.

Molecular compounds are electrically neutral compounds that are comprised of atoms covalently bonded into a discrete unit.

Molecular polarity is a measure of the net dipole moment of a molecule. If a molecule is polar, it must have polar covalent bonds that are arranged in space so that the bond dipole vectors do not cancel out.

Monosaccharides are carbohydrates that are not broken down into simpler sugars by the action of aqueous acid.

Newton (N) is a unit of force. A one newton force is required to accelerate a one kilogram mass at a rate of 1 m/s^2.

Newton's Law of Universal Gravitation states the gravitational attractive force (F) between two objects that have mass is directly proportional to the masses of the two objects (m_1 and

m_2) and inversely proportional to the square of the distance (r) between them. $F = G\dfrac{m_1 m_2}{r^2}$.

The constant G is the universal gravitational constant and has a value of $6.67 \times 10^{-11} \dfrac{N \cdot m^2}{kg^2}$.

Newton's Laws of Motion state: (1) Objects at rest or in uniform linear motion remain in that state until a net external force acts upon them; (2) The force (F) required to accelerate an object is equal to the product of the mass (m) and the acceleration (a), $F = ma$; and (3) For every action there is an equal and opposite reaction.

Nucleic acids are biomolecules that are used to direct protein synthesis. DNA is located in the cell nucleus and serves as a "recipe" box for proteins. Messenger RNA transcribes DNA and carries the message to the ribosomes where protein synthesis takes place. Transfer RNA matches up to the messenger RNA strand to deliver amino acids in the correct order for synthesis of a given protein.

Nucleons are residents of the atomic nucleus and include protons and neutrons.

Nucleosides are building blocks of nucleic acids that consist of one of the nitrogenous bases bonded to a five-carbon sugar.

Nucleotides are nucleosides in which the C5 alcohol functional group has been phosphorylated.

Nucleus is the center of an atom that contains the protons and neutrons.

Nuclide is a general term used to describe an atomic species defined by its atomic number, number of neutrons, and the energy state of the nucleus.

Octet rule states atoms try to gain access to a total of eight valence electrons so as to be isoelectric with a noble gas. This arrangement maximizes attraction among the electrons for the atomic nuclei while minimizing repulsions among the electrons.

Ohm's Law relates the electrical potential (voltage, V), the resistance (R), and the current (I) in an electric circuit. Symbolically, Ohm's Law is $V = IR$.

Optical activity is the phenomenon of chiral molecules interacting with plane-polarized light.

Osmotic pressure (Π) results from two solutions of unequal tonicity that are separated by a semipermeable membrane. The osmotic pressure depends on the molarity of the solution (M) times the ideal gas constant (R) times the Kelvin temperature, or $\Pi = MRT$.

Oxidation and reduction reactions involve the transfer of electrons between reacting species. Oxidation is the loss of electrons, and reduction is the gain of electrons.

Parts per million (ppm) is used to express the concentration of very dilute solutions and is equal to grams of solute contained in one million grams of solution, ppm = g solute/1 x 10^6 g solution. For very dilute aqueous solutions, parts per million is approximately equal to milligrams of solute divided by liters of solution, ppm $\approx$ mg solute/L solution.

Pascal's Principle states the force applied to a noncompressible fluid in a closed container is transmitted uniformly in all directions throughout the fluid.

Peptide bonds join amino acids into proteins. A peptide bond is an amide-type covalent bond between the amino functional group of one amino acid to the carboxylic acid functional group of another amino acid.

Percent concentrations of a solution are defined as the amount of solute per 100 units of solution. Percent by weight [% (*w/w*)] is defined as grams of solute per 100 g of solution. Percent weight-to-volume [% (*w/v*)] is defined as grams of solute per 100 mL of solution. Percent by volume is unreliable since volumes are not always additive.

Periodic Law states the physical and chemical properties of the elements repeat in a periodic and predictable way.

pH function defines the hydrogen ion concentration, pH = $-\log[H^+]$.

Phase boundaries are sharp demarcations in a heterogeneous system where the physical or chemical properties of the mixture change.

Phase diagrams show the states of a substance as a function of temperature and pressure. Phase diagrams include the triple point (where all three phases can coexist in equilibrium), the critical point, the solid–liquid phase boundary line, the solid–gas phase boundary line, and the liquid–gas phase boundary line.

Phenols are organic functional groups characterized by a hydroxyl group on an aromatic ring.

Photons are quanta or bundles of electromagnetic radiation that exhibit particle-like properties. Albert Einstein proposed that a photon's energy is directly related to the frequency of electromagnetic radiation. Thus electromagnetic radiation exhibits both wave-like and particle-like properties.

Physical changes convert a substance into a different state or geometry, but do not change the chemical identity of the substance.

Physical properties can be observed without changing the chemical identity of the substance.

Physics is the general study of nature in an attempt to ascertain the fundamental laws that govern the universe.

Planck's constant (h) relates the energy of a photon (E) to its frequency (f); $E = hf$. The value of h is 6.626 x 10^{-34} J·s.

Plasma is a state of matter observed at very high temperatures. In the plasma state, electrons are dissociated from the atomic nuclei so that the positively charged nuclei are floating in a sea of negatively charged electrons.

Poiseuille's equation describes the flow rate of a fluid through a tube as a function of a pressure difference, ($P_1 - P_2$), the radius of the tube (r), the length of the tube (L), and the viscosity of the fluid (η). Symbolically, Poiseuille's equation is: flow rate $= \dfrac{(P_1 - P_2)\pi r^4}{8\eta L}$.

Polar covalent bonds result from sharing a pair of electrons between atoms of differing electronegativity. The electrons spend more time on the more electronegative atom, resulting in the development of partial charges.

Polypeptides are polymers of amino acids. Proteins are polypeptides.

Positron decay, or beta-plus decay, results from decay of a proton in a radioactive nucleus into a positively charged electron (aka a positron) and a neutron. Positrons are an example of antimatter. When a positron collides with an electron, the result is mutual annihilation; all of the mass is converted into energy, which is carried away by two gamma ray photons.

Potential energy is stored energy and is a property of the system. Potential energy can be created as a result of a gravitational or electric field, from compression of a spring or a gas, or it can be stored in chemical bonds.

Power (P) is the rate of doing work. Numerically, power equals work (W) divided by time, $P = W/t$. Power is measured in watts.

Precision is the agreement among multiple measurements of the same quantity.

Pressure (P) is equal to force (F) divided by area (A), $P = F/A$. Pressure is measured in pascals (Pa), torr, mmHg, or bar.

Primary structure of proteins is the sequence of amino acids in the polypeptide chain.

Protons are positively charged particles that have a mass of approximately 1 amu. In normal matter, protons are found in the atomic nucleus. In acid–base chemistry, the term *proton* refers to a hydrogen ion (H^+) that is transferred from a Brønsted acid to a Brønsted base.

Pyramidal geometry is observed around atoms having three ligands and one lone pair of electrons, AX_3E.

Quantum mechanics describes the particles in the world as wave functions and restricts the energy of the particles to specific states (rather than a continuum of possibilities). By applying the proper operator to the wave function, a specific property can be observed. Quantum theory does not allow for causality, or being able to predict all properties of the system simultaneously.

Racemic mixtures contain equal amounts of two enantiomers. Racemic mixtures are not optically active. Medicines that are racemic are labeled +/– or D/L.

Radioactive decay is the spontaneous process by which an unstable nucleus goes from a level of higher energy to one of lower energy.

Relative humidity expresses the level of saturation of air with water vapor. Relative humidity is equal to the partial pressure of water in the air divided by the vapor pressure of water at that temperature.

Representative elements are the A-group elements (under the old numbering system for the periodic table). Representative elements have a number of valence electrons equal to the group number.

Resistance (electrical resistance, R) describes the ease (or lack thereof) with which an electrical current can pass through an object. Materials with low resistance conduct electricity more easily than materials with higher resistance. Resistance is measured in ohms, Ω.

Reversible work on a gas occurs when a gas is expanded or compressed at a constant pressure at an infinitely slow rate. When these conditions are met, the work (W) is equal to the product of pressure (P) and the change in volume (ΔV), $W = P \cdot \Delta V$.

Saponification is the hydrolysis of an ester by an aqueous base into an alcohol and the conjugate base of a carboxylic acid. Saponification of the ester bonds in a triglyceride affords glycerin and three moles of the conjugate base of three fatty acids, which is commonly called "soap." Soaps are surfactants.

Second Law of Thermodynamics states the entropy of the universe is always increasing.

Secondary structure of proteins includes organization of a polypeptide chain into an alpha helix or lined up as a beta pleated sheet. Secondary structure is largely maintained by hydrogen bonding.

Self-ionization of water is the phenomenon where a tiny fraction of water molecules in liquid water ionize into a hydrogen ion and a hydroxide ion, $H_2O = H^+ + OH^-$.

Semiconductors are prepared by doping crystals of silicon. An n-type semiconductor is doped with a small amount of a group V element (e.g., arsenic or phosphorus), which has more valence electrons than silicon, and introduces an excess of negative charge into the crystalline lattice. p-type semiconductors are doped with a group III element (e.g., boron or aluminum), which has fewer valence electrons than silicon and introduces a deficiency of electrons into the crystal lattice, which are called holes.

Solids are characterized by both a definite volume and a definite shape. In the solid state, intermolecular forces fix the molecules in place with respect to each other.

Solute is the component of a solution that is dispersed at a molecular or ionic level throughout a medium called the solvent.

Solution is a homogeneous mixture comprised of a solvent (present in the largest amount) and one or more solutes.

Solvation is the process of surrounding a solute with solvent molecules.

Solvent is the component of a solution that is present in the largest quantity, and it is the medium throughout which the solutes are dispersed.

Specific gravity (sg) is a dimensionless quantity that is equal to the density of a substance divided by the density of water, $sg = \dfrac{\text{density of substance}}{\text{density of water}}$. Both densities must be expressed in the same units.

Specific heat (c) is an intensive physical property that describes the amount of heat (Q) required to raise the temperature of a mass of material (m) by 1 degree Celsius. Symbolically, these quantities are related as: $Q = (m)(c)(\Delta T)$. Specific heat is normally expressed in J/g/degree or cal/g/degree. The heat capacity of an object depends on the mass of the object; the specific heat does not. The specific heat is equal to the heat capacity divided by the mass of the sample.

Spectroscopy studies the interaction of light (electromagnetic radiation) and matter and provides both qualitative and quantitative information.

Standard conditions in thermodynamics include a temperature of 298 K, pressure of 1 bar, and concentration of 1 M.

Standard temperature and pressure (STP) for gases is now defined as 273 K and 1 bar of pressure.

State functions depend only on the initial and final states of the system, but do not depend on how the system was changed in order to move from one state to the other. Work and heat are not state functions, but internal energy is a state function. The ideal gas law is an equation of state made up of three state functions (P, V, and T).

Stereocenters are atoms that have four different ligands. Stereocenters were once known as chiral centers, chiral carbons, or asymmetric centers.

Stereochemistry is the study of the three-dimensional nature of molecules.

Sublimation is a phase transition of a solid going directly to a gas.

Surface tension is created by the unbalanced intermolecular attractions among the molecules near the surface of a liquid.

Surfactants (surface active agents) include soaps and detergents. Surfactants interrupt the surface tension at the surface of a liquid. Surfactant molecules have a polar (hydrophilic) head and a non-polar (hydrophobic) tail.

Temperature is a measure of the hotness or coldness of a sample. At a kinetic-molecular level, the temperature is related to the average kinetic energy of the molecules in the sample.

Tertiary structure of proteins describes how a protein is folded into a compact globular structure. Tertiary structure is largely maintained by hydrogen bonding and London forces.

Tetrahedral geometry is observed in atoms having four ligands and no lone pairs of electrons, AX_4.

Thermodynamics studies the processes through which energy is transferred into and out of a system.

Third Law of Thermodynamics states the entropy of a perfect crystal at absolute zero is zero.

Tonicity describes the relative concentration of solute particles between two solutions. Two solutions are isotonic if they have equal concentrations of solute particles. A hypertonic solution has the greater concentration, while the hypotonic solution has the lesser concentration. In medical settings, tonicity is implicitly compared to normal saline, 0.89% NaCl in water.

Transcription is the process of creating a strand of messenger RNA from DNA.

Translation is the process of matching transfer RNA to a strand of messenger RNA.

Trigonal planar geometry is observed around atoms having three ligands and no lone pairs of electrons, AX_3.

Triodes or transistors are comprised of three semiconductors bonded together. A pnp transistor consists of an n-type semiconductor sandwiched between two p-type semiconductors. An npn transistor consists of a p-type semiconductor sandwiched between two n-type semiconductors. Transistors are used to amplify electrical signals.

Turbulent flow in a fluid is characterized by chaotic flow.

Tyndall effect is the scattering of light that is observed when light passes through a colloid.

Uncertainty Principle states that certain quantities, such as energy and time, for an object cannot be exactly known at the same time.

Valence electrons are the electrons used to form chemical bonds. These are the highest energy electrons in the ground state of an atom.

Van der Waals equation attempts to compensate for the nonzero volume of real gas molecules and the attraction that exists between real gas molecules.

Vapor pressure is the pressure exerted by molecules that escape from a solid or liquid sample into the gas phase.

Vaporization is the equilibrium process of a liquid converting into a gas. Vaporization is the opposite of condensation.

Vectors are quantities that have both magnitude and direction. Scalar quantities have size but no direction. Vectors are represented in boldface or with an arrow over the symbol.

Velocity (v) is a vector quantity that is defined as displacement (d) divided by time (t), $v = d/t$. Velocity is measured in meters/second (m/s) and includes a direction. Speed is the scalar component of velocity and is expressed simply as m/s.

Viscosity is the resistance of a fluid to flow.

VSEPR Theory (Valence Shell Electron Pair Repulsion Theory) is based on the mutual repulsion of bonding electrons and lone pairs of electrons around an atom. VSPER Theory predicts the geometry around an atom.

Weak acids are only partially dissociated into a hydrogen ion and its conjugate base when it is dissolved in water. To determine the pH of a weak acid, you need to use an equilibrium constant calculation.

Weak bases are only partially dissociated into a hydroxide ion and its conjugate acid when it is dissolved in water. To determine the pH of a weak base, you need to use an equilibrium constant calculation.

Weight (W) is a force created by the acceleration of gravity (g) acting on a mass (m). The magnitude of the weight can be calculated using Newton's Second Law: $W = mg$. Weight is measured in newtons.

Work is the expenditure of energy. Applying a force (F) that moves an object some distance (d) accomplishes an amount of work equal to $W = Fd$, as long as the object moves in the same direction as the applied force. Changing the volume of a gas (ΔV) against a constant pressure (P) also does work, $W = P\Delta V$.

Work–Energy Theorem states that the total work done on a system is equal to the change in the kinetic energy of the system.

Zwitterions are electrically neutral species that have an internal positive charge and an internal negative charge. Amino acids at their isoelectric pH exist as zwitterions.

Supplemental References

The following materials served as references for this textbook. If you want to further explore any of the concepts discussed in this text, the following books may be of interest.

ANESTHESIA

Barash, P. G., Cullen, B. F., Stoelting, R. K., Cahalan, M. K., & Stock, M. C. (2009). *Clinical anesthesia* (6th ed.). Philadelphia, PA: Lippincott Williams & Wilkins.

Davis, P. D., Parbrook, G. D., & Kenny, G. N. C. (1995). *Basic physics and measurement in anaesthesia* (4th ed.). Boston, MA: Butterworth-Heinemann.

Miller, R. D. (2000). *Anesthesia* (Vol. 2, 5th ed.). Philadelphia, PA: Churchill Livingstone.

Nagelhout, J. J., & Plaus, K. L. (2010). *Nurse anesthesia* (4th ed.). St. Louis, MO: Saunders.

PHYSICS

Barcelona, S. L., Vilich, F., & Cote, C. J. (2003). A comparison of flow rates and warming capabilities of the level 1 and rapid infusion system with various-size intravenous catheters. *Anesthesia & Analgesia*, 97, 358–363.

Fishbane, P. M., Gasiorowicz, S. G., & Thornton, S. T. (2005). *Physics for scientists and engineers with modern physics* (3rd ed.). Upper Saddle River, NJ: Pearson Education.

Guyton, A. C., & Hall, J. E. (2006). *Textbook of medical physiology* (11th ed.). Philadelphia, PA: Elsevier.

Lazzaro, G. M., & Huey, R. J. (1991). *Oscillometric blood pressure device*. United States Patent 5,054,494.

Nave, C. R., & Nave, B. C. (1985). *Physics for the health sciences* (3rd ed.). Philadelphia, PA: Saunders.

Pickering, T. G., et al. (2005). *Recommendations for blood pressure measurements in humans: BP measurement methods*. Retrieved March 12, 2012 from http://www.medscape.com/viewarticle/499972_3.

Sears, F. W., Zemansky, M. W., & Young, H. D. (1991). *College physics* (7th ed.). Reading, MA: Addison-Wesley.

Serway, R. A., & Faughn, J. S. (2003). *College physics* (6th ed.). Pacific Grove, CA: Thomson Learning.

Serway, R. A., & Jewett, J. W., Jr. (2004). *Physics for scientists and engineers* (6th ed.) Pacific Grove, CA: Brooks/Cole-Thompson.

Serway, R. A., & Vuille, C. (2007). *Essentials of college physics.* Pacific Grove, CA: Brooks/Cole-Thomson.

Shipman, J. T., Wilson, J. D., & Todd, A. W. (2006). *An introduction to physical science* (11th ed.). Boston, MA: Houghton Mifflin.

Skoog, D. A., Holler, F. J., & Crouch, S. R. (2007). *Principles of instrumental analysis* (6th ed.). Pacific Grove, CA: Brooks/Cole-Thomson.

Steven Engineering, Inc. Hospital Isolated Power Systems, Class 4800 (4800CT9801R4/08).

The American Heritage College Dictionary (4th ed.). (2004). Boston, MA: Houghton Mifflin.

The complete guide to home wiring: A comprehensive manual, from basic repairs to advanced projects. (1998). Minnetonka, MN: Creative Publishing International.

Tipler, P. A., & Mosca, G. (2008). *Physics for scientists and engineers* (6th ed., Standard Version). New York, NY: Freeman.

Walker, J. S. (2010). *Physics* (4th ed.). Upper Saddle River, NJ: Pearson Education.

Wilkins, R. L., Stoller, J. K., & Kacmarek, R. M. (2009). *Egan's fundamentals of respiratory care* (9th ed.). St. Louis, MO: Mosby.

Wilson, J. D., Buffa, A. J., & Lou, B. (2003). *College physics* (5th ed.). Upper Saddle River, NJ: Pearson Education.

CHEMISTRY

Alberty, R. A., & Silbey, R. J. (1992). *Physical chemistry.* New York, NY: Wiley.

Brown, T. L., LeMay, H. E., Jr., Bursten, B. E., Murphy, C. J., & Woodward, P. M. (2012). *Chemistry: The central science* (12th ed.). Upper Saddle River, NJ: Pearson Education.

Campbell, M., & Farrell, S. (2006). *Biochemistry* (5th ed.). Pacific Grove, CA: Thomson, Brooks/Cole.

Horton, H. R., Moran, L., Scrimgeour, K. G., Perry, M., & Rawn, J. D. (2006). *Principles of biochemistry* (4th ed.). Englewood Cliffs, NJ: Prentice Hall.

Interactive Chart of the Nuclides. (2012). *National nuclear data center.* Brookhaven National Laboratory. Retrieved August 3, 2012 from http://www.nndc.bnl.gov/chart

IUPAC Gold Book. (2012). Retrieved August 17, 2012 from http://goldbook.iupac.org/N04257.html

IUPAC Periodic Table of the Elements. (2012). Retrieved August 6, 2012 from http://www.iupac.org/reports/periodic_table

Levine, I. N. (2009). *Physical chemistry* (6th ed.). New York, NY: McGraw-Hill.

Petrucci, R. H., Harwood, W. S., Herring, F. G., & Madura, J. D. (2007). *General chemistry: principles & modern applications* (9th ed.). Upper Saddle River, NJ: Pearson Education.

Shubert, D. (2008). *Organic chemistry: A multi-media approach.* (Private Publication).

Solomons, T. W. G., & Fryhle, C. (2000). *Organic chemistry* (7th ed.). New York, NY: Wiley.

Tro, N. J. (2008). *Chemistry: A molecular approach.* Upper Saddle River, NJ: Pearson Education.

Tuli, J. K. (2011). *Nuclear wallet cards* (8th ed.). National Nuclear Data Center. Brookhaven National Laboratory. Upton, NY: U.S. Department of Energy.

Zumdahl, S. S., & Zumdahl, S. A. (2007). *Chemistry* (7th ed.). Boston, MA: Houghton Mifflin.

RADIOACTIVITY

Brown, T. L., LeMay, H. E., Jr., Bursten, B., & Murphy, C. J. (2006). *Chemistry: The central science* (10th ed.). Upper Saddle River, NJ: Pearson Education.

Cember, H. (1996). *Introduction to health physics* (3rd ed.). New York, NY: McGraw-Hill.

Chaisson, E., & McMillan, S. (2007). *Astronomy: A beginner's guide to the universe* (5th ed.). Upper Saddle River, NJ: Pearson Education.

Choppin, G. R., Liljenzin, J.-O., & Rydberg, J. (1995). *Radiochemistry and nuclear chemistry* (2nd ed.). London, UK: Butterworth-Heinemann.

Christian, P. E., & Waterstram-Rich, K. M. (Eds.) (2007). *Nuclear medicine and PET/CT: Technology and techniques* (6th ed.). St. Louis, MO: Mosby.

Friedlander, G., Kennedy, J. W., Macias, E. S., & Miller, J. M. (1981). *Nuclear and radiochemistry* (3rd ed.). New York, NY: Wiley.

GE Healthcare. (2011). *Where your nuclear medicine future begins: Brivo NM615 brochure.* Waukesha, GE.

Hoag Hospital. (2008). Retrieved October 2, 2008 from http://www.hoaghospital.org/gammaknife/WhatisGammaKnife.aspx

Knight, R. D. (2008). *Physics for scientists and engineers: A strategic approach* (2nd ed.). Upper Saddle River, NJ: Pearson Education.

Knoll, G. F. (2000). *Radiation detection and measurement* (3rd ed.). New York, NY: Wiley.

Lars Leksell Center for Gamma Knife Surgery. (2008). Retrieved October 2, 2008 from http://www.healthsystem.virginia.edu/internet/gammaknife/overview.cfm.

Loveland, W., Morrissey, D. J., & Seaborg, G. T. (2006). *Modern nuclear chemistry.* New York, NY: Wiley.

Meyers, W. G. (1979). The Anger scintillation camera becomes of age. *Journal of Nuclear Medicine, 20,* 565–567.

Nuclear News, September 2008.

Philips Medical Systems. (2003). *SKYLightTM nuclear camera platform: Setting new standards in molecular imaging brochure.* Pays-Bas: Phillips.

Shipman, J. T., Wilson, J. D., & Todd, A. W. (2009). *An introduction to physical science* (12th ed.). Boston, MA: Houghton Mifflin.

Siemens Medical Solutions USA, Inc. (2006). *e.cam Signature series brochure.* Knoxville, TN: Siemens AG.

Skoog, D. A., Holler, F. J., & Nieman, T. A. (1998). *Principles of instrumental analysis* (5th ed.). Pacific Grove, CA: Thomson Learning.

The General Electric Company. (1984). *Chart of the nuclides* (13th ed.).

Walker, J. S. (2007). *Physics* (3rd ed.). Upper Saddle River, NJ: Pearson Education.

Index